KV-482-317

Biomonitoring of Polluted Water

- Reviews on Actual Topics -

Editor:

A. Gerhardt

ttp **TRANS TECH PUBLICATIONS LTD**
Switzerland • Germany • UK • USA

Cover Illustration:
Arrangement: A. Gerhardt
Photo left: A. Gerhardt
Photo right: L. Janssens de Bisthoven

Copyright © 2000 Trans Tech Publications Ltd, Switzerland
ISBN 0-87849-845-1

Volume 9 of
Environmental Research Forum
ISSN 1421-0274

Distributed *in the Americas by*

Trans Tech Publications Inc
PO Box 699, May Street
Enfield, New Hampshire 03748
USA
Phone: (603) 632-7377
Fax: (603) 632-5611
e-mail: ttp@ttp.net
Web: http://www.ttp.net

and worldwide by

Trans Tech Publications Ltd
Brandrain 6
CH-8707 Uetikon-Zuerich
Switzerland
Fax: +41 (1) 922 10 33
e-mail: ttp@ttp.net
Web: http://www.ttp.net

Printed in the United Kingdom
by Hobbs the Printers Ltd,
Totton, Hampshire SO40 3WX

UNIVERSITY OF PLYMOUTH
Item No. 9 005081730
Date 1 3 MAY 2002 S
Class No. 574.583 BIO
Cont. No. ✓
PLYMOUTH LIBRARY

90 0508173 0

WITHDRAWN
FROM
UNIVERSITY OF PLYMOUTH
LIBRARY SE

Biomonitoring of
Polluted Water

Dedication

This book is dedicated to Sara and all the children of the blue planet for a healthy environment.

Preface: Why another book about biomonitoring?

Biomonitoring of water pollution grew out of disciplines such as aquatic ecology and (eco)toxicology. Biomonitoring has become a scientific tool during the last few decades and is now part of administrative practice. A variety of methods has been described and reviewed in several excellent books and reports [*e.g.*: 1 - 16]. All of these contributions mark the considerable progress made in biomonitoring in recent times, *e.g.* the development of rapid assessment methods, the use of multivariate statistical techniques, the incorporation of models into biomonitoring programs, paleolimnological biomonitoring and the use of laboratory techniques in the field; such as online biomonitors. The field is so huge, and spread over such different disciplines, *e.g.* zoology, botany, microbiology, chemistry, physics, as well as various types of ecosystems (such as groundwater, marine and freshwater ecosystems) and different methodological approaches (*e.g.* toxicology, ecology, mathematics, geography etc.) that it is impossible to summarise all of the relevant topics in one book.

The aim of this book is to present some selected examples of new or odd methods and aspects of aquatic biomonitoring. It also gives space to subjects which receive little attention, are somewhat "forgotten" or are overshadowed by more "fashionable" topics. The book is thus meant to be "out of the rut" of usual ideas and thinking. Due to the restricted length of the book, a selection had to be made and some interesting themes could not be covered. It is hoped that the book will stimulate critical discussion and new research directions. Perhaps, it will also help to get some of the described methods and concepts more widely accepted by environmental managers or politicians.

The book begins with an introductory chapter which consists of a short general discussion of the genesis and *status quo* of biomonitoring and its various methodological approaches. This article by the editor is not intended to be an all-embracing review, but rather a short summary and reminder of a few important ideas, definitions and trends in aquatic biomonitoring. Roux gives the theoretical background to how to structure biomonitoring in a design for resource management, citing the example of the South African River Health Programme, which is under development and is trying to incorporate concepts from other international monitoring programs. Other monitoring programs have been extensively described elsewhere, whereas the South African approach is not so well known. Following the new trend, the new policy goals emphasize the need to protect rather than to exploit the ability of ecosystems to recover from disturbances [17]. Particularly in developing countries, environmental monitoring needs to consider equally the social, economic and ecological aspects. The South African River Health Programme is based upon a broad interdisciplinary monitoring approach; resulting in the assessment of the "ecological condition" of running water as compared to "reference conditions". In a case study of the Eland River, the application of the River Integrity Classification Scheme (RICS) is described.

A second chapter deals with some biomonitoring methods applied at different biological organisation levels. Biomarker responses should be included in routine biomonitoring since information from all biological organisation levels form a sound basis for an integrated environmental monitoring approach. Schramm *et al.* describe the use of cellular, histological and biochemical biomarkers in fish for the purpose of running-water biomonitoring within the framework of the VALIMAR project in Germany. The use of heat-shock proteins, metabolic and biotransformation enzymes, and ultrastructural biomarkers are discussed with regard to their ability to provide evidence of change, at the individual and population levels, which are caused by pollution. Only biomarkers with a proven relationship to ecologically relevant parameters *e.g.* growth, reproduction, fish migration etc. can be used for environmental monitoring.

The use of morphological deformities as environmental indicators arises from studies of fish and has now been extended to several groups of aquatic invertebrates. In some countries, this method is already integrated into biomonitoring programs. The article by Janssens de Bisthoven has been chosen so as to encourage the use of morphological deformities, especially of *Chironomidae*, for sediment toxicity assessment of suspected contaminated sites. Another recent development is the continuous biomonitoring of water quality along rivers which receive toxic effluents. Automated biotests have been described since the seventies, especially in the USA, and their use in routine biomonitoring in Europe was triggered by the Sandoz accident in 1986. The article by the editor briefly describes and compares the available biomonitors, with special emphasis being placed on recent developments based upon video-techniques and impedance conversion. There is a trend towards using other species, including benthic insects, rather than the typical standard toxicity organisms, in a multi-species biomonitoring approach in order to obtain locally relevant ecological information. Unfortunately, the general opinion concerning online biomonitors is rather negative, because (1) the water quality of several rivers has improved and (2) many biomonitors give only partialy convincing results. The article stresses that the development of optimal methods should continue, as these instruments have the best deterrent effect upon polluters, and may have contributed to the improved water quality observed in some rivers. Moreover, online biomonitors should be placed at the effluents discharges of industries and waste-water purification plants, rather than being scattered along the river. Online biomonitoring of marine water should be more widely encouraged.

The third chapter discusses a "forgotten" but, for human health, extremely important aquatic ecosystem: groundwater. Groundwater biomonitoring has been slow in development due to its highly complex and heterogeneous environmental characteristics and the difficulty of access. The article of Mösslacher and Notenboom provides an introduction to groundwater ecosystems, and a critical assessment of the difficulties of groundwater biomonitoring.

The last chapter is devoted to selected bioindicators. Pratt and Bowers discuss the use of Protozoa, mainly *Ciliata*, as indicator species for biomonitoring. As this difficult group needs accurate taxonomy, other methods have evolved: (1) changes in community structure at the taxa level, (2) changes in functional groups and (3) changes in biomass or metabolic activity. Protozoa are very useful, especially for rapid toxicity assessment, due to their rapid response, easy storage in cysts and easy culture. They should therefore be incorporated into an integrated biomonitoring approach.

Rinderhagen *et al.* describe the biological characteristics and life history pecularities of Crustacea as bioindicators for pollution biomonitoring. The article concentrates on freshwater Crustaceans: mainly amphipods and isopods. These groups are important links in many freshwater and marine ecosystems due to their abundance and feeding habits. However, they have not yet been incorporated into routine biomonitoring in Europe. The use of these groups has probably been overshadowed by the well-established traditional use of daphnids in (eco)toxicology.

Cleveland *et al.* discuss the use of fish in biomonitoring under three headings: chemical registration, regulatory compliance and natural resource damage assessment. Fish have been used for a long time in toxicology. The use of laboratory toxicity tests with different end-points, as well as field approaches, are described. In recent times, ethical aspects have contributed to a negative attitude towards fish toxicity tests and fish biomonitors. However, as often top-predators, fish are invaluable surrogates for human beings in the evaluation of bioaccumulation and toxic effects. The recent development of a range of population and community end-points, combined with statistical methods and models, focusses on the assessment of the health and status of free-ranging fish populations.

Any integrated biomonitoring program should also consider aquatic vegetation. Benthic algae and bryophytes have seldom been included in trend biomonitoring. Tremp discusses the use of bryophytes in the field for population- and community-level biomonitoring. Submerged *bryophyta* seem to be useful indicators of stress araising from waste heat, acidification and eutrophication.

Two articles concerning the use of aquatic macrophytes have been included in the book in order to encourage strongly this approach to integrated environmental biomonitoring; especially in conjunction with GIS methods. Scientific opinion concerning the value of phytoassessment in biomonitoring is split, and further research and convincing evidence are needed before phytoassessment will receive the same attention as algae and animal bioassessment methods. Lewis and Wang provide an introduction to phytoassessment, its importance, state-of-the-art, biomonitoring methods and indicator species, followed by two important examples of eutrophication and wetlands for bioremediation of water quality. Stewart *et al.* focus on the use of phytotoxicity and plant community studies for biomonitoring purposes. The use of plant communities is not restricted to the description of water-quality conditions, but also handles trophic status and habitat destruction. The development of the floristic quality index and the coefficient of conservatism are pioneering efforts towards developing an index of biotic integrity.

References

[1] D. Pascoe and R. W. Edwards (eds): Freshwater Biological Monitoring. Pergamon Press, Oxford (1984).

[2] D. Gruber and J. Diamond (eds): Automated Biomonitoring: Living Sensors as Environmental Monitors. Ellis Horwood LtD, Chichester, USA (1988), 206 pp.

[3] K. J. M. Kramer and J. Botterweg: Aquatic Biological Early Warning Systems: An Overview. In: D. W. Jeffrey and B. Madden (eds): Bioindicators and Environmental Management. Academic Press, London (1991), 95-126.

[4] J. Cairns, Jr. and T. V. Crawford (eds.): Integrated Environmental Management. Chelesea, Michigan, Lewis Publishers (1991).

[5] D. M. Rosenberg and V. H. Resh (eds.) Freshwater Biomonitoring and Benthic Macroinvertebrates. Chapman & Hall, New York (1993), 488 pp.

[6] P. Calow (ed.): Handbook of Ecotoxicology. Vol. 1. Blackwell Scientific Publications LTD, Oxford, UK (1993).

[7] K. J. M. Kramer (ed.): Biomonitoring of Coastal Waters and Estuaries. CRC Press, Boca Raton (1994), 362 pp.

[8] S. L. Loeb and A. Spacie (eds): Biological Monitoring of Aquatic Ecosystems. Chelesea, Michigan, Lewis Publishers (1994).

[9] S. M. Haslam (ed.): River Pollution. J. Wiley. & Sons Ltd., Chichester, UK (1992).

[10] G. Holmes, L. Theodore and B. R. Singh (eds.): Handbook of Environmental Management and Technology. J. Wiley & Sons Ltd., Chichester, UK (1993).

[11] G. Gunkel (ed.): Bioindikation in aquatischen Ökosystemen. Bioindikation in limnischen und küstennahen Ökosystemen. Grundlagen, Verfahren und Methoden. Gustav Fischer Verlag, Jena (1994).

[12] D. de Zwart: Monitoring Water Quality in the Future, Vol. 3: Biomonitoring. Ministry of Housing, Spatial Planning and the Environment (ed.) Bilthoven, The Netherlands (1995).

[13] B. C. Rana (ed): Pollution and Biomonitoring. Tata McGraw-Hill Publ. Company LtD, New Delhi (1995).

[14] J. Cairns, Jr. and R. M. Harrison (eds.): Methods of Environmental Data Analysis. Chapman & Hall, London (1995), 309 pp.

[15] J. Cairns, Jr. and B. R. Niederlehner (eds.): Ecological Tixicity Testing. Scale, Complexity and Relevance. Lewis Publ., Boca Raton, USA (1995).

[16] W. S. Davies and T. P. Simon (eds.): Biological Assessment and Criteria. Tools for Water Resource Planing and Decision Making. Lewis Publ., Boca Raton, USA (1995).

[17] D. Roux, P. L. Kempster, C. J. Kleynhans, H. R. van Vliet and H. H. du Preez: Integrating Stressor and Response Monitoring into a Resource-Based Water Quality Assessment Framework. Environm. Management Vol. 23, No. 1 (1999), 15-30.

Acknowledgments

Herewith I would like to thank TTP for giving me the opportunity to publish this book. Moreover, I wish to thank all colleagues for a fruitful cooperation in either contributing with an article or reviewing a chapter.

Table of Contents

Preface   v

List of Authors   xi

Introduction

Biomonitoring for the 21st Century ✳
   A. Gerhardt   1

Design of a National Programme for Monitoring and Assessing the Health of Aquatic Ecosystems, with Specific Reference to the South African River Health Programme
   D.J. Roux   13

Selected Methods

Cellular, Histological and Biochemical Biomarkers
   M. Schramm, A. Behrens, T. Braunbeck, H. Eckwert, H.-R. Köhler, J. Konradt, E. Müller, M. Pawert, J. Schwaiger, H. Segner and R. Triebskorn   33

Biomonitoring with Morphological Deformities in Aquatic Organisms
   L. Janssens de Bisthoven   65

Recent Trends in Online Biomonitoring for Water Quality Control
   A. Gerhardt   95

Aquatic Ecosystems

Groundwater Biomonitoring
   F. Mösslacher and J. Notenboom   119

Selected Bioindicators

Protozoa in Polluted Water Biomonitoring
   J.R. Pratt and N.J. Bowers   141

Crustaceans as Bioindicators
   M. Rinderhagen, J. Ritterhoff and G.-P. Zauke   161

Biomonitoring and Ecotoxicology: Fish as Indicators of ✳ Pollution-Induced Stress in Aquatic Systems
   L. Cleveland, J.F. Fairchild and E.E. Little   195

Submerged Bryophytes in Running Waters, Ecological Characteristics and their Use in Biomonitoring
    H. Tremp   233

Biomonitoring using Aquatic Vegetation
    M.A. Lewis and W. Wang   243

The Use of Aquatic Macrophytes in Monitoring and in Assessment of Biological Integrity
    P.M. Stewart, R.W. Scribailo and T.P. Simon   275

Author Index   303

Keyword Index   305

Alphabetic list of authors:

A. Behrens
UFZ Umweltforschungszentrum
Abt. Chemische Ökotoxikologie
Permoserstr. 15
D-04318 Leipzig

N.J. Bowers
Environmental Sciences & Resources
Portland State University
PO Box 751
Portland, OR, 97207, U.S.A

T. Braunbeck
Zoologisches Institut
Abt. Zellbiologie
Universität Tübingen
Auf der Morgenstelle 28
D-72076 Tübingen

L. Cleveland
Columbia Environmental Research Center
Biological Resources Division
U.S. Geological Survey
4200 New Haven Road
Columbia, MO, 65201, U.S.A

H. Eckwert
Universität Heidelberg
Abt. Zoologie 1
Im Neuenheimer Feld 230
D-69120 Heidelberg

J. F. Fairchild
Columbia Environmental Research Center
Biological Resources Division
U.S. Geological Survey
4200 New Haven Road
Columbia, MO, 65201, U.S.A

A. Gerhardt
'LimCo International'
An der Aa 5
D-49477 Ibbenbüren

L. Janssens de Bisthoven
'LimCo International'
An der Aa 5
D-49477 Ibbenbüren

H.-R. Köhler
Universität Heidelberg
Abt. Zoologie 1
Im Neuenheimer Feld 230
D-69120 Heidelberg

J. Konradt
Zoologisches Institut
Abt. Zellbiologie
Universität Tübingen
Auf der Morgenstelle 28
D-72076 Tübingen

M. A. Lewis
U.S. Environmental Protection Agency
Natl. Health and Environm. Effects Research Laboratory
Gulf Ecology Division
1 Sabine Island Drive
Gulf Breeze, FL 32561, U.S.A

E. E. Little
Columbia Environmental Research Center
Biological Resources Division
U.S. Geological Survey
4200 New Haven Road
Columbia, MO, 65201, U.S.A

F. Mösslacher
Institut für Limnologie
Österr. Akademie der Wissenschaften
Gaisberg 116
A-5310 Mondsee, Austria

E. Müller
Zoologisches Institut
Physiol. Ökologie
Universität Tübingen
Auf der Morgenstelle 28
D-72076 Tübingen

J. Notenboom
Natl. Institute of Public Health and the Environment
Laboratory for Ecotoxicology
PO Box 1
NL-3720 BA Bilthoven

M. Pawert
Zoologisches Institut
Physiol. Ökologie
Universität Tübingen
Auf der Morgenstelle 28
D-72076 Tübingen

J. R. Pratt
Environmental Sciences & Resources
Portland State University
PO Box 751
Portland, OR, 97207, U.S.A

M. Rinderhagen
Fachbereich Biologie (ICBM)
Carl von Ossietzky Universität Oldenburg
Postf. 2503
D-26111 Oldenburg

J. Ritterhoff
Fachbereich Biologie (ICBM)
Carl von Ossietzky Universität Oldenburg
Postf. 2503
D-26111 Oldenburg

D.J. Roux
Division of Water, Environment & Forestry Technology, CSIR
PO Box 395
Pretoria 0001, S-Africa

M. Schramm
Zoologisches Institut
Physiol. Ökologie
Universität Tübingen
Auf der Morgenstelle 28
D-72076 Tübingen

J. Schwaiger
Labor für Fischpathologie
Steinseestr. 32
D-81671 München

R. W. Scribailo
Biological Sciences and Chemistry Section
Purdue University North Central
1401 S. U.S. 421
Westville IN, 46391 U.S.A

H. Segner
UFZ Umweltforschungszentrum
Abt. Chemische Ökotoxikologie
Permoserstr. 15
D-04318 Leipzig

T. P. Simon
U. S. Environmental Protection Agency
Water Division
77 W. Jackson Street
WW-16J, Chicago
IL 60604 U.S.A

P. M. Stewart
Lake Michigan Ecological Research Station
Biological Resources Division
U. S. Geological Survey
1100 N. Mineral Springs Road
Porter, Indiana 46304, U.S.A

H. Tremp
Universität Hohenheim
Institut für Landschafts- und Pflanzenökologie (320)
Schloss Mittelbau (West)
D-70599 Stuttgart

R. Triebskorn
Zoologisches Institut
Physiol. Ökologie
Universität Tübingen
Auf der Morgenstelle 28
D-72076 Tübingen

W. Wang
U.S. Geological Survey
Water Resources Division
720 Gracern Road
Columbia, South Carolina 29210, U.S.A

G. P. Zauke
Fachbereich Biologie (ICBM)
Carl von Ossietzky Universität Oldenburg
Postf. 2503
D-26111 Oldenburg

Advisory editorial board:

J. Gibert
Université Claude Bernard Lyon 1
ESA/ CNRS 5023
Ecologie des Eaux Douces et des Grands Fleuves
43 Bd du 11 Novembre 1918
F-69622 Villeurbanne cedex

S. Molander
Teknisk miljöplanering
Chalmers tekniska högskola
S-41296 Göteborg

Dr. H. Tremp
Universität Hohenheim
Institut für Landschafts- und Pflanzenökologie (320)
Schloss Mittelbau (West)
D-70599 Stuttgart

Environmental Science Forum Vol.96 (1999) pp. 1-12
© 1999 Trans Tech Publications, Switzerland

Biomonitoring for the 21st Century

A. Gerhardt

'Limco International', An der Aa 5, DE-49477 Ibbenbüren, Germany

Keywords: Biomonitoring, Toxic Effects, Bioaccumulation, WET, Integrated Environmental Monitoring, Biological Indices, Community Response Biomonitoring, Exposure Monitoring, Bioindicator

Abstract

This article provides a synthesis of important facts, trends and ideas about freshwater biomonitoring. After some definitions, biomonitoring methods based on the three subdisciplines of ecotoxicology are described. New trends for biomonitoring of water quality are increasing ecological realism in (eco)toxicological test methods, *e.g.* multispecies tests, *in situ* tests, use of indigenous species, consideration of uncertainties and links between the different methods within the triad approach. The emphasis shifts from the dose-response approach towards the stress-response approach, which is more relevant to protect aquatic ecosystem health. Comprehensive and time-consuming ecological bioassessment methods are being replaced by rapid assessment methods combined with geographical/hydrological background data, which aim at classification of aquatic ecosystems. The development of predictive models based on exposure data and (eco)toxicological results is booming. As a conclusion, a general proposition of an integrated aquatic biomonitoring concept is outlined and discussed.

1) What is biomonitoring?

Biomonitoring uses biological variables to survey the environment and is a complement to chemical monitoring. Even if continuous chemical monitoring with ion-specific probes and data loggers is possible and cost-effective, life is the ultimate monitor of environmental quality. Aquatic organisms integrate all of the influential biotic and abiotic parameters in their habitat, thus providing a continuous record of environmental quality [1]. Biological variables include bioaccumulation (measurement of pollutant concentrations in biological material), toxicity (responses to chemicals at different biological organisation levels by means of bioassays and biological early warning systems) and ecosystem responses (assessment of ecosystem integrity). There is a variety of methods, about 120 different laboratory toxicity tests are presented and ca. 100 different variables to describe community effects in the field are presented in international literature [2]. According to different aims and strategies (control, alarm, prediction), several types of biomonitoring can be distinguished: Some use "autoecological" parameters, mainly in (eco)toxicological biomonitoring in the laboratory, *in situ* or online biomonitoring, and some use of "synecological" parameters, mostly in long-term trend biomonitoring in the field [3].

2) Definitions

A **stressor** is any physical, chemical or biological entity or process that can induce adverse effects on individuals, populations, communities and ecosystems [4]. The concept of indicator species is a crucial basis for trend biomonitoring. Ideally, the indicator has narrow and specific environmental tolerances and its presence is a reflection of the quality of its environment [5]. A "bioindicator" can be a suborganismal parameter (**biomarker**), an organism (**bioindicator, sentinel, biomonitor,** test species) or a community (**ecological indicator**), reacting to pollution with changes in their life functions. Different types of bioindicators are distinguished: **Bioindicators** *sensu strictu* are species which indicate a certain situation in the field, **test species** are standardised species which are used in **biotests**, and **biomonitor species** are species which serve to survey the environmental situation [6] and can be used to describe the temporal and/or geographical variations of bioavailability of contaminants [7]. Sometimes, a distinction between "response-bioindicator" and "accumulation-bioindicator" is done [3]. An **ecological indicator** is a variable which describes patterns and processes in ecosystems (*e.g.* species diversity). A **biomarker** is a biochemical, cellular, physiological, morphological or behavioral response at the organism level to sublethal exposure to chemicals. Sometimes, even ecological indicators are included in a broad definition of biomarkers [8].

3) Why biomonitoring?

The world-wide loss of biodiversity, both through the extinction of species by destruction of habitats, and through the anthropogenic selection of resistance against various chemicals, is one of the main environmental problems facing society today [9]. In order to achieve the vision of a "society living in harmony with healthy natural systems", the following milestones are (1) pollution prevention by reduction of consumption of resources, recycling of materials, controlling runoff from agricultural and urban lands, (2) watershed planning and management over political boundaries and (3) increased individual and collective responsability [10]. Environmental monitoring is an activity that is essential to maintain human quality of life, *e.g.* storage, purification and distribution of water [11].

Biological monitoring information has two primary functions: 1) to control, signal and predict calamities and accidental spills as early warning system and 2) to document long-term trends, both natural and deleterious [12]. However, a warning of an early warning system needs to be coupled with a sufficiently high appreciation of the dependence of human quality on ecosystem services [11]. The management of aquatic ecosystems has to integrate social, economic and ecological aspects in order to achieve a successful, broadly accepted and sustainable protection of the resource water.

4) Biomonitoring: genesis and status quo

The field of biological monitoring has its genesis in the protection of humans, and subsequently organisms living in natural systems, against deleterious effects clearly evident to the general public [12]. In order to protect humans, single species fish toxicity testing has been developed and applied, where fish were seen as surrogates for humans in natural systems. Aquatic toxicology was born as an illegitime child of classical mammalian toxicology [13]. Fish toxicity testing has mostly been used for the classification of toxicity of chemicals. In recent times, the three subdisciplines of ecotoxicology, such as toxicology, ecology and environmental chemistry are equally integrated in different aspects of biomonitoring (Fig. 1). Aquatic toxicology first concentrated on bioaccumulation and toxic effects in fish as surrogates for humans. Standard toxicity tests for other aquatic organisms such as *Daphnia* sp. and chironomids were developed in the following steps. Environmental chemistry has been used to measure toxicant concentrations in different compartments of the aquatic ecosystem, such as sediment, water column, selected biota and food webs. Whole effluent toxicity testing (WET) combines toxicology and water chemistry, as dilution series of the whole "toxic cocktail" are tested, in contrast to single substance testing in the laboratory. Thus, chemical interaction between different toxic compounds and their degradation products are considered. Aquatic ecology offers the methods for ecological evaluation of aquatic communities. A combination of aquatic ecology and water chemistry or toxicology is the approach of *in situ* experiments and bioassays, *e.g.* online biomonitors. For the assessment of water quality, each of the three pilars of ecotoxicology contributes with their specific methods, *e.g.* biosensors and biomonitors in biological early warning biomonitoring, biotic indices in trend biomonitoring and physical/chemical models in predicting water chemistry. The "overlapping zones" of the three approaches of biomonitoring are the most interesting ones, leading to an integrated environmental monitoring. In the following text, the three different approaches based on the triad of (eco)toxicology are discussed in their actual and future use for environmental assessment and monitoring.

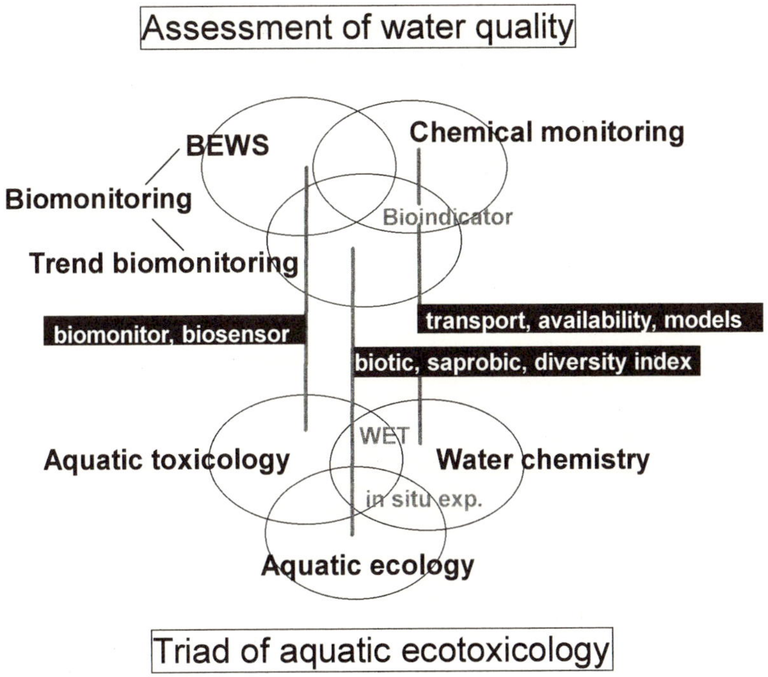

Assessment of water quality

Fig. 1: Methods for the assessment and monitoring of water quality resulting from the triad of ecotoxicology.

4.1) Trends in (eco)toxicological biomonitoring

1) Laboratory toxicity tests remain valuable in deriving and assessing water quality criteria, screening and ranking chemicals and predicting their hazard and risk, establishing dilution levels of chemicals or effluents prior to discharge into water bodies, determining cause-effect relationships and calibrating field bioindicators [14].

2) The paradigm of the "most sensitive species" as the "best environmental indicator" is controversial, because species sensitivity is a function of a number of interrelated factors (*e.g.* environmental conditions, state of health, competitive interactions), therefore, predictive capacity rather than sensitivity may be the most important issue to consider when assessing the ultimate utility of any bioassay method [15].

3) Most test designs lack chemical, physical and biological realism. Considering the "Toxicity Bioassay Continuum", ecological realism can be achieved by planning experimental designs, which move from the centre of the bioassay continuum outwards on the respective axes (Fig. 2).

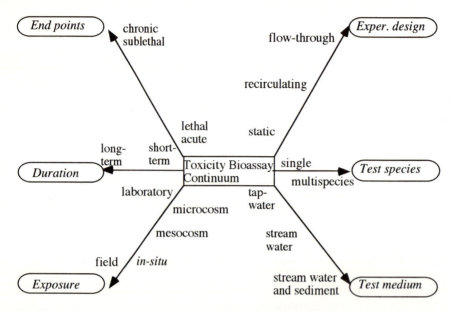

Fig. 2: The "Toxicity Bioassay Continuum": Each test criterium (duration, end points, exposure, species, medium, design) can be adapted to ecologically relevant conditions by moving from the centre in a centrifugal way on the respective axis of the diagram.

For example, <u>chemical realism</u> would consider biotransformation, degradation, adsorption, fluctuation in water chemistry and interactions among chemicals in the test medium. Whole effluent toxicity testing (WET) has been incorporated in monitoring programs, as it overcomes analytical problems and uncertainties resulting from mixture toxicity effects. However, often WET tests do not relate to in-stream impairment, sometimes a time lag between WET results and in-stream effects was found [16]. <u>Biological realism</u> would consider factors like acclimation, avoidance, resistance, predation, competition and recovery. The lack of ecological realism in single species toxicity tests has been overcome by the development of microcosm and mesocosm designs, *in situ* testing and by the use of community response parameters, *e.g.* the ratio between photosysnthesis to respiration [17]. The use of indigenous species for toxicity testing will gradually replace *e.g.* the standardised (*Cerio*)*Daphnia* sp., fathead minnow and *Selenastrum* sp. toxicity tests. Applied for waste discharges into aquatic ecosystems, where these species are not ecologically important or even do not occur, the ecological information from these tests lacks precision, even if the tests are inexpensive. However, if the cost of poor management decisions based on imprecise information is considered, then individual tests are not so inexpensive [18].

4) <u>Community toxicity testing</u> (ecotoxicology *sensu strictu*) has been proven not to be of high variability and of high costs, both previous arguments for single species toxicity testing instead of ecologically relevant testing on the community level [19]. In an attempt to bridge the gap between toxicity testing and ecological reality, Calow *et al.* [20] developed a five-step approach that uses simplified life-history models to make ecologically relevant links between toxicity test results and their implications for population dynamics. This model is still untested [21].

5) The emerging field of ecosystem health prefers the <u>stress-response approach</u> to the dose-response approach for two reasons: First, organisms in natural systems are simultaneously exposed to multiple stressors, some of natural origin, some of anthropogenic origin. Focussing on the aggregate of stress achieves a more accurate and integrative measure of ecosystem condition. Second, the stress-response approach emphasizes ecosystem condition and health much more than the dose-response approach, which emphasizes thresholds of *e.g.* lethality, growth etc. In the latter approach, exposure is considered acceptable if no clinical predefined symptoms of malfunction are present. However,

absence of observable deleterious effects will not be sufficient for making complex decisions about ecosystem health [18].

6) The concept of risk management for chemicals requires that the consequences of release of chemicals can be described or predicted as well-defined events. However, in complex environmental systems many ecological changes cannot be described like that. Therefore, a concept based on the harmlessness of chemicals rather than their ecological noxiousness should be preferred [22].

7) Increasing interest in predictive toxicology with consideration of uncertainties has evolved in environmental monitoring research. The risk quotient (RQ), derived out of the ratio between "predicted environmental concentrations"(PEC) to "no observed effect concentrations"(NOEC), takes uncertainty into account as both parameters represent probability distributions [23]. The problem of differentiation between natural and acceptable variation from unnatural and unacceptable variation/changes in biological processes is difficult to solve, even after long-term data collection, which in itsself is time-consuming and expensive, so that such a monitoring might only be performed at a few selected ecosystems [2].

8) Integrated environmental monitoring should focus on the combination of chemical, toxicological and ecological field appraoches. For example, whole effluent toxicity testing below a point pollution source should be backed up with carefully designed field biomonitoring, to ensure, that the permitted discharge levels have their desired effect. This combination is better than creating artificial hybrides, *e.g.* mesocosm experiments [24].

4.2) Trends in community response biomonitoring

1) Like (eco)toxicological biomonitoring, aquatic ecology can contribute enormously to ecologically relevant biomonitoring. Trend biomonitoring is based on a classical set of methods, which experience a come back with growing interest [25]. Biological indices, which incorporate concepts such as diversity and integrity, provide important measures of ecosystem health. However, construction of biological indices requires considerable efforts, and they may have problems such as unclear identification of habitats, differences in sampling intensity, lack of both pre- and post-impact data and reliance on only one "unimpaired" reference site. Furthermore, the distinction between an anthropogenic cause or natural variation as cause for an observed trend is rather difficult [26]. Several types of biotic indices have widely been used in order to monitor water pollution [25]. They all assign numerical values to individual taxa (species, genera, families) that reflect their inferred sensitivities or tolerances, then summed up or averaged for all taxa in a sample [27]. As individual taxa might not be equally sensitive to all types of anthropogenic disturbance, biotic indices targeted to specific types of disturbances have been developed [28]. Calculating values of each index for a study site, it might be possible, not only to obtain an indication of the degree of impact, but also a diagnosis of the particular type of impact [27]. However, often a lack of specifity in response or sites with a mixture of different pollution types may prevent the development of adequate diagnostic tools [29].

2) Ecosystem response monitoring is based on the monitoring of biological or ecological indicators. The focus is to protect ecological integrity and the measurement end points are structural and functional attributes of biological communities in terms of resilience and elasticity. According to the "bubble-of stability" concept, an ecosystem has several states of equilibria, shifts between them is possible without loss of stability, *e.g.* maintaining a species. If, however, due to a severe stress, the equilibrium point moves out of the bubble, the stability is lost. Measurements of stability include the following issues: temporal scale, spatial scale, level of taxonomic resolution and numerical resolution (abundance) [30].

4.3) Trends in exposure monitoring

1) Surface waters contain a cocktail of many chemical substances and their degradation products. The receiving water's assimilative capacity, expressed as "total maximum daily load", is the capacity of a water body to receive disturbances without resulting in water quality changes according to the

predefined use. Assimilative capacity should be considered in biomonitoring studies. It relies on the following processes in the aquatic ecosystem: dilution, chemical break down of complex components, metabolisation, physical removal, biological removal (e.g.bioaccumulation, emergence) [30].

2) The development of <u>models</u>, extrapolation and estimation methods is crucial, because analysis and online monitoring of the huge number of chemical substances in the water is technically impossible, expensive and time consuming. Moreover, data gaps constitute the main bottle-neck in risk management of chemicals. For at least 95 % of all existing chemicals, short-term toxicity data on fish, *Daphnia* sp. and algae are not available [31]. Water quality models are often simplified abstractions of physical, chemical and biological processes in the aquatic ecosystem *e.g.* for lakes, complete mixing, steady state conditions and mass balance equations are assumed [32]. Another one-dimensional model to predict water quality in lakes combines a biogeochemical model with long-term time series of meteorological data [33]. QSAR (Quantitative Structure Activity Relationship) and QSPR (Quantitative Structure Property Relationship) are often used to relate chemical and physical properties of chemicals to biological parameters, *e.g.* bioaccumulation, toxicity [34]. They have been developed for the response of a single species and may be of limited value to predict complex ecosystems [35]. The PER-system (Potential Ecological Risk) and the ELS model (Effect-Load Sensitivity analysis) are used to structure, analyse and rank chemical threats to aquatic ecosystems [36]. Other used fate models are TOXSWA (TOXic substances in Surface WAter) for pesticides, or SREMOTOX for evaluation of ecotoxicological risk of toxic substances in the North Sea. The aim of GREAT-ER (Geography-referenced Regional Exposure Assessment Tool for European Rivers), a set of coupled exposure models, is to develop a powerful tool for predicting concentrations of chemicals in European rivers [37].

4.4) Integrated environmental monitoring

Environmental management and sustainable development are related to our capacity to produce scientifically sound decision-making tools. These decisions can be arranged in five keywords: diagnosis, control, monitoring, prediction and restoration [38]. <u>Predictive</u> issues were mainly based on ecotoxicological bioassays, most aspects regarding <u>control</u>, <u>monitoring</u>, <u>restoration</u> and even <u>diagnosis</u> were mostly based on chemical analysis and the (eco)toxicological interpretation of the analytical results according to available toxicological information. A monitoring program needs three steps: establishing baseline conditions, developing a predictive model and validating these predictions in the natural system [12]. While most of the classical biomonitoring programs are rather descriptive, a new approach towards predictive power through experiments is needed. The measurement end points should be in close relation to the assessment end points [19]. Not all biological variables are equally fit for serving in every monitoring program. Their suitability can be evaluated in each case by checking a set of requirements related to scientific and fundamental aspects, efficiency, costs, logistic and policy aspects [2]. Some important national biomonitoring programs are RIVPACS (UK), Australian National River Health Program, EMAP (USA) *etc*. An overview on how biomonitoring programs for effluent assessment, ambient toxicity monitoring including online biomonitoring and ecosystem monitoring are designed in different European countries as well as in the USA and Canada is given by de Zwart [2]. A general scheme of integrated environmental biomonitoring should consider the following steps and methods (Fig. 3). Starting point of any biomonitoring program is the intended usage of the water body, here always seen together with the corresponding sediment, under consideration that there are different criteria for water quality according to the different types of usage, *e.g.* agri/aquacultural, domestic, industrial and recreational. The aims of biomonitoring are either to control a toxic effluent, water and sediment quality, or status quo diagnosis and recording of long-term changes. Whereas the first aim mostly corresponds to a toxicological biomonitoring approach, the second aim has often been met by a set of ecological methods.

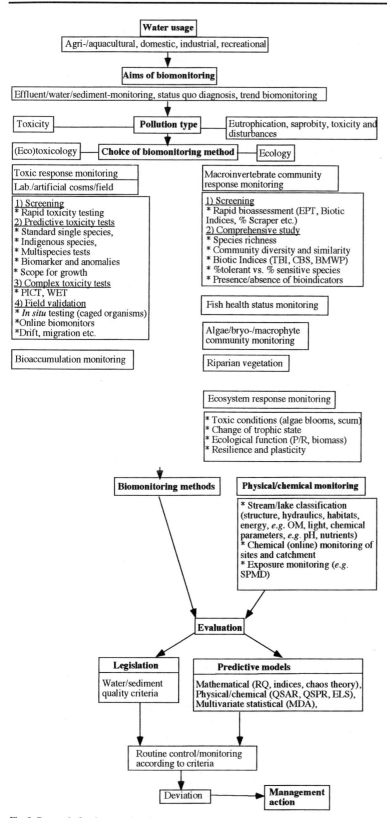

Fig. 3: Proposal of an integrated environmental monitoring.

In case, the (eco)toxicological approach is chosen, toxic effect biomonitoring should be performed at different biological organisation levels: (1) at the suborganismal level by use of biochemical biomarkers with strong relation to other toxic effect parameters, (2) at the individual level by use of acute or chronic single species toxicity bioassays for water and sediment, using standard organisms such as *Daphnia* sp., fish, chironomids, algae and bacteria and indigenous species in static bioassays and flow-through online biotests, and 3) at the community level by use of multispecies toxicity tests and rapid toxicity testing with PICT."Pollution Induced Community Tolerance"(PICT) has often successfully been applied for the determination and quantification of site-specific effects of contaminants on microorganisms [39]. Toxic effects should be studied in a gradient of increasing ecological realism from laboratory bioassays to artificial cosms to *in situ* toxicity testing of caged organisms or in online biotests (Fig. 2). Already Pontash and Cairns [40] recommended the use of indigenous species in multispecies toxicity tests, as they are better predictors of ecosystem level effects than single species tests with standard organisms. In order to support results from such *in situ* ecotoxicity tests, total invertebrate drift might be a parameter to be monitored at sites below effluents with varying toxicant discharges at least additionally to chemical sampling in a water quality alarm situation,which has been indicated by one of the above mentioned tests. WET (Whole Effluent Toxicity) testing comprises studies about toxicity, degradability, persistence, mutagenicity and bioaccumulation of a water containing a mixture of toxic compounds. Wherever possible, simplistic toxicity tests should be reserved to preliminary screening and ranking of chemical substances or sites to be further investigated.

Ecological methods have mostly been used for long-term trend biomonitoring, however, they could also be performed at effluents in addition to WET testing, *e.g.* as above/below comparisons or as a gradient analysis. Freshwater biomonitoring has mostly used benthic macroinvertebrates as bioindicators. Community diversity, eveness and similarity, species richness, enumerations as well as the presence/ absence of selected pollution bioindicators have been used to classify aquatic ecosystems. Several pollution indices have been developed [25]. The problem of such studies is always the clear correlation of the observed community status to pollution as the only cause. Also different types of pollution, *e.g.* toxicants, increased eutrophication, increased saprobity may affect the observed stream simultaneously and lead to confusion and contradiction in the evaluation of water quality. Habitat degradation, floods and other physical disturbances as well as adult insect mortality, migration, colonisation may be other reasons for a distorted community structure. As this approach demands comprehensive and time consuming studies, often by specialists for the determination of species, rapid bioassessment methods, analogous to rapid toxicity testing methods, have been developed [41-43]. The most promising methods are the number of Ephemeroptera, Plecoptera, Trichoptera taxa (EPT), the percentage of scrapers and some biotic indices [42]. However, rapid bioassessment methods rely on one single sample per site, rough determination to family or taxa level, and physical/chemical site classifications [42]. These methods are invaluable to screen and rank sites to be further investigated.

Aquatic trend biomonitoring should also include bioassessment methods for the health status of fish populations and communities. Algae-, bryo- and macrophyte communities offer additional invaluable information about the ecological status of a site and should be incorporated in an integrated biomonitoring program, especially in case of eutrophication problems or habitat destruction. The consideration of riparian vegtation as important food in form of allochthonous detritus and habitat structure in the ecotone between the aquatic and the terrestric environment of small streams or streams with backwaters should be taken into account.

A useful approach for routine monitoring of ecosystem stability parameters, *e.g.* resilience and plasticity, might be the measurement of ecosystem functions. Ecosystem responses to environmental disturbances can be measured as P/R (Production/ Respiration) ratio, biomass, PPR (Primary PRoduction) and change of trophic state. Parameters to measure the resilience (ability to readjust in order to neutralize the effect) and plasticity (the speed with which the system returns to its original state) need to be developed in order to better understand and judge observed changes on the community level as "acceptable" or as "detrimental" for the ecosystem functions. This approach demands less time and specialist work than the community response approach and should be encouraged.

Irrespective of the choice of biomonitoring methods, all biomonitoring efforts in the respective programs need to be combined with physical/chemical monitoring methods, in order to link and evaluate the observed biological responses to water quality parameters and exposure situations in the water body, and to generate data bases for predictive modelling. For example, stream/lake classification systems based on geographic and hydraulic parameters together with habitat structure information are a necessary basis for rapid bioassessment screening of different types of water bodies. Chemical online monitoring of summation and group parameters (*e.g.* conductivity, buffer capacity, TOC, CSB, AOX etc.) complements online biotests. Bioaccumulation studies might be backed up with exposure monitoring by use of SPMD (SemiPermeable Membrane Devices).

After data collection from the different complementary monitoring methods, evaluation methods based on univariate or multivariate statistics are necessary to surpass the descriptive step towards predictive statements, from which quality criteria for water, sediment, effluents and chemical substances can be developed [43]. In case no deviation from the predictive model is found during routine environmental monitoring of a site, this preventive (protective) biomonitoring is still useful and also serves as field validation of the model. In case of a severe deviation from the model, management actions *e.g.* rehabilitation may be needed.

Conclusions

In the next century environmental monitoring should in a cost and time effective way concentrate on the integration of the various existing methods and bridge the gap between (eco)toxicological, ecological and chemical approaches. Rapid toxicity testing methods in combination with rapid bioassessment methods are useful for ranking and mapping sites in combination with GIS. Toxic response monitoring should always be performed together with bioaccumulation and exposure monitoring at selected end-of-pipe sites, which are of toxicological concern for human and ecosystem health. A comprehensive ecological response monitoring using macroinvertebrate, fish and plant communities, identified to species level should be restricted to a few typical sites according to physical/chemical water body classification systems, especially where different types of pollution and habitat degradation overlap. However, as long as there is no clear distinction between community responses due to water quality changes compared to those due to other factors and as long as the potential of resilience is not known, such time and cost intensive approaches are not useful for large scale routine biomonitoring. The gap between traditional toxicology and ecology is already closing due to new approaches *e.g.* WET, *in situ* testing, multispecies testing, use of indigenous test species and communities and online biomonitoring. Predictive biomonitoring models should consider uncertainties, maybe even make use of chaos theory, in order to better understand the variability of biological parameters.

References

[1] R. L. Lowe and Y. Pan: Benthic Algal Communities as Biological Monitors. Algal Ecology 22 (1996), 705-739.
[2] D. de Zwart: Monitoring Water Quality in the Future, Vol. 3: Biomonitoring. Ministry of Housing, Spatial Planning and the Environment (ed.) Bilthoven, The Netherlands (1995).
[3] U. Arndt and A. Formin: Wissenschaftliche Perspektiven der ökotoxikologischen Bioindikation. UWSF-Z Umweltchemie. Ökotoxikologie 5 (1) (1993), 19-26.
[4] D. Roux: Design of a National Programme for Monitoring and Assessing the Health of Aquatic Ecosystems With Specific Reference to the South African River Health Programme. In: A. Gerhardt (ed): Biomonitoring of Polluted Water. Reviews on actual topics (1999) TTP Zürich, Switzerland.
[5] R. K. Johnson, T. Wiederholm and D. M. Rosenberg: Freshwater Biomonitoring Using Individual Organisms, Populations, and Species Assemblages of Benthic Macroinvertebrates. In: D. M. Rosenberg and V. H. Resh (eds): Freshwater Biomonitoring and Benthic Macroinvertebrates (1993), 40-159, Chapman and Hall, New York.
[6] U. Obst: Wirkungsbezogene Umweltanalytik. 1: Grenzen der chemischen und biologischen Analysenverfahren. Bioeffects-related environmental analytics. 1. Limits of chemical and biological analytics. GIT Labor-Fachzeitschrift 9 (1998), 905-908.

[7] P. S. Rainbow and D. J. H. Phillips: Cosmopolitan biomonitors of trace metals. Review. Marine Pollution Bulletin 26 (11) (1993), 593-601.

[8] M. H. Depledge and M. C. Fossi: The role of biomarkers in environmental assessment. 2. Invertebrates. Ecotoxicology 3 (1998), 161-172.

[9] N. M. van Straalen: Biodiversity of ecotoxicological responses in animals. Netherlands Journal of Zoology 44 (1-2) (1994), 112-129.

[10] P. H. Woodruff: Water Quality for the 21st Century. Water Environment and Technology (1993), March, 64-67.

[11] J. Cairns, Jr.: Environmental Monitoring For Sustainable Use Of The Planet. Animals on Earth 14 (2) (1996), 16-18.

[12] J. Cairns, Jr.: The genesis of biomonitoring in aquatic ecosystems. The Environmental Professional 12 (1990), 169-176.

[13] W. Sloof: Biological Effects of Chemical Pollutants in the Aquatic Environment and their Indicative Value. PhD thesis (1983). De Rijksuniversiteit te Utrecht, the Netherlands.

[14] J. C. Chapman: The role of ecotoxicity testing in assessing water quality. Austr. J. Ecology (1995) 20, 20-27.

[15] J. Cairns, Jr: The myth of the most sensitive species. Bioscience 36 (1986), 670-672.

[16] C. Geradi and J. Diamond: Defining the Relationship between Whole Effluent Toxicity Testing and Instream Toxicity. Abstract. SETAC 19th Annual Meeting 15 - 19 Nov. 1998, Charlotte, NC.

[17] A. L. Buikema, B. R. Niederlehner and J. Cairns, Jr.: Biological Monitoring Part 4- Toxicity Testing. Water Research 16 (1982), 239-262.

[18] J. Cairns, Jr. 1993. Environmental Science and Resource Management in the 21st Century: Scientific Perspective. Environmental Toxicology and Chemistry 12 (1993), 1321-1329.

[19] J. Cairns, Jr., J. R. Bidwell and M. E. Arnegard: Toxicity Testing with Communities: Microcosms, Mesocosms and Whole-system Manipulations. Reviews of Environmental Contamination and Toxicology, Vol. 147 (1996), Springer-Verlag, New York.

[20] P. Calow, R. M. Sibly and V. Forbes: Risk assessment of the basis of simplified life-history scenarios. Environ. Toxicol. Chem. 16 (1997).

[21] J. J. Cura: Ecological risk assessment. Water Environment Research 70(4) (1998), 968-971.

[22] K. Mathes: For a new paradigm in ecological risk assessment and regulation of chemicals. Abstract. SETAC 8th Annual Meeting 25-29 May 1999, Leipzig, Germany

[23] P. Calow (ed.): Handbook of Ecotoxicology. Vol. 1. Blackwell Scientific Publications LTD (1993), Oxford, UK 478 pp.

[24] R. O. Brinkhurst: Future Directions in Freshwater Biomonitoring Using Benthic Macroinvertebrates. In: D. M. Rosenberg and V. H. Resh (eds): Freshwater Biomonitoring and Benthic Macroinvertebrates (1993), 442-461, Chapman and Hall, New York.

[25] V. H. Resh and J. K. Jackson: Rapid Assessment Approaches to Biomonitoring Using Benthic Macroinvertebrates. in: D. M. Rosenberg and V. H. Resh (eds.): Freshwater Biomonitoring and Benthic Macroinvertebrates (1993), 195-234, Chapman & Hall, New York.

[26] B. Wallace, J. W. Grubaugh and M. R. Whiles: Biotic indices and stream ecosystem processes: Results from an experimental study. Ecological Applications 6 (1) (1996), 140-151.

[27] B. C. Chessman and P. K. McEvoy: Towards diagnostic biotic indices for river macroinvertebrates. Hydrobiologia 364 (1998), 169-182.

[28] R. Johnson: Spatiotemporal variability of temperate lake macroinvertebrate communities: detection of impact. Ecological Applications 8 (1) (1998), 61-70.

[29] E. Castella, M. Bickerton, P. D. Armitage and G. E. Petts: The effects of water abstractions on invertebrate communities in U.K. streams. Hydrobiologia 308 (1995), 167-182.

[30] D. J. Roux, P. L. Kempster, C. J. Kleynhans, H. R. Van Vliet and H. H. du Preez: Integrating Stressor and Response Monitoring into a Resource-Based Water-Quality Assessment Framework. Environmental Management 23 (1) (1999), 15-30.

[31] K. van Leeuwen: Ecotoxicological Effects Assessment in the Netherlands: Recent Developments. Environmental Management 14 (6) (1990), 779-792.

[32] D. S. Mahamah: Simplifying assumptions in water quality modeling. Ecological Modelling 109 (1998), 295-300.

[33] B. Tessin and V. Leite: Forecasting of Water Quality in Lakes: A Predictive Use of a One-dimensional Model. Application to Lake Bourget (Savoie, France). Hydrobiologia 373/374, (1998) 47-60.

[34] D. Calamari and M. Vighi: Quantitative Structure Activity Relationships in Ecotoxicology: Value and Limitations. In: Hodson E. (ed.): Reviews in Environmental Toxicology 4. Toxicology Communications (1990), Raleigh, North Carolina, USA.

[35] P. Donkin: Quantitative Structure-Activity Relationships. In: P. Calow (ed.): Handbook of Ecotoxicology. Vol. 2, 321-348, Blackwell Scientific Publ., Oxford, UK.

[36] L. Håkansson (ed): Water Pollution - methods and criteria to rank, model and remediate chemical threats to aquatic ecosystems (1999), 267 pp, Backhuys Publishers, Leiden, the Netherlands.

[37] F. Koormann, J.- O. Wagner and M. Matthies: GREAT-ER application as tool for data quality check and monitoring design. Abstract, SETAC 9th Annual Meeting 25-29 May 1999, Leipzig, Germany

[38] M. V. Pablos, P. Sanchez, G. Carbonell and J. V. Tarazona: Use of Ecotoxicity Tests in Environmental Management: Facts and Fantasies. Abstract, SETAC 9th Annual Meeting 25-29 May 1999, Leipzig, Germany.

[39] A. M. Breure, L. Posthuma and M. Rutgers: Pollution-induced community tolerance; concept for quantifying effects of contaminants in the field. Abstract, SETAC 9th Annual Meeting 25-29 May 1999, Leipzig, Germany.

[40] K. W. Pontasch and J. Cairns, Jr.: Multispecies Toxicity Tests Unsing Indigenous Organisms: Predicting the Effects of Complex Effluents in Streams. Arch. Environ. Contam. Toxicol. 20 (1991), 103-112.

[41] B. C. Chessman: Rapid assessment of rivers using macroinvertebrates: A procedure based on habitat-specific sampling, family level identification and a biotic index. Austr. J. Ecology 20 (1995), 122-129.

[42] V. H. Resh, R. H. Norris and M. T. Barbour: Design and implementation of rapid assessment approaches for water resource monitoring using benthic macroinvertebrates. Austr. J. Ecology 20 (1995), 108-121.

[43] R. H. Norris and K. R. Norris: The need for biological assessment of water quality: Australian perspective. Austr. J. Ecology 20 (1995), 1-6.

Environmental Science Forum Vol.96 (1999) pp. 13-32
© 1999 Trans Tech Publications, Switzerland

Design of a National Programme for Monitoring and Assessing the Health of Aquatic Ecosystems, with Specific Reference to the South African River Health Programme

D.J. Roux

Division of Water, Environment and Forestry Technology, CSIR,
P.O. Box 395, Pretoria 0001, South Africa

Keywords: Biomonitoring, Ecological Integrity, Environmental Assessment, Monitoring Programme Design, South African River Health Programme

ABSTRACT

Globally environmental policies increasingly emphasise the need to protect, rather than to use, the ability of ecosystems to recover from disturbances. This necessitates the adoption of response measurements to quantify ecological condition and monitor ecological change. Response monitoring focuses on properties that are essential to the sustainability of the ecosystem. These monitoring tools can be used to establish natural ranges of ecological change within ecosystems, as well as to quantify conceptually acceptable and unacceptable ranges of change.

Following a world-wide trend, a monitoring programme is being developed for assessing the ecological condition of rivers in South Africa. The approach followed for the design of the River Health Programme (RHP) consists of several phases. The main considerations that influence the design of this programme, as well as the rationale that led to its initiation, are discussed in this paper.

INTRODUCTION

Water is the basic resource upon which society relies for the quality of its life, including its health and recreation. It is also the primary resource upon which social and economic developments are based and sustained [1]. Aquatic ecosystems must, therefore, be effectively protected and managed to ensure that they retain their inherent vitality and remain fit for domestic, industrial, agricultural and recreational uses, for present and future generations. However, effective decision-making, and thus resource management, are entirely dependent on the information provided by appropriate and proper resource monitoring. Therefore, the development and application of monitoring techniques play a critical role in the ongoing process of harmonising economic development, human welfare and environmental protection.

Traditionally, information gathered to assist the management of water resources was predominantly non-ecological in nature. Monitoring actions focused largely on chemical and physical water quality variables, and regulatory efforts were aimed at controlling individual physico-chemical stressors. The presumption was that measurable improvements in water quality would result in an improvement in ecosystem condition.

However, the measurement of only physical and chemical water quality variables cannot provide an accurate account of the overall condition of an aquatic ecosystem. Chemical monitoring alone is insufficient to detect, for example, the cumulative and/or synergistic effects on aquatic ecosystems resulting from multiple stressors [2]. Many factors other than chemical water quality may have an influence on the ecological state of an ecosystem. Some examples include habitat alteration, creation of barriers that alter stream flow, water abstraction and the introduction of exotic species. Effective management of aquatic ecosystems must therefore address the cumulative effect of all these changes.

A recent development worldwide is the introduction of in-stream biological effects or response monitoring in water resources management. This type of response monitoring, commonly referred to as biomonitoring, is increasingly being recognised as an important component in the overall monitoring and assessment of water resources. The use of biological field assessments of, for example, fish or macroinvertebrate communities, provides an integrated and sensitive measurement of environmental problems and represents progress in the assessment of ecological impacts, and hence in the management of water resources [e.g. 3, 4].

In South Africa the historical lack of ecological indicators in monitoring activities is at present being addressed through the design of the River Health Programme (RHP), as a sub-programme of a proposed National Aquatic Ecosystem Biomonitoring Programme [5]. Several local advances in applied aquatic science provide a basis for integrating *in situ* biological assessment into the country's surface water monitoring and assessment strategy. These advances include the development and standardisation of rapid bio-assessment techniques and the delineation of homogeneous ecological regions, which provides a spatial framework for selecting reference and monitoring sites within the biomonitoring context.

This paper provides an overview of how biomonitoring can be structured into a monitoring design that supports water resources management. The paper draws mainly from the South African experience and the results of a test application of the South African RHP are presented.

MANAGING AQUATIC ECOSYSTEMS

An ecological system or ecosystem can be defined as "any unit that includes all of the organisms (i.e. the community) in a given area interacting with the physical environment so that a flow of energy leads to clearly defined trophic structure, biotic diversity and material cycles (i.e. exchange of material between living and non-living parts) within the system" [6]. Ecosystems thus include the physical and chemical (abiotic) environments in addition to biological components. Aquatic ecosystems are those environments that provide a medium for habitation by aquatic organisms and sustain aquatic ecological processes. These ecosystems also provide drinking water for wildlife and water for maintaining riparian biota and processes.

Social, economic and ecological factors must be considered in their inter-related nature when managing aquatic ecosystems. The social element includes the concepts of beauty, value, history and relevance. These concepts must be defined by the beholder, and are derived from cultural norms and expectations as they relate to natural systems [7]. The economic element includes aspects such as resource use, manufacturing, distribution and consumption [8]. The ecological element of an ecosystem includes factors such as species distribution and abundance, the structure, stability and productivity of ecosystems and the ability of ecosystems to self-organise and evolve.

The objective of the South African Department of Water Affairs and Forestry (DWAF), as the mandated authority responsible for managing the country's water resources, is "to manage the quantity, quality and reliability of the nation's water resources in such a way as to achieve optimum, long-term, environmentally sustainable social and economic benefit for society from their use" [9]. This objective incorporates all three (social, economic and ecological) elements of ecosystems (Fig. 1). The social element is dependent on the ecological element, and the economic element is dependent on the social and ecological elements [8]. Following from this, the goals of societies must reflect the constraints and boundaries inherent to natural ecosystems. Thus, resource management should, as a first priority, not focus on how the resource can be used, but on the ecological state in which the resource should be maintained and how it should be protected to allow sustainable utilisation. It follows that integrated ecosystem management requires a proactive planning approach in which ecological well being is the governing factor and the permissible level of economic activity is the dependent variable (Fig. 1) [10].

Just as human value judgements are an integral part of assessing health, ecosystem health is based on perception and individual judgements rather than universally accepted measurements. In practice there is a need to define, then quantify what people expect and government does about ecosystem health. However, a composite indexing system for measuring **ecosystem health** is, at the current level of ecosystem science, not available. The ecosystem concept is, therefore, often broken down to its three basic components (ecological, social and economic) for separate measurement and evaluation.

In order to find a balance that will sustain ecosystem health, management decisions regarding ecosystem health rely upon expert input from each of the three ecosystem components. The collection of appropriate and adequate data, of dependable quality, is essential to generate the kinds of information that will effectively guide decision-making in the ecosystem arena. The basis, adopted by DWAF, for measuring and assessing the ecological component of aquatic ecosystems, is ecological integrity.

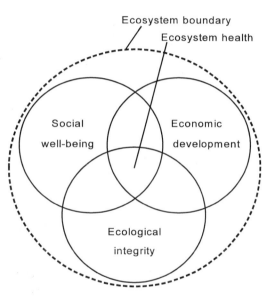

Figure 1:   The   inter-relatedness   of   the ecological,   social   and   economic elements of an ecosystem.

INDICATORS FOR MEASURING ECOLOGICAL INTEGRITY

What is ecological integrity?

Integrity generally refers to a condition of being unimpaired, i.e. corresponding with an original condition. **Biological integrity** has been defined as the ability of an aquatic ecosystem to support and maintain a balanced, integrated, adaptive community of organisms having a species composition, diversity and functional organisation comparable to that of the natural habitats within a region [11].

Similarly, **habitat integrity** has been defined as the existence of a balanced, integrated composition of physico-chemical and habitat characteristics on a temporal and spatial scale that are comparable to the characteristics of natural habitats within a region [12]. Essentially the habitat (physical and chemical) integrity of a river provides the template for a certain level of biological integrity to be realised. It follows that habitat integrity and biological integrity together constitutes ecological integrity.

In terms of the above definitions, **ecological integrity** of a river can be defined as the ability of the river to support and maintain a balanced, integrated composition of physico-chemical habitat characteristics, as well as biotic components, on a temporal and spatial scale, that are comparable to the natural characteristics of ecosystems of the region.

Ecological indicators

Ecological indicators are characteristics of the environment, both biotic and abiotic, that can provide quantitative information on the condition of ecological resources [13]. Such indicators can be used to measure and quantify ecological changes in an ecosystem.

There are five major classes of environmental factors that may affect the ecological condition or integrity of aquatic ecosystems. These are chemical variables, flow regime, habitat structure, biotic interactions and energy source [4]. Alterations to the physical, chemical or biological processes associated with these factors can adversely affect the ecological integrity of the water body. Fig. 2 illustrates how the alteration of the dynamic character of any of these factors, as a result of natural events or anthropogenic activities, can have an impact on the ecological integrity of an aquatic ecosystem. Therefore, a suite of indicators ideally needs to be considered in the assessment of overall ecological integrity.

Because of resource realities, it is impossible to measure and monitor all possible contributors to overall ecological integrity. Efforts to assess ecological integrity thus need to focus on indicators that will identify perturbations in an integrated manner. Since resident aquatic communities integrate and reflect the effects of chemical and physical impacts, occurring over extended periods of time, they are regarded as good indicators of overall ecological integrity. The in-stream biological condition of a river ecosystem is, for example, determined by the nature of geomorphological characteristics, hydrological and hydraulic regimes, chemical and physical water quality, riparian vegetation and other factors. Employing such a broad-based monitoring approach on a national scale is more likely to be cost-effective and also provide the pertinent ecological information to water resource managers.

When designing a monitoring programme, attention should be given to aquatic community components that are representative of the larger ecosystem and are practical to measure. In determining the taxonomic group(s) appropriate for a particular biomonitoring situation, the advantages of each group must be considered along with the objectives of the programme. The taxonomic groups may also vary depending on the type of aquatic ecosystem being assessed. For example, benthic macro-invertebrates and fish are often used as taxonomic groups to assess flowing waters, while plants are used in wetlands and algae and zooplankton in lakes and estuaries. The design of a biomonitoring programme should be tailored for the particular type of water-bodies assessed (e.g. wetland, lake, stream, river or estuary) [14].

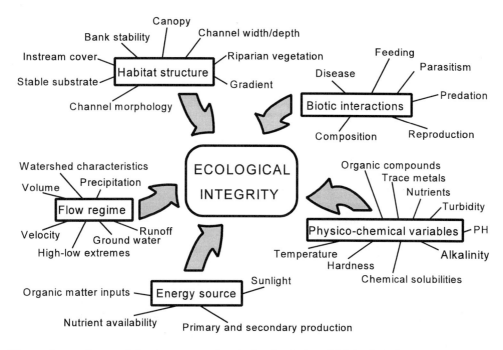

Figure 2:   Some of the important chemical, physical and biological factors that influence ecological integrity (modified from [4]).

The above rationale for focussing on biological indicators does not mean that other ecological indicators should be ignored. Information derived from non-biological indicators often support interpretation of biological results. Furthermore, protecting ecological integrity requires the monitoring and protection of the physical and chemical habitats that shape the structural and functional attributes of biota. For this purpose, qualitative and quantitative information on habitat characteristics is required.

Ecological indices

For the purpose of disseminating results of a monitoring programme, the information resulting from measuring ecological indicators should be simplified to a point where it can be of use to resource managers, conservationists and the general public. This can be done with an ecological index which integrates and summarises ecological data within a particular indicator group. Ecological indices are used to quantify the condition of aquatic ecosystems, and the output format of the resulting information is usually numeric. Appropriate indicators, for example selected fish community attributes, need to be tested and justified, and linked to measuring units (metrics) that can be used to index ecological condition.

ASSESSING THE ECOLOGICAL STATE OF AQUATIC ECOSYSTEMS

Measurement versus assessment end-points

Ecological indicators can be used to **measure** changes in ecosystems, and these measurements can in turn be used to **assess** the implications (or consequences) and acceptability of such change. For the purpose of designing a monitoring programme it is, therefore, important to distinguish between measurement end-points and assessment end-points, where:

- A **measurement** end-point is the result of an actual measurement of some ecological response to a stressor(s). Measurement end-points can be seen as characteristics of an ecological indicator, for example the mortality of a fish population, that may be affected by exposure to a stressor [15]. The values generated through indices of water quality are further examples of measurement end-points.

- An **assessment** end-point is the result of an interpretation (assessment) of measured data, often in conjunction with other related data. Assessment end-points are explicit expressions of an actual environmental value which bears direct relation to the management of ecological resources. An example is where measured indicator values for impacted and unimpacted sites are assessed to express the degree and/or acceptability of impairment at the impacted site.

Environmental assessment provides a synthesis and interpretation of scientific information, and can often be linked with policy or regulatory questions and issues. Environmental assessment is usually characterised by a value added perspective, ranging from a formal, quantitative cost/benefit analysis of all alternatives to a qualitative improvement in our understanding of potential impacts or effects [13].

The measurement and assessment concepts have important implications for a monitoring programme, which must:

- reflect and describe the relationship between measurement and assessment end-points,
- describe in sufficient detail the assessment process so that different people using the same measured information will consistently arrive at the same assessment, and
- recognise that for the purpose of management decision-making, information has to be reported in the form of assessment rather than measurement end-points [13].

Area-specific benchmarks for assessment

Ecosystems are naturally dynamic, and their evolutionary histories and capabilities are never static in either structure or function. For example, hydrological regimes include variability on many time scales, and include not only the "normal" range of conditions at a site, but also the "extremes" of floods and other infrequent conditions. From an ecological point of view there is, however, nothing abnormal about these extremes. These occurrences are a natural and often crucial part of ecosystem dynamics, especially over the long-term.

When interpreting or assessing the results from an ecosystem monitoring programme, the challenge lies in distinguishing between natural and unnatural ranges of change in measured ecological values. Managers will benefit from the knowledge that an ecosystem is responding in some way that is outside its natural range of variation. This would allow remedial steps to be taken before such change becomes permanent. One way of distinguishing between natural and unnatural ranges of variation, is to establish a "natural" benchmark or reference condition with which similar monitoring sites can be compared.

In general, quality assessment requires a procedure for comparing the state of an ecosystem with a reference condition. This means that both the state of the ecosystem to be assessed and the reference conditions have to be made explicit [16].

In South Africa, establishing reference conditions is complicated by a large range of ecosystem types. The variability among natural surface waters, resulting from vast climatic, landform, land cover (vegetation), soil type and other geographic differences, favours the use of area-specific reference conditions rather than national reference conditions. Such area-specific reference conditions should describe, within the relevant geographic area, the characteristics of river segments least impaired by human activities in order to define attainable biological or habitat conditions [14]. The development of area-specific reference conditions will allow environmental conditions at any site(s) under investigation to be compared with conditions found or expected in undisturbed streams or rivers, of similar size and habitat type, and located in the same area.

As completely undisturbed environments are virtually nonexistent, and even remote waters are impacted by factors such as atmospheric pollution, "minimally impacted" sites have been used (for example in the United States) to define the "best attainable reference condition" [17]. However, care should be taken in cases where the best sites in a specific area are already considerably modified. In such cases expert knowledge and extrapolation techniques may be required to construct a hypothetical "best attainable" condition, which can be used as an area-specific reference.

Once appropriate reference conditions have been set for a particular area, standardised measurements of ecological integrity can be used and the resulting data can be compared against these reference conditions. Fig. 3 shows how the results obtained at reference sites can be used to calibrate biological indices. Whereas the reference condition represents the top end of such a calibrated scale, an almost sterile system will represent the lowest possible state [8]. An area-specific calibration of ecological state will enable the assessment of the current ecological state of any site or reach within that area. The current state for a particular site can be anywhere between the reference condition (100%) and the lowest possible state (0%).

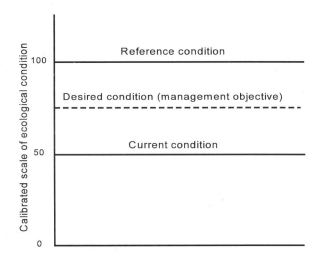

Figure 3:   A conceptual model for assessing the ecological condition of an aquatic ecosystem.

Setting Resource Quality Objectives (RQOs)

The availability of quantitative information on the reference biological integrity as well as the current biological integrity of a river will contribute towards setting realistic and ecologically sound resource management goals. A third critical condition in goal setting is the future condition that the various stakeholders desire for the river; this would typically include an assessment of the social-cultural, economic and ecological importance of the resource. Once consensus is reached on a management goal for a particular river, and if this goal can be expressed in terms of a specific integrity parameters, then measurable Resource Quality Objectives (RQOs) can be allocated per ecological indicator group. In other words, the range of index scores coinciding with the desired integrity class, for each biological indicator group, become measurable and auditable RQOs. However, making choices about the RQOs entails more than the assessment of measured data, and requires input from all stakeholders [e.g. 18].

It is clear that the monitoring of ecological responses can be used to indicate the effects of changing ecological conditions. However, assessment of the monitored data is required to determine the significance of such change in terms of (a) the degree of deviation from the hypothetical "natural" ecological condition, or (b) an ecological management objective. The objective must, in turn, reflect sustainable levels of ecosystem structure, function and processes as well as the expectations of stakeholders.

THE SOUTH AFRICAN RIVER HEALTH PROGRAMME

Existing international programmes

Aquatic biomonitoring programmes are developed for various purposes, including the following:
*   surveillance of the general ecological state of aquatic ecosystems;
*   assessment of impacts (before and after an impact or upstream and downstream of an impact, both for diffuse and point-source impacts);
*   audit of compliance with ecological objectives or regulatory standards; and
*   detection of long-term trends in the environment as a result of any number of perturbations.

National approaches to the design and implementation of aquatic biomonitoring programmes have been followed over the world. The most noteworthy of existing programmes are:
*   the British River Invertebrate Prediction and Classification (RIVPACS) methodology [19];
*   the Australian National River Health Programme [20]; and
*   the Rapid Bioassessment Protocols For Use in Streams and Rivers of the United States [21].

The programme that is being designed for South Africa incorporates appropriate concepts from these international models, yet is tailored to reflect the environmental conditions and resource realities specific to the country.

RHP design process

A monitoring programme is usually developed in response to a need for information. The programme design *per se* will, however, not provide the required information. The design needs to be implemented, and the programme must be maintained and modified through ongoing learning, to match our evolving information needs. The design will consist of tools, protocols and methodologies which will be needed in the implementation, and which will make the programme functional. Furthermore, the selection of these tools will be guided by an overall vision and the specific objectives of the programme. Finally, when the programme has been designed, many individuals and organisations may play a role in turning the design into an operational programme which will produce the information for which it has been designed.

A phased approach was adopted for the design of the RHP, to facilitate the formulation of a design framework, the conceptual development and testing, demonstration and eventual full-scale implementation of the programme [22]. The main design phases are shown in Fig. 4.

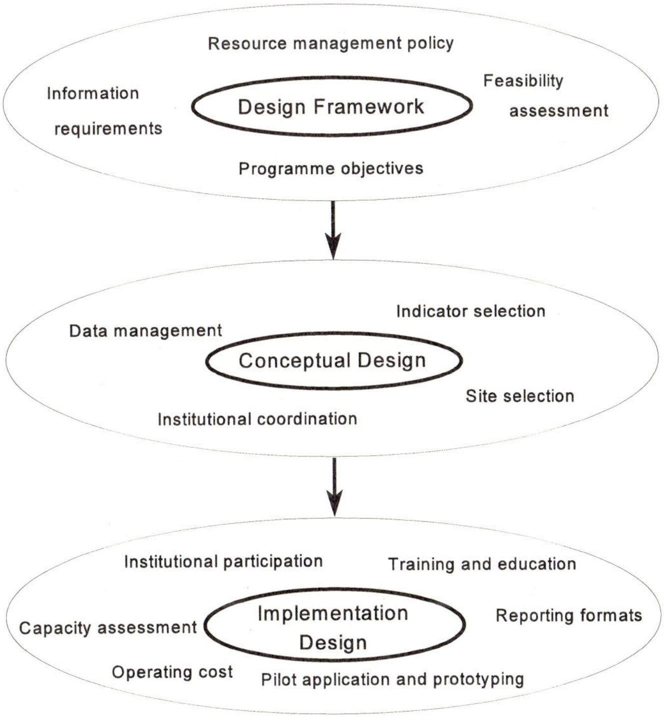

Design framework formulation

The RHP framework was based on two issues, namely:

- the qualitative and quantitative information requirements related to the management of aquatic ecosystems, as expressed by aquatic resource managers, and
- the ability of a national-scale biomonitoring programme, at the current level of scientific development, to deliver the required information.

Figure 4:   Phased design of the River Health Programme.

An important outcome of the framework design was a definition of the objectives of the programme as well as the scope and specifications to which the rest of the design phases must adhere. The design specifications can be summarised as follows [5]:

- The RHP is being designed primarily as a *management information system*. The approach of designing programmes as management information systems recognises that the ultimate purpose of a monitoring programme is to produce information for a specific objective. In the case of the RHP the information must support the management of water resources.
- It is unlikely that one biomonitoring programme will meet all information needs, for example as expressed by resource managers with national, provincial, catchment or local interests and

responsibilities. As the RHP is required to provide information on a national level, its design must be specified accordingly. As such, it was agreed by managers that the primary focus of the RHP should be on the **state of health** of aquatic ecosystems.

•   Although the monitoring focus of the RHP will be on biological indicators, relevant non-biological indicators should also be incorporated to have a suite of ecological indicators for assessing aquatic ecosystems.

•   Models for coordination and co-participation among relevant organisations will have to be investigated. It will be necessary to pool and optimise available resources and capabilities in order to successfully implement and maintain a national programme of the complexity and specialised nature of the RHP.

The main objectives of the RHP are to (1) measure, assess and report on the ecological state of riverine ecosystems, (2) detect and report on spatial and temporal trends in the ecological state of riverine ecosystems, and (3) identify and report on emerging problems regarding the ecological state of riverine ecosystems in South Africa. Each of these objectives are discussed in more detail below [from 5].

Ecological state reporting
The level of information which could be reported is determined by the breadth and detail of the data that are collected. "Breadth" depends on the number of ecosystem processes and components (indicators) that are included in the data. "Detail" refers to the degree to which each ecosystem indicator is measured and analysed. The more detailed the available data, the better the insight that can be obtained about the functioning of the ecosystem, i.e. the interrelations among ecological components as well as their relationship to stressors [16].

Current ecological knowledge does not allow for obtaining a complete picture of ecosystem components and all the processes associated with them. Therefore, a compromise has to be made between the breadth of the information and the degree of detail. Breadth is often at the expense of detail. A broad approach can be sensitive to all kinds of stressors, however, subtle responses may not be detected. Similarly, detail is at the expense of breadth. Although diagnostic capacity depends on the detail of information, the evaluation may become too narrowly focused, with an increasing risk that important effects on other ecosystem components can be overlooked. Also, there comes a point at which too much detail can cloud the issue or make analysis unnecessarily complex [16].

Ultimately, the breadth and detail of monitoring specifications need to be tailored according to resource realities. On a national scale, the programme will be designed to measure and assess the general state and annual changes over river reaches, rather than to provide day-to-day operational answers or for measuring exact river conditions at any specific site.

Trend detection
Natural ecological variation will complicate direct comparison of monitoring results between sites. However, through the development of a spatial classification scheme, it is hoped that geographical areas could be delineated within which it is valid to compare data from different sites. Normalising the conditions at each site, relative to a reference condition for the particular geographical area, will allow direct comparison and the detection of spatial trends in the ecological state among sites.

Once the programme has been in existence for a few years, the detection of temporal trends should provide an ability to quantify changes (whether an improvement or deterioration), or to qualitatively predict ecosystem degradation.

Problem identification

The data collected through a national programme are unlikely to be sufficient to establish causal relationships with a high level of confidence, i.e. specific detail on impairment due to habitat degradation, hydrological alteration or chemical water quality deterioration. Therefore, to address questions related to emerging problems, the national programme needs to feed into regional or site-specific bio-assessment initiatives, tailored for the particular problem experienced. More detailed and frequent monitoring can be instituted to provide answers to specific questions as part of such specific studies (Fig. 5). An example of such a question may be the extent to which the quality of an effluent discharge must be improved in order to achieve a specified in-stream ecological objective.

Whereas national ecological indicator surveys should allow the detection of unacceptable change, regional detailed surveys would be required to link, with a significant level of confidence, specific causes to the change. National assessment would thus allow limited resources to prioritise regional activities and create focus on specific problem areas. Although regional biomonitoring activities will not be addressed as part of the national programme design, regional bio-assessment will be essential to complement the national information and hence to optimise decision-making competence. Provision must, therefore, be made for linking national and regional bio-assessment programmes.

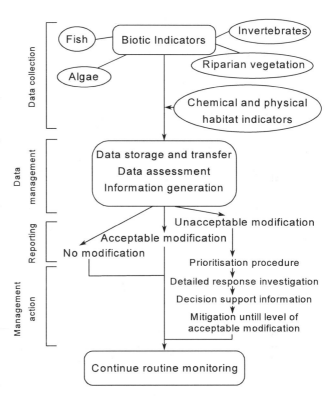

Figure 5: Flow diagram indicating the different components of a biomonitoring programme and how the results of the programme may influence management actions.

Conceptual programme design

During the conceptual design phase, aspects which were addressed included the development of:
- a spatial classification scheme, which would allow the delineation of geographical areas within which it would be valid to compare biological data from different sites [23];
- a protocol for selecting reference and monitoring sites to support state-of-the-environment (SOE) reporting [24];

- protocols for selecting and using biological and other ecological indicators to measure the health of river systems [25];
- procedures for the transfer, storage and retrieval of data resulting from the RHP [5];
- mechanisms and structures for institutional coordination, which is essential for the long-term maintenance of any national environmental monitoring programme [5].

Spatial classification scheme

A regional approach to defining reference conditions has been proposed for South Africa. Following on the outcome of a National Workshop, additional work led to the development of a three-tiered hierarchical classification scheme, as follows [from 23]:

Level 1 - Bioregional Classification: a modification of an existing biogeographic classification based on the broad historical distribution patterns of riverine macroinvertebrates, fish and riparian vegetation. This level produced 18 bioregions for South Africa.

Level 2 - Sub-regional Classification: based on patterns of river zonation within bioregions. It was envisaged that this would reflect agreement between broad geomorphological characteristics (e.g. landform, lithology, soils, hydrology, climate, basin relief, river profile morphology) and distribution patterns of fish, macroinvertebrates and, to a lesser extent, riparian vegetation.

Level 3 - River Types: which were to account for variation between rivers within a sub-region. It was envisaged that this level of the hierarchy would account for differences in factors such as river size, hydrological pattern (e.g. perennial or intermittent flow) and other geomorphological or chemical characteristics.

Reference and monitoring sites selection protocol

The RHP is being designed to allow comparison between reference and monitoring sites, where:

- **Reference sites** are relatively unimpacted sites that can be used to define the best physical habitat, water quality and biological parameters for each kind of river.
- **Monitoring sites** are commonly those sites identified as important in assessing the condition of a river or reach known or thought to be experiencing an impact on water quality or habitat degradation. In the case of SOE reporting, however, monitoring sites are randomly selected impacted or unimpacted sites that reveal the range of conditions in their types of rivers.

A detailed discussion of criteria for the selection of reference sites and a proposed protocol for selecting reference sites has been published [24].

Indicator selection protocols

A key element in the design of a biomonitoring programme is the decision as to which indicators to measure, and which indices to select to represent these indicators. While biological indicators are the main focus of the RHP, the inclusion of physical and chemical indicators will substantially increase the long-term information value of the programme.

The biological indicators most commonly used in biomonitoring are aquatic macroinvertebrates and fish. Riparian vegetation serves to link the in-stream aquatic ecosystem to the surrounding terrestrial ecosystem which, in turn, influences river processes and patterns. Although this component of aquatic ecosystems is often overlooked, riparian vegetation is considered a vital element in determining the state of aquatic ecosystems.

The state of riverine biota is a reflection of the chemical and physical habitat conditions in that river. In order to interpret the meaning of the biological index values accurately, it is necessary to gather information about the chemical and physical environment of the river. These physical and chemical indicators provide a framework within which to interpret the biological results. As an example, the community of fish or invertebrates will be very different in a river with high habitat diversity compared to one with low habitat diversity, or before and after a prolonged drought.

The non-biological indicators considered to provide the most comprehensive support framework for the interpretation of biological data are aspects of physical habitat, hydrology, water chemistry and geomorphology.

To accommodate a range of regional requirements, capabilities and the availability of resources in the implementation of the RHP, five alternative biomonitoring protocols (BPs) are being proposed (BP1 to BP5). The options range from the use of a single biological index and an associated habitat index at a site, to the use of several biological and non-biological indices. The latter option provides a comprehensive assessment of the state of the riverine communities and their environmental conditions.

Once a biomonitoring initiative has started, it would be possible to scale up or down on the BP adopted for a certain catchment, province or the country. Such a decision would depend on the resolution of information required, available resources and expertise, and the possible prioritisation of particular rivers or sites.

Data management procedures
Options for data capturing include filing and distributing hard-copy data sheets, updating and maintaining local databases, and sharing a centralised database. The last option is preferred for its long-term data security and accessibility advantages. However, the feasibility of this option would depend on a uniform data structure and reliable high-speed data links. Both of these qualifying aspects are currently receiving attention from the DWAF.

Information derived from the RHP will potentially be utilised by a very wide spectrum of users. To focus the communication of information, these audiences can broadly be divided into three levels, namely political or administrative, operational and grassroots levels.

To a large extent the success of the RHP will be determined by the effectiveness of communicating results to the different target audiences. While raw data or index values may be sufficient for the specialists familiar with interpreting biological results, these formats may be meaningless to anyone who does not have an understanding of the derivation of the index, how it reflects deviation from natural or best attainable conditions, or how to correctly interpret the value of the index. For such audiences the information reflected in the index may need to be reported in simple graphical formats.

More detailed or more generalised presentations can be made of the same data, according to the preference of the target audience. The critical factor is that the source data must be reliable, and based on scientifically acceptable and standardised collection protocols. The data assessment process must also be described in sufficient detail so that different people using the same measured data will consistently arrive at the same assessment information.

Institutional structures and coordination
The design of the RHP started as a national initiative driven by the DWAF. However, the DWAF does not have the required regional infrastructure and resources to implement and maintain the programme on a national basis. Also, the geographic framework for decision-making in South Africa is moving from the national scale down to provincial and more local scales. The information generated from the RHP will also provide decision-making support for the resource managers with a more local interest.

Therefore, to ensure the successful implementation and long-term maintenance of the programme, it will be necessary to involve regional stakeholders. As such, a model of national coordination (quality control, standardisation of procedures, etc.) and regional implementation and maintenance (data collection and ownership, coordination among regional stake holders, etc.) is currently being pursued.

Implementation of the RHP would thus require the development, within the broad RHP Implementation Design, and establishment of procedures which cater for coordination among and the specific needs of the national, provincial and regional stakeholders. Apart from technical and coordination issues, the implementation phase also needs to address the vast educational needs, capacity building, reporting formats for key audiences and funding opportunities and options in order to ensure long-term maintenance of the monitoring programme

For the purpose of implementing the RHP, it might be necessary to identify "Regional Lead Agencies". Such lead agencies may include Provincial Governments, water boards, university departments or consulting firms. While keeping within the broad design framework, each lead agency may decide independently on site selection, who would do the monitoring and be responsible for data transfer to a central body. Lead agencies may take on their work themselves, or appoint research groups with appropriate expertise.

After implementation, the role of scientists in maintaining the programme will remain significant. Although the physical aspects of conducting biomonitoring need not be done by specialists, it is essential that specialists be responsible for the interpretation of collected data. However, involvement of administrators, regulators and engineers, who must translate reported results into real world, everyday activities such as resource management and development, will increase.

Testing and implementation phase

The implementation design phase is about matching the ideals of the conceptual design with the realities of the real world, in order to create a feasible platform for implementing and maintaining the RHP. During the conceptual design phase developments are largely theoretical and substantial testing, modification, demonstration and integration are still required to mould all the concepts into an operational programme. Pilot testing also allows small-scale demonstration of the programme through reporting the generated information to key target audiences. Additional research and developmental needs will also be identified during this phase.

The final implementation design must provide the information required to implement and maintain the programme successfully. It must, therefore, address aspects such as start-up cost, operating cost, human resource requirements, training needs, institutional participation, equipment needs and maintenance requirements. As the programme is intended for national application, survey methods should not be too resource intensive, nor so complex that only specialists can conduct monitoring. Similarly, site selection should balance the realities of resource availability with obtaining sufficient data to comply with the objectives of the programme.

THE ELANDS RIVER, A CASE STUDY

For some of the indicators proposed for the RHP, indices have been developed and applied in South Africa. For the majority of the indicators, however, indices are in the early phases of conceptualisation and still need practical development and testing [5].

An approach of applying the latest developmental prototypes of the above technologies, in the context of case studies, is being followed. By doing so, a high degree of alignment and synergy between programme components can be encouraged. This also applies to a systematic and adaptive procedure for linking monitoring, assessment and management outputs. This case study demonstrates a prototype framework for linking biological response data, as generated by the RHP, through a systematic approach to river management. The relevant concepts are broadly demonstrated with the aid of the case study.

Study area and biological survey

The results of fish and invertebrate sampling on the Elands River, Mpumalanga, were assessed by means of a prototype Fish Community Integrity Index (FCII) [26] and the fourth version of the South African Scoring System (SASS4) [27], respectively. The sampling surveys took place during the second half of 1996.

The Elands River (Fig. 6) was divided into reaches, based on physical characteristics which determine habitat suitability for fish. These reaches were refined by checking them against historical fish distribution patterns. In other words, each reach represents a segment of the river in which the fish community would, under unimpaired conditions, remain generally homogenous due to the relative uniform nature of the physical habitat [26]. Each fish-based reach was assumed also to support a homogenous community of benthic invertebrates. Based on site suitability, accessibility and representativeness, surveys were conducted at between three and five sampling sites per reach.

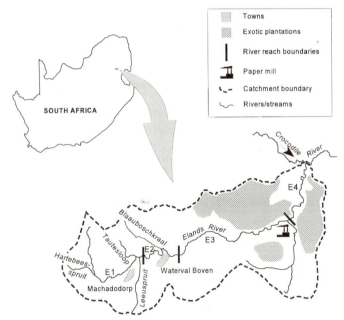

Figure 6: Location and characteristics of the study area.

Assessment framework

Monitoring results were assessed by comparing the collected data against reference conditions derived for each reach. These reference conditions approximate the best attainable biological condition for a reach, in the absence of impact from human activities. A combination of historical data and expert opinion for fish, and relatively unimpaired reference sites and expert opinion for invertebrates, was used to define the reference conditions.

To provide a management perspective to the assessment of the monitoring data, a provisional River Integrity Classification Scheme (RICS) was followed. This is shown in Table 1.

Table 1.   The provisional River Integrity Classification Scheme (RICS).

River Integrity Class	Biological community characteristics (FCII and SASS4)	% of reference index score
Class A: Unmodified	Community characteristics approximate natural conditions.	>85%
Class B: Moderately Modified	Moderate change to community characteristics (lower abundances and possible loss of some intolerant species); basic ecosystem functions remain predominantly unchanged.	61-85%
Class C: Considerably Modified	Considerable modification of community characteristics and basic ecosystem functions have occurred; several intolerant species have been lost or occur only in low numbers.	40-60%
Class D: Severely Modified	Community characteristics have been seriously modified with an extensive loss of basic ecosystem functions; tendency towards domination by a few tolerant species.	<40%

Table 2 indicates the outcome when the results of the biological survey were applied to the RICS.

Table 2.   The current biological integrity classes for each river reach, according to assessment of fish and invertebrate communities.

River reach (altitude in m)	River/Stream	Biological condition class	
		FCII	SASS4
E1 (>1500)	Tautesloop	A	A
	Hartbeesspruit	B	A
	Leeuspruit	D	A
E2 (1200-1500)	Elands River	A	B
	Blaauboschkraal	B	A
E3 (900-1200)	Elands River	A	A
E4 (800-900)	Elands River	A	A

Resource quality objectives and management actions

An exercise involving stakeholders to set management goals for the Elands River has not yet been undertaken. However, to take this exercise further, we assume the following hypothetical desired conditions: a) No indicator group should deteriorate from its current integrity class and b) the whole of the Elands River should at least be maintained at Integrity Class B - given the social (e.g. recreational trout angling in upper reaches) and economic (e.g. trout aquaculture and forestry) importance of the Elands River catchment, a goal of Class A for the entire river may be unrealistic and impossible to achieve. According to the above rules for goal setting, it is only the Leeuspruit in reach E1 for which the fish community needs to improve from Integrity Class D to Integrity Class B. The current low integrity class according to the FCII is associated with the presence of exotic black bass in the Leeuspruit.

The RICS will provide the range of index values for the FCII in order to comply with a goal of Class B. Based on expert and system-specific knowledge, management options could be suggested for improving the fish community characteristics accordingly. An example of a management option for the Leeuspruit is the removal of exotic black bass and/or trout and the reintroduction of the appropriate indigenous fish species. Various management options could be rated on the basis of their political and technical feasibility and perceived efficacy [e.g. 28], in order to prioritise and guide management action.

CONCLUDING REMARKS

The RHP is not intended to replace any water quality monitoring approach, but rather to expand on the approaches currently in use. Implementation of the RHP would, however, require a substantial broadening of the traditional water quality monitoring and assessment focus. It would require a far more integrated collection and analysis of data, as well as the assessment of new types and combinations of data.

In essence, biomonitoring is a scientific procedure, which can be used to provide resource information. The principal role of this monitoring information is to drive and direct the processes of decision making and management. These processes include:
* assessing information and identifying problems;
* drafting regional and national policies, regulations and eventually legislation;
* establishing criteria, standards and management objectives for combatting deteriorating environmental conditions, and
* demonstrating results in the environment [29].

In broad terms, the RHP has to contribute to science-based management of aquatic ecosystems, in support of national and regional mandates to manage the water resources of South Africa. If the nation's rivers (and estuaries, seas and impoundments) are maintained at an appropriate level of ecological integrity, then the efforts of resource managers will have been successful. The information generated by the RHP will assist in identifying those areas where water resource managers have been successful and those areas where they need to focus their attention. Through the monitoring of structural and functional attributes of ecosystems, the RHP will also provide an information base for managing the chemical, physical and biological processes that shape ecological integrity.

A systematic process which involves the collection and assessment of biological data, setting goals and quantifiable objectives for managing the biological integrity of rivers, predicting how various management options will affect components of the ecosystem, and monitoring responses to the chosen management actions, will close the loop between monitoring, assessment and management. By following this iterative cycle and improving the individual components, the balance between water resource protection and utilisation can be optimised.

ACKNOWLEDGEMENTS

The Institute for Water Quality Studies of the Department of Water Affairs and Forestry initiated and is funding the design of the RHP. The case study results were obtained as part of a study supported by the Water Research Commission.

REFERENCES

[1]   Department of Water Affairs and Forestry (1994): Water Supply and Sanitation Policy: Water - an Indivisible Asset. White Paper of the Department of Water Affairs and Forestry, Pretoria, South Africa.

[2]   S. Jackson and W. Davis (1994): Meeting the goal of biological integrity in water-resource programs in the US Environmental Protection Agency. Journal of the North American Benthological Society, 13, 592-597.

[3]   J.R. Karr, K.D. Fausch, P.L. Angermeier, P.R. Yant and I.J. Schlosser (1986): Assessing biological integrity in running waters - a method and its rationale. Illinois Natural History Survey, Special Publication 5. Champaign, Illinois.

[4]   J.R. Karr and E.W. Chu (1997): Biological Monitoring and Assessment: Using Multimetric Indexes Effectively. EPA 235-R97-001. University of Washington, Seattle.

[5]   D.J. Roux (1997): National Biomonitoring Programme for Riverine Ecosystems: Overview of the design process and guidelines for implementation. NAEBP Report Series No 6. Institute for Water Quality Studies, Pretoria.

[6]   E.P. Odum (1971): *Fundamentals of Ecology*. Third Edition. W.B. Saunders Co. Philadelphia.

[7]   R.J. Steedman (1994): Ecosystem health as a management goal. Journal of the North American Benthological Society, 13, 605-610.

[8]   C.K. Minns (1995): Approaches to assessing and managing cumulative ecosystem change, with the Bay of Quinte as a case study: an essay. Journal of Aquatic Ecosystem Health, 4, 1-24.

[9]   Department of Water Affairs and Forestry (1997): White Paper on a National Water Policy for South Africa, Pretoria.

[10]   C. Cocklin, S. Parker and J. Hay (1992): Notes on the cumulative environmental change I: Concepts and issues. Journal of Environmental Management, 35, 31-49.

[11]   J.R. Karr and D.R. Dudley (1981): Ecological perspectives on water quality goals. Environmental Management, 5, 55-68.

[12]   C.J. Kleynhans (1996): A qualitative procedure for the assessment of the habitat integrity status of the Levuvhu River (Limpopo System, South Africa). Journal of Aquatic Ecosystem Health, 5, 1-14.

[13]   K.W. Thornton, G.E. Saul and D.E. Hyatt (1994): Environmental Monitoring and Assessment Program Assessment Framework. Report No. EPA/620/R-94/016. U.S. Environmental Protection Agency, Research Triangle Park, NC 27711.

[14]   U.S. Environmental Protection Agency (1990): Biological Criteria: National Programme Guidance for Surface Waters. Criteria and Standards Division, Office of Water Regulations and Standards, U S Environmental Protection Agency, Washington DC.

[15]   G.W. Suter II (1990): Endpoints for regional ecological risk assessments. Environmental Management, 14, 9-23.

[16]   M.I. NIP and H.A.U. De Haes (1995): Ecosystem approaches to environmental quality assessment. Environmental Management, 19, 135-145.

[17]   T. Oberdorff, R.M. Hughes (1992): Modification of an index of biotic integrity based on fish assemblages to characterise rivers of the Seine Basin, France. Hydrobiologia, 228, 117-130.

[18]   C. Breen, N. Quinn, and A. Deacon (1994): A Description of the Kruger Park Rivers Research Programme (Second Phase). Foundation for Research Development, Pretoria, South Africa.

[19]   J.F. Wright, M.T. Furse and P.D. Armitage (1993): RIVPACS - a technique for evaluating the biological quality of rivers in the UK. Eur. Wat. Poll. Contr., 3, 15-25.

[20]   J.M. King (1995): Enquiries regarding a national biomonitoring initiative, made during a trip to Australia in June - July 1995. Unpublished Southern Waters' report to the Institute for Water Quality Studies, Pretoria.

[21]   Barbour, M.T., Gerritsen, J., Snyder, B.D. and Stribling, J.B. (1997): Revision to Rapid Bioassessment Protocols For Use in Streams and Rivers: Periphyton, Benthic, Macroinvertebrates, and Fish. EPA 841-D-97-002, Washington, DC.

[22]   D.R. Hohls (1996): National Biomonitoring Programme for Riverine Ecosystems: Framework document for the programme. NBP Report Series No 1. Institute for Water Quality Studies, Pretoria.

[23]   C.A. Brown, S. Eekhout and J.M. King (1996): National Biomonitoring Programme for Riverine Ecosystems: Proceedings of spatial framework workshop. NBP Report Series No 2, Institute for Water Quality Studies, Pretoria.

[24]   S. Eekhout, C.A. Brown and J.M. King (1996): National Biomonitoring Programme for Riverine Ecosystems: Technical considerations and protocol for the selection of reference and monitoring sites. NBP Report Series No 3. Institute for Water Quality Studies, Pretoria.

[25]   M.C. Uys, P-A. Goetsch, and J.H. O'Keeffe (1996): National Biomonitoring Programme for Riverine Ecosystems: Ecological indicators, a review and recommendations. NBP Report Series No 4. Institute for Water Quality Studies, Pretoria.

[26]   C.J. Kleynhans (1998): The instream biological integrity of the Crocodile River, Mpumalanga, as based on the assessment of fish and invertebrate communities. Draft report of the Institute for Water Quality Studies, Pretoria.

[27]   H.F. Dallas (1997): A preliminary evaluation of aspects of SASS (South African Scoring System) for the rapid bioassessment of water quality in rivers, with particular reference to the incorporation of SASS in a national biomonitoring programme. Southern African Journal of Aquatic Sciences, 23, 79-94.

[28]   A. Haney and R.L. Power (1996): Adaptive management for sound ecosystem management. Environmental Management, 20, 879-886.

[29]   C.O. Yoder (1994): Towards improved collaboration among local, State, and Federal agencies engaged in monitoring and assessment. Journal of the North American Benthological Society, 13, 391-398.

Environmental Science Forum Vol.96 (1999) pp. 33-64
© 1999 Trans Tech Publications, Switzerland

Cellular, Histological and Biochemical Biomarkers

M. Schramm[1], A. Behrens[2], T. Braunbeck[3], H. Eckwert[4], H.-R. Köhler[4], J. Konradt[3], E. Müller[1], M. Pawert[1], J. Schwaiger[5], H. Segner[2] and R. Triebskorn[1]

[1] Department of Physiological Ecology of Animals, Zoological Institute, University of Tübingen,
Auf der Morgenstelle 28, DE-72076 Tübingen, Germany

[2] UFZ Centre of Environmental Research, Department of Chemical Ecotoxicology,
Permoserstr. 15, DE-04318 Leipzig, Germany

[3] Department of Cell Biology, Zoological Institute, University of Tübingen,
Auf der Morgenstelle 28, DE-72076 Tübingen, Germany

[4] Department of Zoology I, University of Heidelberg,
Im Neuenheimer Feld 230, DE-69120 Heidelberg, Germany

[5] Laboratory for Fish Pathology, Steinseestr. 32, DE-81671 München, Germany

Keywords: Biomarker, Monitoring, Stress Protein, Metabolic Enzymes, Biotransformation Enzymes, Cellular Pathology, Ultrastructure, Liver, Gills, Kidney, Trout, *Salmo Trutta f. Fario*, Fish

INTRODUCTION:

Biomarkers are biological responses to environmental chemicals at the "below individual" level, measured either inside an organism or in its products [1]. Furthermore, biomarkers give a measure of exposure of toxic effects, or of suspectibility [2, 3, 4]. So biomarkers represent useful tools with which to characterise the state of health of important members of e.g. the aquatic systems [5].

Biomarkers can be applied directly in the field, but also to organisms exposed to test water by "active monitoring" [6]. Chemical analyses alone often - only insufficiently - indicate water pollution. They often represent only the results obtained at a given time point, decribed by McCarthy & Shugart [7] as "a snapshot". The use of biomonitor organisms, in which biomarker reactions can be measured, have the advantage over chemical analyses, that biological reactions of organisms integrate over a long time period, as long as the environmental impact is in a sublethal range.

In biological systems, biomarker reactions take place in a sequential order of responses to pollutant stress. Effects at higher hierarchical levels (i.e. cellular, tissue, organism) are always preceeded by changes in earlier biological processes (i.e. molecular, subcellular). This relation enables the prediction of effects at higher levels on the basis of early biomarker reactions [8]. These connections of biomarkers to effects at higher organisation levels mentioned above, are very important. Several factors, e.g. respond in a dose or time-dependent manner, variability to seasons, temperature, sex, handling, have to be taken into consideration for the selection and application of biomarkers [9]. This means that biomarker reactions can potentially provide strong ecological evidence, if the biological implications of their responses can be correlated to reactions at higher levels of biological organisation (e.g. growth or reproduction). Furthermore, basic knowledge of the biochemical, physiological, and ecological constraints is required in order for biomarker-based ecotoxicological statements to be made[10].

Generally, fish are regarded as to be suitable organisms with which to monitor environmental pollution. They are located at the top of the aquatic food chain, and are known to accumulate toxicants [11, 12]. Secondly, they are in direct contact with polluted water via their gills and their body surface [6].

In the following, several of the most common biochemical and cellular biomarkers in fish (i.e. cell ultrastructure of the gill and liver, activities of metabolic enzymes, biotransformation enzymes and heat shock proteins) will be presented. These biomarkers were examined as a part of a combined ecotoxicological project of the German Ministry of Education, Science, Research and Technology (BMBF, 070TX21-25) [13, 14, 15]. This project "Valimar" aims to compare sensitivities and response patterns of a series of biomarkers, their capacity to reflect chemical and limnological data, and their suitability to indicate impact on the population level of indigenous fish species as well as on biological and structural features of the aquatic ecosystem. Two fish species, brown trout (*Salmo trutta* f. *fario*) and loach (*Barbatula barbatula*) were exemplarily chosen as test organisms which were exposed to the waters of two streams (Körsch and Krähenbach), in the south-west of Germany which are differently polluted with respect to pesticides, PAHs and heavy metals. In the following, particular emphasis is laid on the results obtained within the frame of this projekt.

Stress proteins (heat shock proteins)

First discovered in the fruit fly, *Drosophila melanogaster*, in response to increased temperature [16], a set of proteins, originally called heat shock proteins, has since been found to occur ubiquitously in different animal (and plant) tissues. The majority of these heat shock proteins are synthesized constitutively and abundantly in the absence of any stress and bind transiently and non-covalently to nascent polypeptides and unfolded or unassembled proteins to block non-productive interactions [17]. Experimental treatment with a variety of chemicals, adverse environmental conditions and response to injury and viral infections [18, 19, 20] resulting in intracellular accumulation of unfolded or malfolded polypeptides, may activate the synthesis of increased concentrations of heat shock proteins [21, 22]. The exact mechanism of induction of most stress proteins remain unclear up to the present. For the 70 kDa heat shock protein (Hsp70), a model for its regulation was proposed and supported by experimental evidence. Induction of *hsp70* gene expression involves binding of a transcription factor (Hsf 1) to promotor elements (hse) which facilitates *hsp70* transcription and lead to increased cytosolic Hsp70 concentrations [summarized in

23, 24]. Due to the broad variety of inducing agents, the term "heat shock proteins" frequently is replaced by "stress proteins" which is now also commonly used for this class of molecules.

Based on this background, the induction of these stress proteins can be used as a suitable molecular biomarker for the assessment of the toxic impact of different environmental stressors on a variety of different biota [2, 25, 26, 27]. The potential of stress proteins to be incorporated into biomonitoring studies is based on their sensitivity in induction (compared to classical LC/LD studies) and their ubiquity in nature. In recent years, research on the potential application of Hsps as biomarkers has been undertaken in the terrestrial [28, 29, 30, 31, 32] as well as in the aquatic field [26, 24]. Regarding the latter, the main body of information is available for fish [summarized in 24], although some invertebrate species, e.g. the rotifer *Brachionus plicatilis* [33] and transgenic nematodes [34, 35, 36] have been regarded as potential biomonitor species for aquatic pollution.

Current knowledge of heat shock proteins in fish recently has been reviewed by Iwama and coworkers [24]. The vast majority of studies are focused on the basic principles of Hsp induction (mostly in reponse to elevated temperature) and, thus, Iwama *et al.* finished their review only with speculations on the applicability of fish Hsps for biomonitoring. Nevertheless, some promising approaches to relate stress protein induction in fish cells to environmentally relevant parameters do exist which will be discussed further.

According to their molecular weight and function, eukaryotic stress proteins can be classified into 5 different protein families. In fish (cell lines), a variety of stress proteins of different molecular weight has been described [summarized in 24] but, in the following text, only the proteins expressed predominantly in response to environmentally relevant chemicals are mentioned.

The low molecular weight stress proteins (LMW): The members of this family are heterogenous with a molecular weight of about 14 to 40 (50) kDa and less evolutionarily constrained [40]. Unlike the other stress protein families, which also are expressed constitutively, the synthesis of LMW stress proteins is strictly stress-induced and also developmentally regulated [19, 26]. A wide variety of different LMWs has been discovered in fish cell lines [24]. In desert fish cells (*Poecilionis lucida*), which inhabit a highly stressful environment, high levels of a 30 kDa protein and a lower level of a 27 kDa protein were synthesized in response to prolonged elevated temperatures [38]. Members of this family were also described in different fish tissues, for instance in gill and muscles of arsenite-exposed fathead minnows (*Pimephales promelas*) [39] or in the gills of colour carp (*Cyprinus carpio*) after a heat shock [39]. Characterization of cDNA clones for stress-27 (Hsp27) and stress-30 (Hsp30) in fish from arid regions (*Poecilionis lucida*) revealed that both are members of the LMW stress family but belong to separate lineages within this family and that there are multiple isoforms of both [37]. In response to zinc chloride or cadmium chloride, permanent fish cell lines synthesized stress proteins with a molecular weight of 14, 28, 46, 50 and 51 kDa [40, 41]. It remains unclear, however, whether the latter three proteins belong to the LMW or the Hsp60 protein class.

The stress-60 (chaperonin, Hsp60) proteins: The stress-60 family, a major stress protein family, was found in bacteria as well as in the mitochondria and chloroplasts of eukaryotes [42]. Their members have a molecular weight of about 40 to 60 kDa. The proteins are involved in the assembly of protein oligomers [43], in folding and assembly of polypeptides translocated into organells [44] and bind unfolded precursors before their re-export to intermembrane space [45]. The expression of this protein family has also been subject to biomarker studies of fish. In different fish cell lines, an increased abundance of members of the stress-60 family was described, for example proteins of 42 and 57 kDa in nuclei of fathead minnows (*Pimephales promelas*) epithelial cell line after heat shock [46], but also a 60 kDa stress protein in the gills of the same species [47].

The stress-70 (Hsp70) proteins: The stress-70 family is the most highly conserved of all Hsp families (molecular weight of about 66 to 78 kDa). The members are involved in the intracellular stabilization of nascent or unfolded protein precursors before assembly in the cytosol or translocation into organelles including mitochondria and the ER [48, 49], in stabilization of newly translocated polypeptides before folding and assembly [50, 51], in rearrangement of protein oligomers [52], in dissolution of protein aggregates [50], and in the degradation of rapidly turned-over cytosolic proteins [53]. Among the different stress protein families, the stress-70 group undoubtedly has been the most extensively studied and the most numerous intensions on the establishment of a biotest have been carried out on this protein family. An increased expression of stress-70 proteins follows heat shock or heavy metal exposure such as $CdCl_2$, $ZnCl_2$ or sodium arsenite in different fish cell lines [40, 54, 55, 56] and tissues, for example gill of rainbow trout (*Oncorhynchus mykiss*) [57]. The Hsp70 level in the liver of rainbow trout (*Oncorhynchus mykiss*), chinook salmon (*Oncorhynchus tshawytscha*) [58] or brown trout (*Salmo trutta* f. *fario*) [59] is also a sensitive indicator of aquatic pollution. Infectious diseases may interfere with the use of stress protein induction in fish as a marker of environmental stress, for instance a bacterial strike of coho salmon with *Renibacterium salmoninarum* resulted in an increased stress-70 level in kidney and liver [60]. Exposure to bleached craft pulp mill effluent resulted in an increased stress-70 level even in ovarian follicular cells from white sucker (*Catostomus commersoni*) [61]. Studies of the isoforms in tropical and desert species of poeciliid fish did not show any variation in constitutively expressed stress-70 but a polymorphic induced stress-70 [62]. Such isoforms can have, for example, a molecular weight of 72 to 76 kDa and are synthesized in various tissues of the eurythermal teleost *Fundulus heteroclitus* [63] or in primary cultures of salmon (*Salmo salar*) hepatocytes [64].

The stress-90 (Hsp90) proteins: Stress-90 proteins (molecular weight of about 87 to 92 kDa) are present in all prokaryotic and eukaryotic organisms so far tested [65]. They function as important molecular chaperones in stabilization of a variety of target proteins, such as cellular signalling molecules and transcription factors, in an inactive or unassembled state [65]. In aquatic systems, several studies have shown a heat shock-induced expression of *hsp90* mRNA in zebrafish embryos [66] and of a member of the stress-90 group (stress-87) in cultured gill cells of colour carp

(*Cyprinus carpio*) [39]. Cho and coworkers [67] have found induction of a 90 kDa stress protein by copper sulphate, cadmium sulphate, or 2-mercaptoethanol in a permanent fish cell line. The specific mechanisms involved in this induction, however, have not been identified [26] and also the suitability of stress-90 as a biomarker is unclear up to the present.

The high molecular weight stress proteins (HMW): The members of this not often reported stress protein family have a molecular weight of about 100 to 110 kDa. Little is known about the function or prevalence of them, but stress-104/stress-110 appear to be induced by a wide variety of stressors in addition to heat shock [26]. After elevated temperatures a 100 kDa stress protein was investigated in cultured cells of rainbow trout (*Salmo gairdnerii*) [54], in gill and muscle tissue of fathead minnow (*Pimephales promelas*) [47] and in arsenite-exposed cultured cells of rainbow trout (*Salmo gairdnerii*) [54].

A number of methods and techniques are available for examining different aspects of the stress response. The techniques most frequently used include (1) one- and two-dimensional SDS-PAGE including autoradiography/flurography, (2) Western-blotting techniques with a stress protein-specific first antibody and a second antibody conjugated with fluorescein isothiocyanate (FITC) or peroxidase, (3) hybridization techniques with stress protein cDNA probes, (4) quantification of the stress protein-encoding mRNA by polymerase chain reaction (PCR), and (5) the use of transgenic animals. Within the field of environmental toxicology, the appropriateness of an ELISA has been sporadically reported in fish [68, 69]. Most of the studies on environmental aspects of stress protein induction, however, used Western-blotting techniques for proteotoxicity assessment, which are actually the widest-spread methods in this field. Also, the use of transgenic animals may be integrated in an "early warning system" but it should always be kept in mind, that these strains as well as permanent cell cultures do not reflect the situation of natural populations in the environment.

Like other biochemical biomarkers which require the potency of intact cellular mechanisms to react to stressors, the stress protein response increases at first with increasing intensity of the stressor, then the ability to react reaches its climax, and finally it decreases above a certain threshold level of exposure ([31], for fish see [58]). This may be explained by the incipient inhibition of processes of transcription and protein expression. A precise assessment of the toxicity of environmental conditions only can be given, when it can be determined whether the resulting stress protein level is on the rising or descending part of the stress protein optimum curve. Therefore, to comprehend elevated but still sublethal levels of hazardous factors, investigations on stress proteins should be combined with cellular biomarkers which also comprise pathological symptoms of exposure [15]. The assessment of toxicity can also become more difficult because a considerable number of stress protein inducers are effective in conditioning the cell/tissue/animal to become physiologically "tolerant" [19, 70, 71, 72]. This acquired tolerance, caused by a chronic exposure, protects the animal from a normally lethal temperature [72] or hazardous chemical impact. Because of the diversity of the stress response in various organisms in response to different stressors, it is difficult to determine which specific stress proteins are involved in the establishment of tolerance. Furthermore, a genetically fixed tolerance ("resistance") based on e.g. constitutively increased biotransformation processes may alter the proteotoxic impact of chemicals and, consequently, the stress protein level.

In vertebrates and, therefore, also in fish, interdependations of (e.g. adrenaline-mediated) physiological stress response and the intracellular Hsp induction should be considered. Although an adrenaline effect on Hsp70 has been demonstrated in rats [73], no work relating physiological stress and cellular Hsp response in fish has been conducted up to the present [24].

A seasonal variation in the stress protein level in different fish species (*Pimephales promelas*, *Salmo trutta*, *Ictalurus natalis* and *Ambloplites rupestris*) under natural conditions has been reported [68, 59, 74] which may additionally complicate stress assessment if the obtained data cannot be related to controls. It could be shown that the stress-70 level is much higher in spring and summer then during the colder seasons autumn and winter (Fig. 1) [68, 59]. Despite the variability of parameters dominating field studies, an attempt to incorporate the stress protein response into a set of biomarkers used for aquatic toxicity assessment was made within the mentioned project "Valimar" [15] using semi-field "bypass" exposure systems (Fig. 1). This study showed that the Hsp70 level in the liver and gill of brown trout (*Salmo trutta*) and loach (*Barbatula barbatula*) could account for differences of chemical impact in water and sediment of two model streams, for differences in susceptibility of the two mentioned fish species, and for differences in the toxic impact of water and sediment dependent on the season [59, 74, 70]. These results formed the basis of the following hypothesis on the interrelationship of a permanent heavy metal, PCB and PAH burden in the rivers sediments, the seasonal application of pesticides, the annual temperature regime and the Hsp70 level in fish liver [59, 74]. During the cold seasons when no pesticides are introduced into the water, the permanent metal, PCB and PAH burden is most probably responsible for a relative induction (and accumulation) of Hsp70 compared to controls. In late spring and summer, however, when the usual level of Hsp70 already is enhanced under control conditions (probably in response to the elevated temperature), the combined influence of pesticides, the permanent sediment burden and the high temperature seem to overcharge the regulative competence of the stress response system. This effect may be responsible for the observed Hsp70 level decrease in polluted streams in summer, which was mirrored by experimentally induced overcharge of the Hsp70 system by high concentrations of a cocktail of environmental contaminants. This overcharge of the cellular response system, however, seems to be transient. Recovery experiments showed that the Hsp70 level decrease following Hsp70 induction at high toxicity levels could be tracked back from low to high Hsp70 levels following recovery [76]. Within the following years, it is aimed to show evidence for this hypothesis under field conditions.

One prerequisite for using stress proteins in environmental monitoring is, that the stress response is not induced by handling of the organisms during sample collection or the experiment. This probably depends on the severity of physiological stress posed by handling of the fish since two studies on the influence of physical handling led to different results [77, 69]. Anyway, handling stress of organisms during sampling should be kept to a minimum. It is also necessary to characterize the level of stress proteins expressed under "natural" conditions prior to environmental monitoring studies. The research carried out over the past years has resulted in a basic understanding of the roles of the major stress proteins in the cell. However, particularly from the view of ecologists and ecotoxicologists, much remains to be understood. The major challenge now lies in elucidating (1) the kinetics of stress protein induction in the natural environment, (2) the influence of confounding factors of toxicity such as e.g. temperature and water pH, (3) the interactions of stress response and tolerance, and (4) the relevance of this biomarker for higher levels of biological organization.

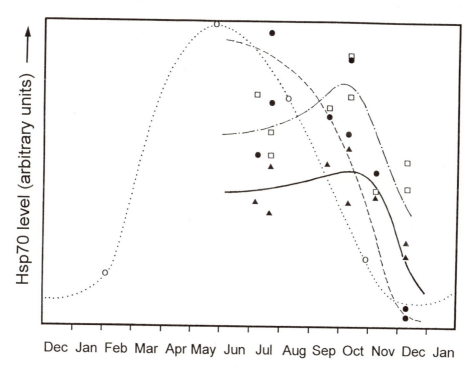

Figure 1: Seasonal variation in the Hsp70 level of the liver in juvenile brown trout, *Salmo trutta*. Data either were taken from Fader and coworkers [69] (open circles, dotted line: field specimens taken from an unpolluted brook, Valley Creek, PA, USA) or represent results obtained in semi-field flow-through exposure systems within the frame of the VALIMAR project (S. Gränzer, N. Kunz and H.-R. Köhler, previously unpublished): filled dots, dashed line: controls; squares, dashed-dotted line: exposure to water from a minor polluted brook (Krähenbach near Tübingen, FRG); triangles, solid line: exposure to water from a brook polluted by sewage plants and surface runoff (Körsch near Stuttgart, FRG).

Metabolic enzymes

Metabolic enzymes are found as constitutive elements in every living cell, and disturbance of their specific catalytic reactions may lead to a misbalance in homeostasis [78]. Since such changes in enzymatic activities may be causally related to exposure to chemicals, changes in the activities of not only biotransformation enzymes, but also metabolic enzymes can be used as biomarkers [7, 79, 80]. The primary targets of xenobiotics at the molecular level [78, 81, 82]; thus, in contrast to parameters at higher levels of biological organization, subcellular systems including enzyme activity modifications should respond more rapidly and sensitively and should be particularly suitable tools to detect the effect of trace amounts of environmental toxicants in the environment.

Since, especially at the beginning of field studies, targets and modes of action of the chemicals involved are not known, central metabolizing organs such as the liver in vertebrates or the midgut gland in invertebrates are usually selected for investigation [83, 84, 85]. In particular the vertebrate liver is of central importance due to its involvement in numerous metabolic pathways [6, 82] and its crucial role in the uptake, accumulation, biotransformation and excretion of xenobiotics [6, 87, 88].

Numerous *in vivo* as well as *in vitro* tests have been performed in laboratories to characterize chemical effects and to understand their mode of action. In contrast to parameters like death, growth or behaviour, the specific reaction of an enzyme may yield discrete information about the site and the functional metabolic effect of the chemical(s) [89]. For example, inhibition of acetylcholine esterase by phosphorodithioates implicates disturbances within the nervous system [90, 91, 92]. Changes in aminolaevulinic acid dehydratase were found to be highly susceptible for lead exposure [93], catalase stimulation was shown after paraquat and menadione exposure [94, 95] and Mayer [96] revealed strong inhibition of succinate dehydrogenase by dieldrin.

Although *in vitro* tests with fish cell cultures represent a special part in laboratory research, alterations in the activities of metabolic enzymes, however, have only rarely been measured. Zahn et al. [97], e.g., determined changes in the activities of lactate dehydrogenase, as well as in the lipid peroxidation rate in isolated hepatocytes of rainbow trout in response to disulfoton exposure in addition to variation in the activity of the biotransformation enzymes glutathione S-transferase and ethoxycoumarin-O-deethylase. In the fibrocytic cell line R1 derived from rainbow trout liver [98], effects of 2, 4-dichlorophenol on lactate dehydrogenase, isocitrate dehydrogenase, glucose-6-phosphate dehydrogenase, alanine aminotransferase, esterase, as well as acid phosphatase, were described [99].

Based on the growing body of information about how toxicants influence the response patterns of biomarkers, increasing efforts have been taken to assess the influence of xenobiotics on enzymes not only in laboratory, but also in field studies. However, most interest has been focused on changes in biotransformation enzymes [e.g., 100, 101, 102, 103, 80].

Contamination of field sites can generally be monitored by different test designs, which comprise collection of sediment or water probes followed by exposure of test organisms in the laboratory, collection of sentinel organisms in field sites, caging of test organisms at contaminated locations [80, 103], or exposure of test organisms in bypass systems [103]. In the frame of the projekt "Valimar" bypass systems, for example, juvenile brown trout and adult loach were exposed to the waters of two differently polluted streams [104]. Among other biomarkers, changes in the activities of acetylcholine esterase, cytochrome c oxidase, succinate dehydrogenase, citrate synthase, esterase, malic enzyme, glucose-6-phosphate dehydrogenase, hexokinase, phosphofructokinase, acid phosphatase, uricase, catalase and alanine amino-transferase were monitored after various periods of exposure. Results indicate appreciable heterogeneity in the reactions of the enzymes measured, which, however, can be correlated with a significant variability in the concentration and composition of the toxicant mixtures in the streams. In addition, in either fish species, the chemically induced enzymatic changes are confounded by season- and temperature-dependent changes.

In the complexly contaminated river Körsch, elevated enzyme activities were predominately found in energy-supplying pathways (hexokinase, glucose-6-phosphate dehydrogenase, phosphofructokinase). Induction was also found for alanine aminotransferase, catalase and malic enzyme. Whereas in loach esterase activities did not show significant changes, activity was strongly inhibited in brown trout at the highly polluted Körsch. No consistent reactions could be demonstrated for uricase and acid phosphatase. As a rule, enzyme activities in brown trout proved to react more sensitively than in loach.

The examples cited document that studies in changes of metabolic enzymes may not only provide insight into the modes of action underlying toxic effects, but may also be used as a general source of biomarkers of environmental pollution. If compared to biotransformation enzymes, however, most metabolic enzymes are even less specific of the substance inducing the effect(s) studied, and due to the embedding into complex metabolic pathways, metabolic enzymes are subject to complicated regulations, which are under the influence of many endogenous and environmental parameters other than xenobiotics, e.g., age, sex, hormonal status, reproductive status, temperature, season, parasitism and others. Profound knowledge on the natural variability of the interaction between the enzymes and confounding factors is, thus, an indispensable prerequisite for the successful use of both biotransformation and metabolic enzymes as biomarkers.

	Loach (Barbatula barbatula)				Brown trout (Salmo trutta f. fario)			
	Krähenbach	Körsch	Krähenbach	Körsch	Krähenbach	Körsch	Krähenbach	Körsch
	related to mg protein		related to g organ		related to mg protein		related to g organ	
Brain:								
Acetylcholin esterase	0	-	-	0	+/-	-	+/-	-
Cytochrome c oxidase	+/-	-	0	-	0	-	+	0
Liver:								
Cytochrome c oxidase	-	-	-	+/-	-	-	+/-	+
Succinate dehydrogenase	+	++	+	++	-	+	+/-	++
Citrate dehydrogenase	0	+/-	+/-	-	-	+/-	+	+/-
Acid phosphatase	+/-	+/-	-	+/-	-	+/-	++	++
Uricase	+/-	+/-	+	+/-	+/-	--	+/-	--
Catalase	+	+/-	+	+	+	-	+	+
Alanine aminotransferase	+	+/-	+	+	++	+/-	+	+
Glucose-6P-dehydrogenase	+	+	+	+	+	++	+	++
Phosphofructokinase	+	+	+	+	+	+	+/-	++
Malic enzyme	+/-	+	+/-	+/-	+	+	+/-	++
Hexokinase	+	+	+	+	+	++	+	++
Esterase	0	0	+	+/-	+/-	--	-	---

Table 1: Average responses of different enzymes of loach (*Barbatula barbatula*) and brown trout (*Salmo trutta f. fario*) after exposure in a bypass system at the differently polluted small rivers Krähenbach and Körsch between 1995 and 1997 in relation to control values derived from fish held under controlled conditions in the laboratory. 0 = no response, - = reduced enzyme activity, -- = strongly reduced enzyme activity, --- = very strongly reduced enzyme activity, + = induced enzyme activity, ++ = strongly induced enzyme a

Biotransformation enzymes

Contamination of our environment with xenobiotics is a widespread phenomenon living organisms have to cope with. Four steps which are uptake, distribution, metabolism and excretion of chemicals are important parameters for bioaccumulation and toxicity of these substances.

Metabolism or biotransformation of anthropogenic compounds converts lipophilic chemicals to more hydrophilic metabolites which can be excreted more easily. Usually, metabolic conversion occurs in two distinct phases of reaction. In phase I, the lipophilic substrate is metabolized mostly by the cytochrome P450 enzyme system (or mixed-function oxygenase (MFO) system), which introduces an oxygen atom into the molecule. In the second phase, the metabolites from phase I are conjugated with endogenous molecules such as glucuronic acid, sulfate, glutathione, etc. [105].

The cytochrome P450 enzyme system includes many isoenzymes. The superfamily P450 is separated into families and subfamilies, whereas an Arabic number describes the family, a letter the subfamily and again an Arabic number the individual isoform coded by a distinct gene. CYP denotes the protein cytochrome P450, so an example of an individual cytochrome P450 form would be CYP1A1.

The cytochrome P4501A1 isoenzyme is of particular toxicological relevance due to its receptor-mediated inducibility by numerous organic xenobiotics, namely polycyclic aromatic hydrocarbons (PAHs), coplanar polychlorinated biphenyls (PCBs), furanes, dioxins and various pesticides [106, 107, 108]. For this reason, the measurement of the catalytic activity of this enzyme system has been used as a biomarker of exposure since the 1970s [109, 110]. The catalytic activity is determined by measuring ethoxyresorufin-O-deethylase (EROD) activity or aryl hydrocarbon hydroxylase (AHH) activity. The frequent use of EROD induction as a biomarker of xenobiotic exposure is due to its easy and cost effective analysis as well as to its proven relationship to chemical inducers [111]. To date, the induction of the P450 enzyme system has been used in a number of field studies. By use of the biomarker EROD, it is possible to discriminate between sites of different pollution conditions [112]. A drawback of EROD measurement is the fact that the enzyme activity does not only depend on exposure to chemical inducers but might be inhibited also by certain chemicals [113] or by physical factors, such as high temperature [114]. The quality of results in environmental monitoring studies is increased if EROD is measured in combination with immunochemical quantification of CYP1A either by Western Blot or ELISA. [115, 116].

In phase II of the biotransformation process, usually metabolites from phase I are conjugated with endogenous molecules to produce even more hydrophilic products. Nevertheless, organic xenobiotics can be also directly handled by phase II enzymes. Two important enzyme families of the phase II enzymes are the membrane bound UDP-glucuronyl transferases (UDPGT) and the cytosolic glutathione S-transferases (GST). UDPGT conjugates UDP-glucuronic acid to organic xenobiotic substrates and GST conjugates them with glutathione, the latter being the most important reaction for electrophilic compounds. Both UDPGT and GST are inducible by organic xenobiotics, but (at least) GST to a lesser extent than phase I enzymes [117, 118]. Nevertheless, the catalytic activities of UDPGT and GST are used as biomarkers, often in combination with catalytic and immunochemical measures of phase I enzymes [119, 120, 121].

In this context, the question arises, whether a biomarker response can be more than a specific tool to demonstrate exposure to environmental pollution. It is known that many xenobiotics, especially PAHs, become even more toxic or carcinogenic upon being converted by the P450 enzyme system [122]. Therefore one could imagine correlating EROD induction with the occurence of carcinoma. In long-term studies on benthic fish populations in the Puget Sound, Washington, USA, correlation between sediment PAH exposure (hepatic EROD inducers) and frequency of hepatic lesions has been demonstrated [123]. However, Van Veld et al. [124] showed, that EROD activity in hepatic lesions of mummichog (*Fundulus heteroclitus*) was 15-77% lower than in non-neoplastic tissue. This may illustrate that, although a sensitive biomarker such as EROD induction may serve as an early warning response of exposure, both interpretation and correlation to toxic effects remain difficult.

Another important toxicological implication of P450 induction is the involvement of the P450 enzyme system in the synthesis and metabolism of steroids. Thus, chemically induced alterations of P450 levels could theoretically result in effects on fish reproduction [125]. Munkittrick et al. [126] showed that white sucker (*Catostomus commersoni*) from Lake Superior, Canada, receiving bleached craft mill effluent showed abnormalities in growth rate, reproductive development, fecundity with age and age to maturation, secondary sexual characteristics, gonadal size, and egg size. These observations were correlated with a decrease in serum levels of estradiol and testosterone, and with increase in the activity of hepatic P450 enzymes. Similar results have been obtained from field studies in Sweden [127, 128]. However, neither causative relationships between these effects and Cytochrome P450 induction nor their impact on fertilization rate and recruitment remain still unclear.

To improve our understanding of the causative mechanisms, *in vitro* and *in vivo* testing is essential. Moreover, much more attention should be paid to the interrelation of endocrine effects - particularly estrogenic/antiestrogenic - with the P450 system [129, 130, 131].

The measurement of biotransformation enzymes is one part of the combined ecotoxicological project introduced above. Analyses of phase I enzymes of fish exposed in a bypass system to river water from Körsch and Krähenbach showed that EROD induction and measurement of the cytochrome P450 protein level by a semi-quantitative ELISA were able to discriminate between these two test streams. Apparently they reflected the different conditions in the experimental sites [132]. EROD induction, in addition, was shown to be able to distinguish between the two different species investigated, the brook trout (*Salmo trutta* f. *fario*) and the loach (*Barbatula barbatula*) (Fig. 2).

Results from investigations on phase II enzymes did neither show any induction of UDPGT (catalytic activity was hardly detectable) nor of GST. This finding agrees with previous observations, that induction of GST occurs less frequently and less consistently than EROD induction [133, 134].

As mentioned above, EROD activity is not only influenced by chemical inducers but does also depend on physical factors such as temperature. It is known, that P450 enzyme activity decreases with increasing temperature [135, 136]. Data on control fish probably reflect the action of this confounding factor: rather low activities were recorded during summer (data not shown) for the polluted Körsch river, however, no influence of temperature were recorded. This might be due to overlapping effects of chemical inducers on one hand and of other environmental parameters on the other hand, and again stresses the need for a careful interpretation of biomarker data, which best should derive from a comprehensive "battery" of marker responses.

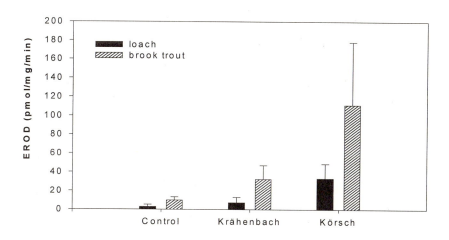

Fig. 2: EROD induction in brook trout (*Salmo trutta* f. *fario*) and loach (*Barbatula barbatula*) in the control and the two model streams Krähenbach and Körsch. Basal activity in loach is lower than in brook trout.

As already mentioned for stress proteins and metabolic enzymes, it might also be possible to correlate the induction of the P450 enzyme system with demographic parameters in the future. Alterations in presence or activity of cytochrom P450 could then be used to assess integrity of free-living populations. In general, however and also for P450, the underlying causal mechanisms for a manifestation of a marker-indicated effect of population development remain to be established at the present state of knowledge.

Histopathological and Ultrastructural Biomarkers

Histopathological and ultrastructural alterations in fish tissues are useful to specific organs or cells as targets for environmental chemicals and to show how they are affected under *in vivo* conditions [137]. Furthermore, cellular biomarkers allow a retrospective as well as a prospective diagnosis regarding the single individuum. Specialy, investigations at the ultrastructural level enable conclusions relating to the mechanisms by which the cell is damaged [138]. Based on this knowledge, it is possible to provide information about potential secondary effects and to predict possible disadvantages for the organisms. Above all, histological indicators have been proved to be indispensable in connection with the determination of NOEC (no effect concentration) values, in short time as well in long time experiments [139].

When using histopathological biomarkers, several sources of errors have to be considered [137]. The detection of chemically-induced injuries depends on proper fixation, processing and staining of tissues. Proper interpretation of histopathologic lesions is also highly dependent on the experience of the investigator. The subjective nature of morphologic studies has made correlations with other, more quantitative physiological and biochemical approaches difficult [140, 141, 142]. This problem can be solved by means of a valuation system of cytopathological changes, which characterises the „state of health" of cells of test fish [143, 144]. A possible method to assess cellular biomarker responses is more closely explained in the next paragraph in context to the description of liver ultrastructure as a biomarker. Generally, the results for the biomarker "histopathology of kidney", "ultrastructure of liver" and "ultrastructure of gill", described in the following text all arise from the projekt „Valimar" which has previously be mentioned [13, 14, 15].

Histopathological Biomarkers:

Kidney: Histopathological investigations represent a sensitive tool to detect toxic effects of environmental pollutants in laboratory experiments (145) and field studies (146). As described previously (147), histopathological studies are important for the interpretation of subcellular and ultrastructural markers, because they have the capacity to differentiate between organ lesions induced by infectious diseases or other environmental factors and those due aquatic toxicants.

Within the frame of the above mentioned research project [13, 14, 15] histopathological investigations were carried out in various organs of fish exposed to water from the teststreams with a different degree of pollution. Findings have been evaluated semi-quantitatively and partly by stereological procedures as published in detail by Schwaiger et al. [147]. As an example to demonstrate the importance of examining the target organs on the level of light microscopy, morphological alterations of the kidney will be summarized briefly within the present article.

The histopathological responses of the kidney clearly differentiated between fish from the two differently contaminated sites. Control fish did not show kidney lesions. In fish exposed to the moderately polluted stream (Krähenbach), kidney alterations consisted in slight degenerative changes and occasional dilation of the kidney tubules (Fig. 3b). Such tubulonephrotic changes are ferequently observed as a consequence due to heavy metal exposure [148, 149, 150] or triazine herbicides [151].

In fish exposed to the more polluted stream (Körsch), these tubulonephrotic changes were overlapped by severe kidney alterations due to *Proliferative Kidney Disease* (PKD) (Fig. 3c), which is induced by a myxosporean parasite [152]. The occurence of this parasitic disease leads to the suggestion, that fish may have been weakened by contaminant stressors rendering them more susceptible to diseases as discussed by Rice et al. [153] and Anderson and Zeeman [154]. Since it is known, that various subcellular markers reflect stress conditions in fish induced by a broad range of environmental factors [137, 155, 156], histopathological investigations - by identifying the aetiology of organ lesions - contribute to a better understanding of stress responses at lower levels of organization.

Fig. 3: Renal tissue of brown trout (exposed for 13 weeks); original magnification 480x, H&E stain
a: Renal tissue of brown trout fro the control group without pathological alterations.
b: Renal tissue of a brown trout exposed to the moderatly polluted river (Krähenbach) showing single cell necrosis of tubular epithelial cells (arrow) and slight dilation of kidney tubules (arrow head).
c: Renal tissue of a brown trout exposed to the highly polluted river (Körsch) showing an almost complete loss of renal tubules and necrotic changes due to PKD (arrow).

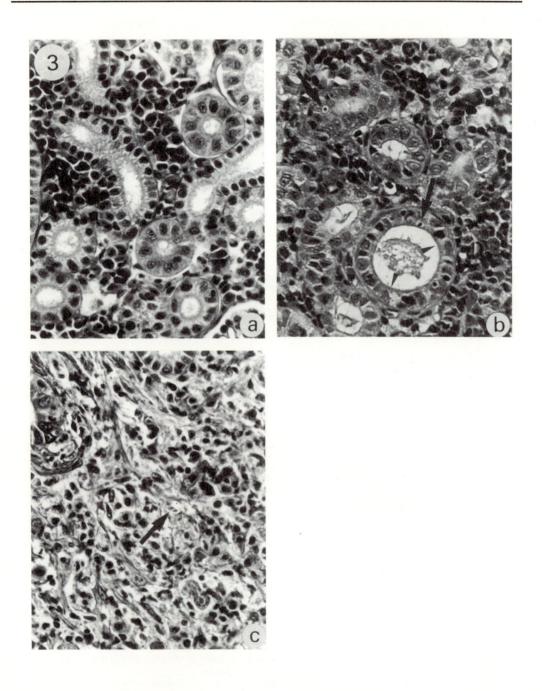

Ultrastructural Biomarkers:

Ultrastructure of the liver: The best established monitor organ in fish is the liver [e.g. 157, 158]. As a major metabolic organ, the liver plays an important role in uptake, accumulation [159, 160], biotransformation [161, 162] and excretion [163, 164] of xenobiotics. Functional changes are known to be reflected in structural changes of hepatocytes [159, 165, 86], which in turn can be used as biomarkers to trace environmental pollution caused by chemicals [7]. Cell alterations caused by nutritional stress were discribed in the hepatopancreas of amphipods by Storch & Burckhardt [66]. These results have been transferred to vertebrates: as well as the midgut gland epithelium of arthropods, the liver cells of fish also respond in characteristic ways to different nutrition [167]. So, confounding issues for histopathological biomarkers include distinguishing changes caused by anthropogenic toxicants from those due to infectious disease, normal physiological variation, and nutritional situation [137].

Within the above mentioned project it became obvious that hepatocellular response patterns in brown trout are usefull biomarkers to discriminate between the two different polluted stream. Fish exposed to highly polluted stream water (Körsch) showed more drastic alterations of the liver ultrastructure than fish exposed to the lightly polluted water (Krähenbach). Qualitative aspectss of those diagnosis are summarized in Fig. 4.

Fig. 4: Brown trout (exposed for 13 weeks)

It means: n = nucleus, g = glykogen, er = rough endoplasmatic reticulum, m = mitochondria, p = peroxisome, mi = mikrovilli, l = lysosome, ge = degranulated endoplasmatic reticulum, m = myelinbody, en = endothel, ko = kollagen, s = sinusoid.

a: Hepatocyte of a **control** with a pronounced cellular compartmentation. Around the centrally located nucleus, few non-fenestrated lamellae of the RER and only few mitochondria can be observed. Large amounts of glycogen are found in the peripheral cell areas.

b: The cellular compartmentation of fish exposed to the **lightly polluted water** (Krähenbach) shows slight changes in comparison to controls. Furthermore, cellular storage products (glycogen) are slightly decreased, cell organelles as e.g. RER show a increase in extent.

c: The cellular effects on fish exposed to the **highly polluted water** (Körsch) are much more pronounced than in the fish of the Krähenbach. The cellular compartmentation is completely lost, a strong augmentation of mitochondria and RER can be noticed. The quantity of glycogen is strongly reduced.

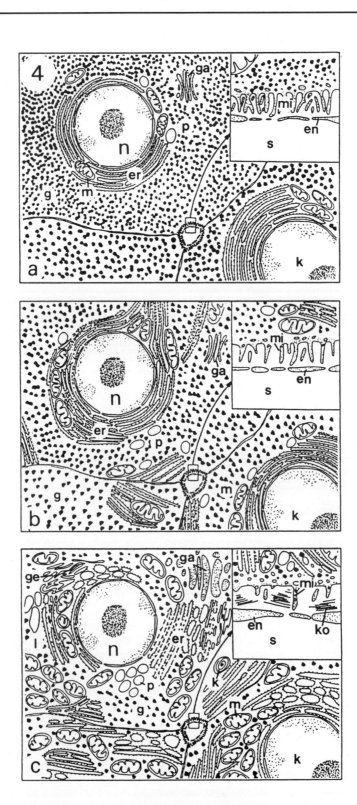

The liver ultrastructure of fish exposed to the lightly polluted stream water did not differ significantly from controls. One obvious reaction was the increase of apical vesicles close to the bile canaliculi. A few fish showed some changes to the cytoplasmic compartmentation and slight reduction in glycogen content of the hepatocytes (Fig. 4b). Occasionally, vesiculation of the rough endoplasmic reticulum (RER) was observed. In these fish, the observed effects were much less pronounced than in fish exposed to the higher polluted Körsch water. In comparison to controls, hepatocytes of brown trout exposed to this appeared more homogeneous due to a loss of the cellular compartmentation (Fig. 4c). Organelles were distributed all over the cytoplasm and only small spots of glycogen could be found. The amount of RER was increased if compared to controls, and rarely arranged in long and parallel cisternae. The short cisternae were distributed all over the cytoplasm. Moreover, the RER showed numerous structural alterations: cisternae were found to be dilated, vesiculated or degranulated. The same was true for qualitative alterations in mitochondria: as in the controls, their size and structure was very flexible. However, there was a pronounced increase in the number of mitochondria, which were distributed throughout the cytoplasm (Fig. 4c). The cisternae of the dictyosomes were partially dilated and surrounded by an increased number of vesicles. Glycogen storage in the periphery of the cell was reduced and the number of cell organelles increased (Fig.4c). Occasionally, large amounts of lipid were found. Close to the bile canaliculi, and to the space of Disse, an increase in the number of apical vesicles was observed in the hepatocytes. The number of microvilli in the space of Disse seemed to be reduced. Macrophages and granulocytes in the bile canaliculi and in the sinusoids increased in number if compared to controls. Myelin bodies only occasionally occurred. The nuclei, however, did not apparently differ between controls and exposed animals.

To characterise the "state of health" of hepatocytes or hepatic parenchyma, the ultrastructural symptoms in liver tissue were classified into three categories: **1**: control state, **2**: slight changes and/or reactions visible, **3**: strong reactions and/or destruction visible. Criteria upon which this classification is based are published by Schramm et al. (1997) [143]. For each test specimen, liver sections were investigated and the conditions of the cell organelles were allocated to one of the three cytological categories. From these values, mean values for each fish were calculated. Mean values for each fish group were calculated from the mean values of the individual fish.

The results of this semi-quantitative evaluation of symptoms in exposed fish are presented in Fig. 5.

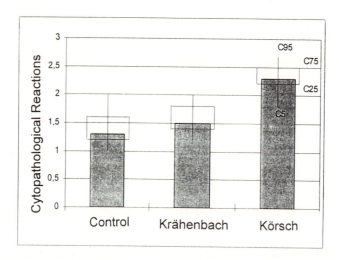

Fig. 5: Semi-quantitative evaluation (median) of cytopathological reaction in hepatocytes of exposed brown trout. Control = control conditions in the lab, Krähenbach = lightly polluted stream, Körsch = highly polluted stream. Y-axis = Cytopathological response scale. 1: control conditions. 2: slight changes and/or reactions are visible. 3: strong reactions and/or destruction are visible. **Median** = 50% percentil. C_{25}-C_{75} = 25% - 75% percentil, C_5-C_{95} = 5% - 95% percentil.

By means of this method, semi-quantitative and less ssubjective data of cytopathological alterations can be obtained, which allow cellular and histological studies to be correlated with quantitative measurements (e.g. metabolic or biotransformation enzymes) [144]. For liver, semiquantitative results can be positively correlated with data obtained by chemical analyses [104]. The cellular reactions can be discussed in relation to functional adaptations/alterations of the metabolism. It is necessary to distinguish between non-specific reactions to a wide variety of environmental changes (such as a loss of cellular compartmentation), which might be due to a chemical attack of the cytoskeleton [168, 169] and alterations, which are connected with, for example intensified biotransformation processes and an increased energy turnover. These might be reflected e.g. by an increased number of mitochondria, dilation of the cisternae of dictyosomes or the RER or the decrease in the amount of hepatic glycogen [161, 164, 170-173, 175-177].

Ultrastructure of the gills: In fish, most of the dissolved and particle-bound pollutants enter the body via the pharynx either by inspiration or by food uptake. The gills represent major sites for respiration; they are always in close contact with the water which makes them important targets for waterborne pollutants. Generally, fish gills comprise more than half of the body surface with an epithelial layer of only a few microns separating the interior of the fish from the external environment [178]. Due to this close association between water and blood, the gills are strongly affected by environmental pollutants [178-183]. From a morphological as well as from a physiological point of view, the gill is a very complex organ involved not only in gas exchange, ion exchange, acid-base balance, but also in nitrogenous waste excretion.

Cellular responses of the gills indicate not only an impairment of the test organism, but also its adaption to the pollution situation. Gills therefore are potentially useful monitor organs to reflect the health of aquatic organisms [184, 185]. Ultrastructural responses are sensitive enough to indicate early effects on the organism of low concentrations of chemicals, which makes them useful biomarkers to assess environmental pollution. In the present study, gills from fish exposed under semi-field conditions to the water and sediment of the two model streams were examined in order to verify the principle suitability of gill ultrastructure as biomarker to assess the pollution of small streams by various chemical compounds. The study focuses on ultrastructural alterations of epithelial, mucus, pillar and chloride cells.

In contrast to controls and to fish exposed to the water of the less polluted Krähenbach, both loach and trout exposed to the water from the highly polluted Körsch clearly showed ultrastructural changes in the gills. The chloride, mucus and epithelial cells of the secondary lamellae as well as the epithelial cells of the multi-layered primary filament showed the greatest effects. The results did not differ significantly with different exposure times. No ultrastructural differences were recorded for the rodlet cells which are distributed over the gill epithelium in controls as well as in exposed fish. However, they seemed to increase in number under polluted conditions.

Epithelial cells of the primary filaments: Controls and fish exposed to the water of the Krähenbach did not show any pathological symptoms in the epithelial cells of the primary filament. These cells bear only few mitochondria, a Golgi apparatus, and parallel-arranged cisternae of rough endoplasmic reticulum. The basal lamina appears homogenous. Cells of the primary filament in fish exposed to the Körsch water showed widening of the intercellular spaces. Large numbers of electron-dense lysosomes were found in macrophages close to these sites. The basal lamina often appeared condensed throughout the primary filament after exposure to the highly polluted stream water.

Epithelial cells of the secondary lamellae: After exposure to the highly polluted stream water, an extense lifting of epithelial cells (intraepithelial „oedema") could be observed in many secondary lamellae. In controls and fish exposed in the lightly polluted water, this effect appeared to a minor extent or not at all. In these cells, the microvilli became longer and irregularly shaped after exposure of fish to the polluted river water. Hyperplasia of epithelial cells occasionally occurred and often led to a clubbing of two neighbouring secondary lamellae.

Chloride cells: In controls, spherically-shaped chloride cells bore large numbers of mitochondria usually located in the apical part of the cells. The nucleus was in the basal part of the cell. Microvilli were either very short or absent (a-cells) or well-shaped (b-cells). The endoplasmic reticulum consisted of small tubules or vesicles (Fig. 6a). In chloride cells of fish exposed in the highly polluted stream, the lumen of the ER often became electron-dense. In a few cases, chloride cells were characterized by destroyed membranes of the ER and dilated vesicles appeared (Fig. 6b). In addition, the intercellular spaces were dilated (Fig. 6c). Occasionally, proliferation of chloride cells could be found.

Mucus cells: Mucus cells of controls bore many mucus-containing vacuoles, which were mostly apically located. The tubules of the rough endoplasmic reticulum were usually in parallel. In animals exposed to water from the highly polluted stream, the mucus-containing granules became more electron-dense. Often, the cells were completely full of these mucus vacuoles and no other organelles were visible. In addition, intercellular spaces were heavily dilated.

Histological staining for mucopolysaccharides also revealed this phenomenon and often showed secondary lamellae to be totally covered by mucus. An important physiological function of mucus secreted by gill cells is the protection of the thin and sensitive gill epithelium from environmental impacts as e.g. by xenobiotics. Hypersecretion of mucus, however, resulting in a complete mucus cover of the gill epithelium may impede gas exchange. A similar hypersecretion of gill mucus has been described by [186] for trout exposed to heavy metals or organic compounds, and by Jagoe (1993) [187] after treatment with beryllium. In fish exposed to the highly polluted Körsch, many of the chloride cells showed alterations of the endoplasmic reticulum (ER) as e.g. dilation, vesiculation or darkening of the lumen (Fig 6a-c). Generally, chloride cells are responsible for the maintenance of the acid-base and the ionic balance e.g. [184, 188, 189]. As a result of damage to these cells, fish could die due to metabolic acidosis or alkalosis. Although unaffected cells might functionally replace those already destroyed, they might themselves be destroyed due to functional overload. Changes observed in the ER could be correlated with increased production of membrane-bound proteins (e.g. ion channels), important for ion-exchange through cell membranes. In addition to their role in regulation of the ion balance, chloride cells are involved in excretory processes. In this context, the proliferation of chloride cells in fish exposed to the polluted stream could be explained by increased excretion or adaptive processes to different ionic environments [190, 191, 192]. Hyperplasia of secondary lamellae could often be observed in animals exposed in the highly polluted stream. This symptom, which often leads to the complete fusion of two neighbouring secondary lamellae, has often been described for fish exposed to specific toxins in the laboratory [193, 186]. Hyperplasia results in an enlargement of the distance between blood and the oxygen-rich water, and finally may lead to an insufficient oxygen supply of the blood. The observed epithelial lifting in the secondary lamellae is thought to be a typical inflammatory reaction [103, 195]. It is so far unclear whether these inflammatory reactions are indicative of a general weakening of animals which may be caused by any kind of stress, or whether they are the result of direct contact of the respective epithelia with a toxin. While in brown trout, ultrastructural responses in the gills were significantly different between fish exposed to the slightly or the highly polluted stream, this was not always true for loach. A possible reason for this might be the fact that loach is a benthic species while trout inhabit the free-water zone. Loach are therefore always in closer contact with the sediment, which is - in both model streams - contaminated with PAHs, PCBs and heavy metals. This two year study showed that ultrastructural responses in fish gills are useful biomarkers to monitor pollution in small streams. To detect pollution in sediments as well as in water, the gills of both bottom- and free water dwelling fish should be examined.

Final remarks

The applied biomarkers applied within the frame of the project of „Valimar" presented in this paper, led to toxicological statements on effects, which stood in line with the results of the corresponding chemical analysis in the two test streams of this project [104, 144, 196]. On principle, ecotoxicological results should be based on profound knowledge of toxicology, environmental chemistry and ecology [10]. Therefore, the final diagnosis based on investigations on heat shock proteins, metabolic and biotransformation enzymes, and (ultra)structural biomarkers for the individual fish can be discussed in view of an affected energy supply of the monitoring organism, which may be caused by detoxification and may be accompained by an affected physiological performance displayed e.g. by a decreased swimming velocity [104].

The subsequent step will be to clarify the relevance of biochemical, cytological and histological alterations to the next biological level beyond the individual, i.e. the population. So, strictly spoken, only those subcellular responses which provide strong evidence for changes at higher levels of biological organization should be assigned „biomarkers" in the sence of early-warning sentinels suitable for *eco*toxicological risk assessment. First results obtained from embryo tests with loach and trout indicate a correlation of the mentioned biomarker responses with an affected hatch of both species in the highly polluted water of the Körsch river [197].

Figures 6: Gills of brown trout
a. Control: Spherical chloride cell in the primary filament (Cc). Numerous mitochondriae (arrows), and a basally located nucleus (N) x 5400.
b. Körsch exposure: Spherical chloride cell in the primary filament (Cc). Dilated vesicles of the ER (arrowheads) and electron-dense lumen of the ER. x 12000.
c. Körsch exposure: Chloride cells (Cc) which have lost contact to the adjacent cells of the primary filament (arrows) due to dilation of the intercellular space (Is). Nucleus (N). x 5400.

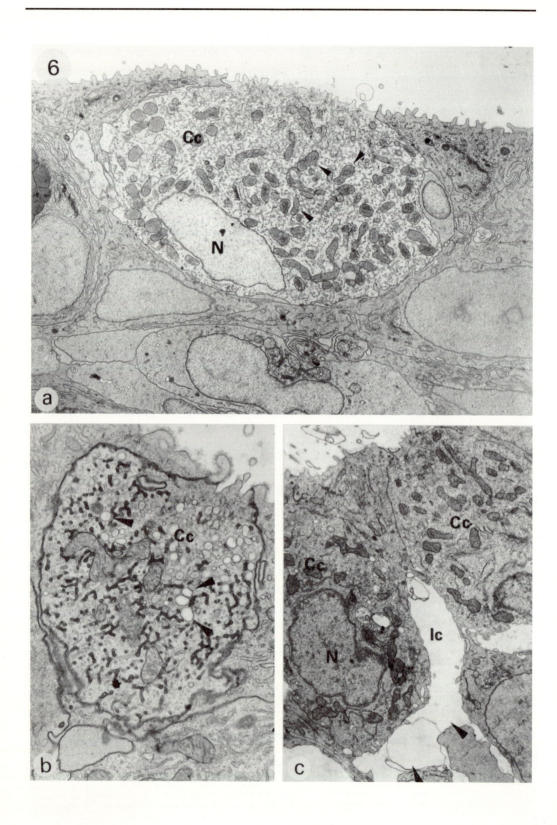

References

[1]Van Gestel, Ecotoxicol. 5 (1995), p. 217.

[2]Peakall, Ecotoxicology 3 (1994), p. 173.

[3]M.H. Depledge, Ecotoxicology 3 (1994), p. 161.

[4]D.A. Holdway, Austr. J. Ecol. 20 (1995), p. 34.

[5]R. Triebskorn (1996), 6. SETAC-Europe Annual Meeting, Taormina.

[6]Arndt, U., Nobel, W., Schweitzer, B. Bioindikatoren: Möglichkeiten, Grenzen und neue Erkenntnisse. Ulmer Verlag, Stuttgart (1987).

[7]McCarthy, J.F., Shugart, L.R. Biomarkers of environmental contamination. Lewis Publishers, Chelsea, MI, USA (1990).

[8]Bayne B.L., Brown, D.A., Burns, K., Dixon, D.R., Ivanovici, A., Livingstone, D.R., Lowe, D.M., Moore, M.N., Stebbing, A.R.D., Widdows, J. The effects of stress and pollution on marine animals. Praeger Publishers, New York (1985), p. 384.

[9]Huggett, R.J., Kimerle, R.A., Mehrle, P.M., Bergman, H.L., Dickson, K.L., Fava, J.A., McCarthy, J.F., Parrish, R., Dorn, P.B., McFarland, V., Lahvis, G. Biomarkers - Biochemical, Physiological, and Histological Markers of Anthropogenic Stress. Proceedings of the Eight Pellston Workshop, SETAC Special Publications Series (Ed.: R.J. Huggett; R.A. Kimerle; P.M. Mehrle; H.L. Bergman), Lewis Publishers, Cheleas, MI, USA (1992).

[10]Fent, K. Ökotoxikologie. Georg Thieme Verlag, Stuttgart (1998).

[11]G. Meyer (1990), Über die Ursachen unterschiedlicher Giftanreicherungen in Fischen, Diploma thesis, Lund University, Sweden.

[12]D.J. Rowan, Great Lakes Research (1992), p. 724.

[13]Triebskorn; W. Honnen; D. Frahne; T. Braunbeck; H.-R. Köhler; G. Schüürmann; J. Schwaiger; H. Segner; A. Oberemm, Erster Zwischenbericht zum BMBF-Verbundprojekt "Validierung und Einsatz biologischer, chemischer und mathematischer Tests und Biomarkerstudien zur Bewertung der Belastung kleiner Fließgewässer mit Umweltchemikalien" (FKZ 070TX21-25) (1996), 323p.

[14]Triebskorn; W. Honnen; D. Frahne; T. Braunbeck; H.-R. Köhler; G. Schüürmann; J. Schwaiger; H. Segner; A. Oberemm, Zweiter Zwischenbericht zum BMBF-Verbundprojekt "Validierung und Einsatz biologischer, chemischer und mathematischer Tests und Biomarkerstudien zur Bewertung der Belastung kleiner Fließgewässer mit Umweltchemikalien" (FKZ 070TX21-25) (1997), 532p.

[15]Triebskorn; W. Honnen; D. Frahne; T. Braunbeck; H.-R. Köhler; G. Schüürmann; J. Schwaiger; H. Segner; A. Oberemm, Dritter Zwischenbericht zum BMBF-Verbundprojekt "Validierung und Einsatz biologischer, chemischer und mathematischer Tests und Biomarkerstudien zur Bewertung der Belastung kleiner Fließgewässer mit Umweltchemikalien" (FKZ 070TX21-25) (1998), 170p.

[16]Tissières, J. Mol. Biol. 84 (1974), p. 389.

[17]Gething, Nature (London) 355 (1992), p. 33.

[18]Schlesinger, M. Ashburner and A. Tissières (eds.). Heat Shock: From Bacteria to Man. Cold Spring Harbour Lab., Cold Spring Harbour (1982), 440 pp.

[19]Nover (ed.). Heat Shock Response of Eukaryotic Cells Springer Verlag, Berlin, Heidelberg, New York, Tokyo (1984), 82 pp.

[20]Schlesinger, M.G. Santoro and E. Garcia (eds.). Stress Proteins. Induction and function. Springer Verlag, Berlin, Heidelberg, New York, Tokyo (1990), 123 pp.

[21]Edington, J. Cell Physiol. 139 (1989), p. 219.

[22]Craig, Trends Biochem. Sci.16 (1991), p. 135.

[23]Morimoto, in: U. Feige, R.I. Morimoto, I. Yahara and B. Polla (eds.). Stress-Inducible Cellular Response. Birkhäuser, Basel (1996), p. 139.

[24]Iwama, Rev. Fish Biol. Fisheries 8 (1998), p. 35.

[25]Bradley, in: W.G. Landis and W.H. van der Schalie (eds.). Aquatic Toxicology and Risk Assessment, Vol. 13 ASTM STP 1096. American Society for Testing and Materials, Philadelphia, PA, (1990), p. 338.

[26]Sanders, Crit. Rev. Toxicol. 23 (1) (1993), p. 49.

[27]Sanders, Sci. Total Environ. 139/140 (1993), p. 459.

[28]Hopkin (ed.). Ecophysiology of Metals in Terrestrial Invertebrates. Elsevier Appl. Sci., London, New York, (1989), 366 pp.

[29]Köhler, Arch. Environ. Contamin. Toxicol. 22 (1992), p. 334.

[30]Dallinger and P.S. Rainbow (eds.). Ecotoxicology of metals in invertebrates. Lewis Publishers, Boca Raton, FL, (1993).

[31]Eckwert, Ecotoxicology 6 (1997), p. 249.

[32]Eckwert, Appl. Soil Ecol. 175 (1997), p. 275.

[33]Cochrane, Comp. Biochem. Physiol. Vol. 98C, No. 2/3 (1991), p. 385.

[34]Stringham, Mol. Cell Biol. 3 (1992), p. 221.

[35]Guven, Aquat. Toxicol. 29 (1994), p. 119.

[36]de Pomerai, Human Experim. Toxicol. 15 (1996), p. 279.

[37]Norris, Mol. Biol. Evol. 14 (10) (1997), p. 1050.

[38]Dyer, Environ. Toxicol. Chem. 12 (1993), p. 1.

[39]Ku, Bull. Inst. Zool. Academia Sinica (Taipei) 30 (4), p. 319.

[40]Heikkila, J. Biol. Chem. 257 No.20 (1982), p. 12000.

[41]Misra, Biochem. Biophys. Acta 1007 (1989), p. 325.

[42]Ellis, Sem. Cell Biol. 1 (1990), p. 1.

[43]Georgopoulos, Semin. Cell. Biol. 1 (1990), p. 19.

[44]Hartl, Science 247 (1990), p. 930.

[45]Bochkareva, Nature 336 (1988), p. 1558.

[46]Sanders, Biochem. J. 297 (1994), p. 21.

[47]Dyer, Can. J. Zool. 69 (1991), p. 2021.

[48]Morimoto, K. Abravaya, G. Mosser and G.T. Williams. In: M.J. Schlesinger, M.G. Santoro and E. Garcia (eds.). Stress Proteins. Induction and function (1990). Springer Verlag, Berlin, Heidelberg, New York, Tokyo (1990).

[49]Beckmann, Science 248 (1990), p. 850.

[50]Pelham, Cell 46 (1986), p. 959.

[51]Gething, Cell 46 (1986), p. 939.

[52]Rothman, Cell 46 (1986), p. 5.

[53]Chiang, Science 246 (1989), p. 382.

[54]Kothary, Can. J. Biochem. 60 (1982), p. 347.

[55]Burgess, Exp. Cell Res. 155 (1984), p. 273.

[56]Kothary, Biochim. Biophys. Acta 783 (1984), p. 137.

[57]Williams, Environ. Toxicol. Chem. 15 (1996), p. 1324.

[58]Vijayan, Aquat. Toxicol. 40 (1998), p. 101.

[59]Triebskorn, J. Aquat. Ecosyst. Stress and Recovery 6 (1997), p. 57.

[60]Forsyth, J. Aquat. Animal Health 9 (1) (1997), p. 18.

[61]Janz, Toxicol. Appl. Pharmacol. 147 (2) (1997), p. 391.

[62]Norris, Mol. Biol. Evol. 12 (6) (1995), p. 1048.

[63]Koban, Mol. Mar. Biol. Biotech. 1 (1) (1991), p. 1.

[64]Grøsvik, Biomarkers 1 (1) (1996), p. 45.

[65]Craig, A. Rev. Genet. 22 (1988), p. 631.

[66]Sass, Mech. Develop. 54 (2) (1996), p. 195.

[67]Cho, Biochem. Biophys. Res. Comm. 233 (1997), p. 316.

[68]Fader, J. therm. Biol. 19 (1994), p. 335.

[69]Vijayan, Life Sci. 61 (1997), p. 117.

[70]Mosser, J. Cell. Physiol. 132 (1987), p. 155.

[71]Mosser, J. Comp. Physiol. B 158 (1988), p. 457.

[72]Spotila, Physiol. Zool. 62 (1989), p. 253.

[73]Udelsman, Surgery 115 (1994), p. 177.

[74]Köhler, in: Validierung und Einsatz biologischer, chemischer und mathematischer Tests und Biomarkerstudien zur Bewertung der Belastung kleiner Fließgewässer mit Umweltchemikalien, Abschlußbericht für das BMBF-Verbundprojekt (1998), p. 323.

[75]Pawert, Zelluläre und proteinchemische Reaktionen von Fischkiemen (Bachschmerlen und Bachforellen) sowie Anwendung limnochemischer Methoden zur Bewertung der Belastung kleiner Fließgewässer mit Umweltchemikalien, Medien Verlag Köhler, Tübingen, FRG, (1998), 212 pp.

[76]Eckwert, unpublished.

[77]Mazur, The heat shock protein response and physiological stress in aquatic organisms. PhD thesis, university of British Columbia, Canada, (1996), 175 pp.

[78]Bridges, Ann. NY Acad. Sci. 407 (1983), p. 42.

[79]Di Giulio, Aquat. Toxicol. 26 (1993), 1.

[80]Vigano. Sci. Total Environ. 151 (1993), p. 37.

[81]Hinson, Annu. Rev. Pharmacol. Toxicol. 32 (1992), p. 471.

[82]Braunbeck, T.. Zelltests in der Ökotoxikologie. Projekt: „Angewandte Ökologie". Veröff. PAÖ 11, (1995), 204 pp.

[83]Johnson, in: Cairns, V. W. Hodson, P. V. Nriagu, J. V. (ed.), Contaminant effects on fisheries. Wiley and Sons, New York critique, (1984), p. 20-36.

[84]Moore, Mar. Pollut. Bull. 16 (1985), p. 134.

[85]Vogt,. Aquaculture. 67 (1987), p. 157.

[86]Arias, I. M., Popper, H., Schachter, D., Schafritz, D. A.. The liver: biology and pathobiology. Raven Press, New York., (1988) 664 pp.

[87]Goksøyr, Mar. Poll. Bull. 33 (1997), p. 36.

[88]Meyers in: Rand, G. M., Petrocelli, S. R. (eds.) Fundaments of aquatic toxicology. Hemisphere Publ. Corp., Washington, (1985) p. 283-323.

[89]Westlake,. Comp. Biochem. Physiol. 76C (1983), p. 15.

[90]Arnold, Aquat. Toxicol. 33 (1995), p. 17.

[91]Thompson, in: Fossi, M.C., Leonzio, C. (eds.), Nondestructive biomarkers in vertebrates. Lewis Publisher, Boca Raton, Florida (1994), p. 37.

[92]Mehrle, in: Rand, G. M., Petrocelli, S. R. (eds.), Fundaments of aquatic toxicology: Methods and application. Hemisphere Publ. Corp., Washington (1985), p. 264.

[93]Hodson,. Water Res. 12 (1978), p. 869.

[94]Livingstone, Funct. Ecol. 4 (1990), p. 415.

[95]Wenning, Aquat. Toxicol. 12 (1988), p. 157.

[96]Mayer, F. L.. Dynamics of dieldrin in rainbow trout and effects of oxygen consumption Ph.D. Utah State Univ. Logan, Utah. Diss. Abstr. Int. 32(1970), p. 527.

[97]Zahn, Exp. Toxicol. Pathol. 48 (1996), p. 47.

[98]Ahne, ZBl. Bakt. Hyg. I. Abt. Orig. B.180 (1985), p. 480.

[99]Braunbeck, T.. Entwicklung von Biotestverfahren mit Zellkulturen aus Fischen zum Nachweis letaler und subletaler Schäden von Organismen durch Umweltschadstoffe im Wasser. Veröff. PAÖ 8 (1994), p. 537.

[100]Beyer, Aquat. Toxicol. 36 (1996), p. 75.

[101]Collier, Environ. Toxicol. Chem. 14 (1995), p. 143.

[102]Jimenez in: McCarthy, J. F., Shugart, L.R. (eds.), Biomarkers of Environmental Contamination. Lewis Publ. (1990), p. 123.

[103]Lindström-Seppä, Ecotoxicol. Environ. Safety 18 (1989), p. 191.

[104]Triebskorn, J. Aquat. Ecosys. Stress Recov. 6 (1997), p. 57.

[105]Andersson, Aquat. Toxicol. 24 (1992), p. 1.

[106]Kobayashi, Bull. Japan. Soc. Sci. Fish. 53 (1987), p 487.

[107]Kobayashi, J. Biochem. 114 (1993), p. 697.

[108]Van Veld, Toxicol. Appl. Pharmacol. 142 (1997), p.348.

[109]Payne, Bull. Envir. Contam. Tox. 14 (1975), p. 112.

[110]Payne, Science (Wash. D.C.) 191 (1976), p. 945.

[111]Kennedy, Anal. Biochem. 222 (1994), p. 217.

[112]van der Oost, Aquat. Toxicol. 39 (1997), p. 45.

[113]Förlin, Aquat. Toxicol. 8 (1986), p. 51.

[114]Monod, Water Res. 25 (1991), p. 173.

[115]Goksoyr, Comp. Biochem. Physiol. 100C, 1/2 (1991), p. 157.

[116]Collier, Mar. Environ. Res. 34 (1992), p. 195.

[117]Jedamski-Grymlas, Ecotoxicol. Environ. Saf. 31 (1995), p. 49.

[118]Otto, Arch. Environ. Contam. Toxicol. 31 (1996), p. 141.

[119]B.K-M Gadagbui, Biomarkers 1 (1996), p. 252.

[120]Machala, Environ. Toxicol. Chem. 16 (1997), p. 1410.

[121]Vigano, Environ. Toxicol. Chem. 17 (1998), p. 404.

[122]Kantoniemi, Ecotoxicol. Environ. Saf. 35 (1996), p. 136.

[123]Horness, Environ. Toxicol. and Chem. 17 (1998), p. 872.

[124]Van Veld, Carcinogenesis 13 (1992), p. 505.

[125]Truscott, Comp. Biochem. Physiol. 75C (1983), p. 121.

[126]Munkittrick, Can. J. Fish. Aquat. Sci. 48 (1991), p. 1371.

[127]Andersson, Can. J. Fish. Aquat. Sci. 45 (1988), p. 1525.

[128]Sandström, Water Sci. Technol. 20 (1988), p. 107.

[129]Sumpter, in: Taylor, E.W. Toxycology of Aquatic Pollution. Phyiological, Molecular and Cellular Approaches. Cambridge University Press. Great Britain (1996), p. 205.

[130]Danzo, Environ. Health Perspect. 105 (1997), p. 294.

[131]Navas, Environ. Sci. Pollut. Res. (in press).

[132]Triebskorn, Journal of Aquatic Exosystem Stress and Recovery 6 (1997), p. 57.

[133]Soimasuo, Aquat. Toxicol. 31 (1995), p. 329.

[134]Novi, Aquat. Toxicol. 41 (1998), p. 63.

[135]Andersson, Aquatic. Toxicol. 8 (1986), p. 85.

[136]Carpenter, Comp. Biochem. Phyiol. 97C (1990), p. 127.

[137]Hinton, in Biomarkers - Biochemical, Physiological, and Histological Markers of Anthropogenic Stress (Ed.: R.J. Huggett; R.A. Kimerle; P.M. Mehrle; H.J. Bergman), Lewis Publishers, Chelsea, MI, USA (1992), p 155.

[138]Braunbeck, Biol. Unserer Zeit 19 (1989), p. 127.

[139]Hofer, R., Lackner, R., Fischtoxikologie. Gustav Fischer Verlag, Stuttgart, (1995).

[140]Weibel, E.R., Stereological methods. Vol. 1, Academic Press, New York, (1979).

[141]Bolender, Annu. Rev. of Pharmacol. Toxicol. 21 (1981), p.549.

[142]Loud, Lab. Invest. 50 (1984), p. 250.

[143]Schramm, Biomarkers 2 (1998), p. 93.

[144]Schramm, Zur Ökotoxikologie kleiner Fließgewässer - Leberultrastruktur und hämatologische Parameter als Biomarker bei Fischen. Dissertation, Abteilung Physiologische Ökologie der Tiere, Universität Tübingen (1998).

[145]P.W. Werster, Comp. Biochem. Physiol. 100C (1991), p. 115.

[146]S.J. Teh, Aquat. Toxicol. 37 (1997), p. 51.

[147]J. Schwaiger, J. Aquat. Ecosytem Stress and Recovery 6 (1997), p. 75.

[148]R.A. Goyer, Current Top. Path. 55 (1971), p. 147.

[149]T.S. Gill, Ecotox. Environ. Safety 18 (1989), p. 165.

[150]P.L. Goering, Toxicol. 85 (1993), p. 25.

[151]T. Fischer-Scherl, Arch. Environ. Contam. Toxicol. 20 (1991), p. 454.

[152]M. El-Matbouli, Annual Rev. of Fish Diseases (1992), p. 367.

[153]C.D. Rice, Ecotoxicol. Eviron. Safety 33 (1996), p. 186.

[154]D.P. Anderson, Am. Fisheries Soc. Symp. 8 (1990), p. 38.

[155]S.M. Adams, Am. Fisheries. Soc. Symp. 8 (1990), p. 8.

[156]B.M. Sanders, Critical Rev. Toxicol. 23 (1993), p. 49.

[157]Braunbeck, Schr.-Reihe Verein WaBoLu 89 (1992), p. 109.

[158]Braunbeck, in Fish in ecotoxicology and ecophysiology (Ed.: T. Braunbeck; W. Hanke; H. Segner), VCH, Weinheim (1993), p. 55.

[159]Couch, in The pathology of fishes (Ed.: W.E. Ribelin; G. Migaki), The University of Wisconsin Press, Madison (1975), p. 559.

[160]Gluth, Ecotoxicol. Environ. Saf. 9 (1985), p. 179.

[161]Achazi, Z. Angew. Zool. 1 (1989), p. 3.

[162]Braunbeck, Ecotox. Environ. Saf. 21 (1991), p. 109.

[163]Pentreath, J. Exp. Mar. Biol. Ecol. (1976), p. 51.

[164]Köhler, Aquat. Toxicol. 16 (1990), p. 271.

[165]Meyers, Mar. Fish. Rev. 44 (1982), p. 1.

[166]Storch, Helgoländer Meeresuntersuchungen 38 (1984), p. 65.

[167]Storch, Marine biology 74 (1983), p. 101.

[168]Réz, Acta Biol. Hung. 37 (1986), p. 31.

[169]Bucher, Arch. Environ. Contamin. Toxicol. 23 (1992), p. 410.

[170]Phillips, M.J., Poucell, S.J., Patterson, J., Valencia, P., The liver: an atlas and text of ultra structural pathology. Raven Press, New York(1987).

[171]Köhler, A. Mar. Environ. Res. 28 (1989), p. 417.

[172]Heining, Exp. Toxic. Pathol. 45 (1993), p. 167.

[173]Klaunig, J. Environ. Pathol. Toxicol 2 (1979), p. 953.

[174]Konradt, Verhandlungen der Deutschen Zoologischen Gesellschaft 159 (1996).

[175]Braunbeck, Ecotox. Environm. Saf. 24 (1992), p. 72.

[176]Segner, in Zweiter Zwischenbericht zum BMBF-Verbundprojekt "Validierung und Einsatz biologischer, chemischer und mathematischer Tests und Biomarkerstudien zur Bewertung der Belastung kleiner Fließgewässer mit Umweltchemikalien" (Ed.: R. Triebskorn; W. Honnen; D. Frahne; T. Braunbeck; H.-R. Köhler; G. Schüürmann; J. Schwaiger; H. Segner; A. Oberemm) /(1997), p. 343.

[177]Negele, R.D., Hoffmann, R.W., Kurz- und Langzeitwirkungen von Atrazin auf Regenbogenforellen (Oncorhynchus mykiss). Klinische, hämatologische und pathomorphologische Untersuchungen. Bayerische Landesanstalt für Wasserforschung, Wielenbach (1991).

[178]Wood, Phys. Zool. 64 (1990), p. 1.

[179]Skidmore, Water Res. 6 (1972), p. 217.

[180]Lemke, Trans. Am. Fish. Soc. 92 (1963), p. 372.

[181]Abel, J. of Fish Biol. 6 (1974), p. 279.

[182]Rombough, Can. J. Zool. 55 (1977), p. 1705.

[183]Smith, J. Fish. R. Board Can. 29 (1972), p. 328.

[184]Laurent, Cell Tiss. Res. 259 (1990), p. 429.

[185]Lacroix, Aquat. Toxicol. 27 (1993), p 373.

[186]Mallatt, Can. J. Fish. Aquat. Sci. 42 (1985), p. 639.

[187]Jagoe, Aquat. Toxicol. 24 (1993), p. 241.

[188]Goss, J. Fish Biol. 45 (1993), p. 709.

[189]Perry, Can. J. Zool. 9, 1775-1786 (1992).

[190]Laurent Can. J. Zool., 67 (1988), p. 3055.

[191]McDonald, Phys. Zool. 64 (1991), p. 124.

[192]Haaparanta, J. Fish Biol., 50 (1997), p. 575.

[193]Kumaraguru, J. Fish Biol. 20 (1982), p. 87.

[194]Bindon, Can. J. Zool., 72 (1993), p. 1395.

[195]Chevalier, Can. J. Zool. 63 (1985), p. 2062.

[196]Schwaiger, J. Aquat. Ecosys. Stress Recov. 6(1997), p. 75.

[197]Luckenbach, Dritte deutschsprachige SETAC-Tagung „Ökosystemare Ansätze in der Ökotoxikologie", Zittau, 18. Und 19. Mai (1998) (in press).

Environmental Science Forum Vol.96 (1999) pp. 65-94
© 1999 Trans Tech Publications, Switzerland

Biomonitoring with Morphological Deformities in Aquatic Organisms

L. Janssens de Bisthoven

'LimCo International', An der Aa 5, DE-49477 Ibbenbüren, Germany

Keywords: Morphological Deformities, Vertebrates, Invertebrates, Chironomidae, Fish, Amphibians, Birds, Biomonitoring

Abstract

This paper focuses on the broad lines of investigation and the actual potential of the use of morphological deformities in aquatic vertebrates and invertebrates for biomonitoring in aquatic ecosystems. The use of morphological deformities in chironomid larvae (Diptera, Chironomidae) for biomonitoring is reviewed.

Introduction

Many aquatic organisms are used as bioindicators for environmental stress in aquatic ecosystems at different functional levels: the ecosystem-, community-, population-, species- and organism-level [1]. The latter level includes biochemical indicators, energy metabolism, enzyme activities, ion regulation, physiological indicators, behavioural responses, life-history responses, use as sentinel organisms (*i.e.* bioaccumulation measurements) - see overview of different methods in Johnson *et al*. [2] - and morphological deformities. The present paper deals with contaminant-induced externally visible morphological (anatomical) deformities and morphometric changes in aquatic animals, and hence mostly excludes chemically induced carcinogenic or mutagenic anomalies such as internal neoplasms in teleost fish and bivalve molluscs. For a review on more than hundred papers on this subject we refer to Couch and Harshbarger [3] and for the underlying genetical mechanisms, to Shugart and Theodorakis [4]. Morphological deformities are discussed here as ecotoxicological endpoints of anthropogenic stress, and hence as possible biomarkers (*sensu latu*) for environmental contamination. Some internal deformations of gills, limbs and vertebrae (*e.g.* skoliosis) in amphibians and fish are common and are often externally visible. Therefore, they are included in the overview and the discussion. Because of the relative scarcity of data on morphological deformities in mammals and reptiles, and their rather marginal role in aquatic ecosystems, they are not discussed here. However, it is clear that marine mammals, with their fat reserves and their position as top predators, are very much affected by pollution. The occurrence of pseudohermaphroditism (to be considered as an internal morphological change) in molluscs is also shortly discussed, as it is an interesting and well-documented response of a particular taxocene on a specific pollutant type in marine environments. A review is given for chironomid larvae (Diptera, Chironomidae), complemented with some original data and new rapid assessment methods, because of their inherent advantages as bioindicators, such as a pronounced morphological response, high densities and a benthic life history in some species. For a precise description and illustrations of the multitude of deformities discussed here, the specialized literature should be consulted.

The morphology of plants and animals is closely associated with their abiotic and biotic environments and is often the expression of phenotypic variation, as is for example the case for marine sponges [5], or for the root morphology of Japanese red cedar saplings, modulated by soil pH and Al content [6]. Deviations from the phenotype under influence of environmental factors may be designated as 'not normal' or as 'anomalies'. 'Deformities' designate morphological malformations, due to the physiological effects of one or more toxicants, and hence exclude damage caused by mechanical stress or wear, ageing and 'anomalies' within the phenotypical plasticity of the species.

An original study along the mediterranean coast of northern Israel [7] revealed the presence of deformed benthic Foraminifera, related to contamination by heavy metals. The bryozoan *Cristatella mucedo* can become deformed due to parasitism by the myxozoan *Tetracapsula bryozoides* [8]. Asymmetry in antler size in roe deer has been used as an index of individual and population conditions [9]. Fluctuating assymetry (FA: the random deviation from perfect symmetry in bilateral traits) reflects environmental (pollution, temperature, food) or/and genetic stress and can be seen as a special type of quantified morphological deformity.

The issue of using morphological deformities in aquatic organisms for biomonitoring purposes merits special attention because of a strong causality to toxic stress and a relatively easy diagnosis. The problematic, however, has a rather disparate character and hence lacks a uniformising theoretical background. Organisms from such different taxa as Bryozoa or aquatic insects are screened for contaminant-induced deformities. A whole range of methods - *e.g.* X-rays in vertebrates, light microscopy, SEM - is applied for organisms living in such different environments as marine and freshwater habitats, and in organs ranging from gills, vertebrae to chitineous mouth parts. Moreover, all kinds of contaminants, both in the sediments and in the water, have been reported to induce these deformities, in all kinds of life stages. Deformities are also designated in the literature as 'anomalies', 'deformations', 'malformations' and 'abnormalities' and they may be of teratogenic, genotoxic/mutagenic, carcinogenic or somatic origin. Besides environmental contamination, other causal factors are mechanical damage, food deficiency and genetic stress. Some deformation processes have been successfully applied as standard bioassays in ecotoxicology, such as the zebra fish embryo test, and the *Xenopus* embryo teratogenesis assay (FETAX, [10]), or as part of a biotic index, such as the Karr index for fish [11]. Different aspects might be investigated, ranging from genetics, histology, morphology, parasitology, biochemistry, up to population ecology. Also, deformities are caused by sub-organismal disturbances, but may have consequences at the population level and beyond, due to depressed fitness, altered behaviour, decreased food uptake, genetical drift and depression, and reproductive disturbances. Many questions remain to be answered as to the effectiveness of using deformities in water quality management and ecotoxicology and the quantification methods employed.

Ecological and genetical background

The fundamental issue behind the occurrence of deformed specimens in a population, due to toxic stress, is how the population will temporally cope with it, in a natural selection perspective. The paper by Holloway [12] is useful in this discussion: the framework is an environment containing sufficiently low levels of toxins which cause morphological damage, without eliminating the species. The damage is sublethal and eventually morphologically visible. The question is whether such a situation can reach an equilibrium, maintaining a state with deformed individuals, or will evolve to a new state due to homeostatic processes or natural selection of susceptible alleles. The crucial question for biomonitoring is at which toxic burden deformities will appear and until which extent the population will survive and still contain deformed individuals. These toxicity thresholds will change in function of several ecological processes, such as differential mortality rates, development rates or reproduction rates in normal and deformed animals. The final outcome is the result of an interaction between population responses and toxic effects. Toxicity may be lethal for part of a population only (the classical ecotoxicology is based on this: a measure is the LC_{50}). In this context, deformities have to be classified as either inheritable or not inheritable. Inheritable deformities will tend to disappear by natural selection from the population over several generations, if they cause lower fitness in the individuals, or if they are the expression of other lower fitness factors. While not inheritable ones will be induced in any generation when the conditions are favourable for induction and hence are a within-generation effect. Crossing of homozygote parents with recessive alleles causing a deformity will produce deformed offspring. However, the occurrence of heterozygoty in these alleles may mask the epigenetic deformity over several generations. That deformed fish juveniles in fish hatcheries may have an epigenetic origin has been elegantly proved [13]. It appeared that these individuals were issued from the fertilisation of aged oocytes, whereby chromosomal abnormalities occurred.

Deformed individuals are the visible part of a complex process of toxicity stress in a population. Depending on the type of contaminants and their combined effects, and the genetic configuration, the population will try to cope with the stress inflicted by contaminant bioaccumulation essentially in two ways: a trade-off between growth and survival rate. This may happen before, during or after the appearance of deformities. Resources may be channeled into detoxification instead of growth, and so achieve relatively low mortality rate, but long development period, or conversely: resources are allocated to growth instead of to detoxification and relatively high mortality rates are achieved. Individuals will respond in one of three ways to a toxin which produces a novel environment with consequent novel gene expression [12]: (1) the new genes expressed may enable some individuals to maintain fitness by continuing to optimize energy allocation, (2) other individuals suffer sublethal effects of the toxin. The new genes expressed give them a chance of survival, but at the expense of other fitness functions. Energy is no longer partitioned optimally and fitnes suffers. (3) The new genes expressed do not enable the individual to survive the toxic effect, resulting in death.

The presence of deformed individuals in a population is due to a type (2) reaction. For Chironomidae larvae (Insecta, Diptera) deformed individuals show higher mortality rates than normal larvae [14]. However, this tendency is strongly population-dependent or site-specific. In some populations no apparent fitness differences were found between both morphotypes [14,15,16]. Two factors may be responsible for that: (1) optimal environmental conditions (food, low predation, low physical disruptions) enable deformed larvae to survive equally well. (2) Genetical adaptation, only possible in constant conditions, plays a role in reaching a stable and relatively fit population, containing both normal and deformed specimens. Adaptation or resistance can result from the production of adaptive enzymes, protein-binding of metals, induction of mixed-function oxydases, population selective action favoring survival of resistant individuals, changes in reproductive strategy [17], higher detoxification rates [18] and the production of metal containing granules [19]. The effect of toxicants on microevolutional events should not be understimated. Toxicant-induced microevolution in lotic Chironomidae is however often disrupted [20], by incoming gene flow of drifting clean upstream larvae. A combination of *in situ* deformity screening and a bioassay involving the assessment of deformities and fluctuating assymetry in parental and first generation animals of the tested species in the suspected sediment was recommended [21]. The FA analysis gives clues about genetic stress, while the deformity rates give indication about the contaminant burden.

Aquatic vertebrates

Most deformities in vertebrates concern bone malformations, easily diagnosed by X-ray analysis, or egg shell thinning in birds. Even in remote areas such as the central North Pacific Ocean, traces of PCBs and DDT are present in albatrosses [22], which were only half the calculated threshold concentration for egg shell thinning or embryo malformation. However, at sufficiently low body concentrations, organochlorine compounds could not be associated with deformities in 4 aquatic bird species, in a bird sanctuary in Texas [23]. Similarly, morphometric changes in embryos of the common tern (*Sterna hirundo*) and the night heron (*Nycticorax nycticorax*) from locations at the Great Lakes, were not related to DDE, other organochlorine pesticides and mercury, while clear correlations were found with PCB concentrations [24]. The dioxin-like planar halogenated hydrocarbons (PHH) were suggested as being the causative agents for congenital deformities in the endangered Forster's tern (*Sterna forsteri*) in Wisconsin, USA [25]. Several species of colonial fish-eating birds nesting in the Great Lakes basin, including herring gulls, common terns and double-crested cormorants, have exhibited chronic impairment of reproduction, probably due to 2, 3, 7, 8-tetrachlorodibenzo-p-dioxin, non-ortho-substituted PCBs and furans. This is characterized by eggshell thinning, high embryonic and chick mortality, edema, growth retardation and deformities, hence the name "the Great Lakes embryo mortality, edema, and deformity syndrome"

(GLEMEDS)[26]. The deformities include bill deformities, club feet, missing eyes and defective feathering. Plausible routes of exposure and sources of these compounds and statistically significant dose-response relationships have been demonstrated. Selenium was suggested as a strong deformity inductor - up to 16% of inspected embryos, indicating a teratogenic action- in American coots (*Fulica americana*) and black-necked stilts (*Himantopus mexicanus*). The deformities included bill, wing, brain, eye, leg and feet defects. Polychlorinated dibenzodioxins (PCDDs) and polychlorinated dibenzofurans (PCDFs) caused intercerebral asymmetry in heron hatchlings in British Columbia, Canada [27]. These deformities were correlated with changed behaviour. Abnormal development increased with polychlorinated aromatic hydrocarbon exposure in eggs of the common snapping turtle *Chelydra serpentina serpentina* [178]. Aquatic reptiles and birds, as top predators in the aquatic food chain, are especially sensitive to waterborne or biomagnified contaminants.

Some examples of fish and amphibian deformity studies are given in Table 1. A measure of ecosystem recovery after the Exxon Valdez oil spill was developed by assessing genetic damage and physical deformities in embryos of exposed Pacific herring [29]. Vertebral deformations in fish are not necessarily direct teratogenic effects. Vertebral deformities in the African catfish (*Clarias gariepinus*), induced by malathion, an organophosphorous acetylcholinesterase inhibitor, were the result of notochord deformities, which for their part resulted from uncontrolled body contractions [30]. More data exist on skeletal lesions in fish, because the high proportion in which they occur is a serious problem in aquaculture [31]. These authors mention the following non-congenital causes of deformities in natural fish populations: high levels of toxins, pesticides, or heavy metals [*e.g.* 32], problems in the gallbladder [33], or bacterial endotoxins [34]. Nutritional deficiencies, high incubation densities and high levels of solar radiation are also possible causes for the occurrence of deformities in cultured fish [31]. Viral infections, a deficiency in polyunsaturated fatty acids or Vitamin A were suggested as possible causal agents in the formation of swimbladder deformities in cultured sea bass [35]. Compounds with mechanisms of action similar to those of beta-naphtoflavone might cause serious neoplasms in fish [36]. Skeletal anomalies are widespread in freshwater and marine fish in Japan [37], due mainly to the effects of agrochemicals. Many internal morphological lesions have similar cellular origins and morphogenesis in fish and mammals [38]. For the fish *Thymalus thymallus* exposed to methylmercury during embryogenesis, no effect on FA or any departure from a normal morphology was found [39].

That deformities in marine fish are effective biomarkers for xenobiotics has been demonstrated in a extensive study [64]: 15 types of 'gross abnormalities' (skin lesions, ocular and branchial abnormalities) were found in samples containing more than 24.000 specimens from more than 300 localities, representing 143 species in the Atlantic and the Gulf coast. Background prevalences were estimated to be less than 0.7%. Demersal fish showed up to three times higher prevalences of deformities than pelagic fish, due to contact with contaminated sediments. Competition and predation may eliminate teratogenically deformed fish larvae, thus masking certain contamination effects [50]. In another study in the Wadden Sea (32 stations, 124.000 fish examined), 0.4 to 8% of the fish, dependent on the species, showed external lesions, such as skeletal deformities, lymphocystis, papillomatosis and infectious ulcerative diseases [65]. The study lacked quantitative data to relate the fish lesions to marine pollution. The authors concluded however, that the effects of a number of pathogenic parasites on fish survival is considered to be more serious than that of deformities and infectious diseases. The question is whether these parasites are also influenced by pollution. *Pleuronectes americanus* and *Micropterus dolomieui* showed higher deformity frequencies in medium-sized fishes, and lower in juveniles and older fish [52,54] . This suggests that these deformities have no teratogenic origin and that deformed fish have a higher mortality rate than unaffected fish. Unexpectedly, levels of pesticides and heavy metals in deformed and normal smallmouth bass were similar [54]. The effects of pulp mill effluents were not lethal for fourhorn sculpin (Table 1), which showed up to 60 % deformity frequency [42]. Migration might have an attenuating influence on the effects of local effluents [42].

Table 1. Examples of deformities in fish and amphibian tadpoles or adults, and their inferred causal agents.

Taxon	Deformity	Causal agent	Reference
FISH			
general	teratogenic deformities in larvae	Selenium, >20μg g^{-1}	[40]
Thymallus thymallus	no effect on FA	Hg, 20μg l^{-1}	[39]
Myoxocephalus quadricornis	scoliosis, deformation/dislocation of vertebrae	heavy metals	[32]
idem	vertebral deformities and changed mechanical properties	tetrachloro-1,2-benzoquinone bleached kraft mill effluents	[41]
Myoxocephalus quadricornis	vertebral deformities 9.4-59.5%	pulp mill effluents	[42]
Pimephales promelas	deformed shape, no scales	Selenium, 10-30 μg/l	[43]
Gambusia holbrooki	gill thickening, loss of lamellae	Hg (II) , 75-300 nM	[44]
Larvae of 10 marine species, Japan	vertebral deformities	herbicide trifluralin	[45]
Perca fluviatilis	opercular deformities	pulp mill effluent	[46]
Rutilus rutilus/Perca fluviatilis	gill anomalies	acid peaks	[47]
4 marine species *	missing/deformed dorsal finrays scale disorientations	sediment PAHs, Cu	[48]
Benthic sculpins (2 species)	fin/tail necrosis	sediment : pulp-paper mill	[49]
Centrarchids	lordosis, kyphosis, scoliosis, edema, exophthalmus, cataract, head-mouth-fin	Selenium	[50]
11 marine species	stunted/missing dorsal spines/rays, depression of dorsal profile, scale orientation	agricultural/industrial/residential	[51]
Pleuronectes americanus	bifurcated gill filaments	contaminated sediments	[52]
Brachydanio rerio	edema, malformation of notochord	TCDD, 5 ng/fish	[53]
Micropterus dolomieui	lordosis, kyphosis, scoliosis up to 30%	unknown	[54]
Gadus morhua, Clupea harengus	severe deformities in embryos	sea-surface microlayer of petroleum	[28]
Ambassis commersoni	vertebral deformities	cadmium	[55]
Seriola quinqueradiata	vertebral deformities	agrochemicals	[56]
Cyprinodon variegatus	skeletal malformations	malathion	[57]
Salmonids	vertebral dysplasia	Herbicide trifluralin	[58]
Colisa fasciatus	gill architecture	Nickel	[59]
AMPHIBIA			
Rana perezi	scoliosis, deformed limbs	carbamate insectic. ZZ-Aphox, 0.25 mg l^{-1}, organophosphate folidol, 1 mg l^{-1}	[60]
Xenopus embryos	skeletal abnormalities, edemas	mixed xylenes, toluene	[61]
Rana perezi	single and multiple malformations in tail, limbs and vertebral column	deficiency in compound diet	[31]
Rana cates beiana	reduced number and deformations in labial papillae	As, Cd, Cr, Cu, Se and other elements	[62]
Rana catesbeiana, X. laevis	gross spinal deformities	aqueous dieldrin, 1.3-25 μg/L	[63]

* species-specific differences were found

That deformities may have consequences for the survival of the species has been elegantly demonstrated with tadpoles [62]: animals with a reduced number of labial papillae had a lower growth rate due to a lower grazing efficiency. In the US, a widespread and dramatic increase of the percentage of deformed amphibians has been noted in the beginning of the nineties [66]. Several hypotheses were put forward: (1) the formation of cysts in the limbs, due to infestation by a trematode parasite, with the formation of extra deformed limbs as a consequence, (2) predation attempts, with missing limbs as a consequence, or (3) chemical stress with the eventual involvement of the disruption of retinoic acid (a metabolite of vitamin A) pathway.

Such massive developmental disruptions in aquatic vertebrates necessarily involve chemicals which are able to strongly interfere with their physiology. One aspect is the endocrine system, which has an strong regulatory function in all aspects of life of an organism. There are a number of well-known xenobiotics which mimic the natural hormone estradiol [67]: organochlorine pesticides, such as DDT and PCBs, and breakdown products of the alkylphenoxy detergents. Kime [67] mentions the following measurable morphological parameters of the reproductive system of fish, which may be used to determine its most sensitive component to a specific pollutant: pituitary morphology, gonad histology, sperm morphology, egg size and larval development.

Invertebrates, excl. Chironomidae

Deformities related to environmental stress have been reported in a number of aquatic insects: Hemiptera [68], Plecoptera Capniidae [69], Plecoptera Perlidae [70], Trichoptera Hydropsychidae [70,71,72,73], and in Oligochaeta [74,75,76]. In one field study, chlorine and crude oil were suspected to be the agents causing deformities in tracheal gills of Plecoptera and Trichoptera [70](Table 2).

Wear in buccal structures of insects as a consequence of prolonged use on hard substrates, has been documented as well: the cuticle of claws and mandibles of a number of forest insects from different families (including Diptera) is hardened by incorporation of some metals [77] , such as zinc and manganese. This in order to provide cutting edges to these wood boring or leaf cutting insects. Mandible length during the final instar of *Heterocampa guttivitta* (Notodontidae) may be reduced by 20%, due to mechanical wearing during the feeding process [78].

Table 2. Examples of deformities in invertebrates, used as biomarkers for environmental contamination (excl. Chironomidae).

Taxon	Deformity	Causal agent	Reference
Phasganophora capitata (Plecoptera) *Cheumatopsyche sp.* (Trichoptera)	deformed gills	chlorine, crude oil	[70]
Gammarus pulex	disorganized peripheral muscle network of ceca	Lindane, copper	[79]
Tubifex tubifex,(Tubificidae) *Ilyodrilus frantzi*	loss of chaetal hair loss of pectinates	pH, salinity, hardness, mercury	[76]
Specaria, Frazeri, Nais communis (Naididae)	no effect	pH, salinity, hardness, mercury	[76]
Potamothrix hammoniensis(Tubificidae)	chaetal hair deformities	mercury, other heavy metals	[74]
Tubificidae	loss of hair setae	sediment contamination	[75]
Horseshoe crabs, 3 species	segment-defective, double, no-posterior embryos /abnormal eye areas	Hg>=organotin>Cr=Cd>Cu>Pb>Zn	[80]
Hydropsyche contubernalis, H. siltalai	damages anal papillae/darkening of abdomen	Cd	[73]
Hydropsyche spp.	abnormal nets	heavy metals/toxic chemical wates	[71]

In order to assess sediment contamination in lakes, Lafont [75] proposes the index $I_O=10.S.T^{-1}$, where S is the total number of oligochaete species and T the relative abundance of Tubificidae without setae in the whole oligochaete community, and the composite index E_0, containing a letter which represents the relative abundance for Tubificidae without hair setae and a suffix which represents species richness. A very specific form of morphological modification is found in Mollusca: under influence of pollution and other abiotic factors in the marine environment, pseudohermaphroditism may occur, mostly in females, in a high number of prosobranch species [81]. These authors state that the occurrence of pseudohermaphroditic females in Mollusca, also called "imposex", has been analysed with increasing intensity since 1970 by different research groups (see more than 70 papers on this subject, [81]). Imposex is induced by leachates from tributyltin (TBT), contained in antifouling paints. Different species show different sensitivities and the occurrence of imposex can vary between 0 and 100 %. Many types of imposex exist, which has lead to several attempts of classification and quantification in function of pollution. The two main indices are the "relative penis size index" (RPS) and the "vas deferens sequence index" (VDS). Histological work or the use of electron microscopy and a specialized knowledge of the sexual apparatus morphology in Mollusca is needed in order to determine these indices. Amongst the genera *Nucella*, *Ocenebra* and *Hinia*, living along the Atlantic French coast, *Nucella* as the most sensitive to TBT, to be used in a concentration range below 2 ng Sn l^{-1} [82]. At higher concentrations, less sensitive genera such as *Hinia* should be assessed for imposex occurrence. Populations of *Ocinebrina aciculata* are declining in France due to TBT pollution [83], a clear example of the deleterious effect of a pollutant at the population level. In 1982, French regulation banned the use of antifouling paints containing TBT on boats less than 25 m long. The reason was that oyster cultures in the neighbourhood of marinas showed a high rate of shell deformities [84]. Anomalies in shell calcification mechanisms appear in *Crassostrea gigas,* exposed to concentrations of 0.05 µg TBT l^{-1} [85].

Gassman *et al.* [48] were not able to find correlations between deformities in the blue crab (*Callinectes sapidus*) and sediment concentrations of aromatic and aliphatic hydrocarbons, polychlorinated biphnyls or 5 heavy metals. This illustrates the inherent difficulty of field studies, where part of the ecosystem remains a black box. The horseshoe crab species show in many parts of the world a population decline and some morphological deformities [80]. These authors not only were able to induce deformities in acute and chronic exposure bioassays with a range of heavy metals and organotin (Table 2), but also tried to elucidate part of the deformity inducing mechanisms of mercury (segment-defective embryos). Probably, mercury inhibits cell movement during morphogenesis.

Chironomidae
Advantages and limitations of using chironomids in biomonitoring
In freshwater biomonitoring, a lot of attention has been focussed on Chironomidae larvae because this insect group offers many advantages as potential bioindicators [86]. A recent reference book on Chironomidae, edited by Armitage, Cranston and Pinder [87] discusses all aspects of their systematics, morphology and ecology. The dipteran family of the Chironomidae is ubiquitous and represents the largest family of aquatic insects. Thanks to their high species diversity, the larvae (4 instars) are found in almost every type of aquatic habitat. Benthic species are in close contact with aquatic sediments and hence with the associated pollutants bound to detrital organics and mineral particles. Most benthic chironomids are filter feeders and/or scrapers (mixed feeders). Consequently they come into close contact with pollutants through their diet. Thanks to their often multivoltine life cycle and their high densities, the larvae are relatively easy to sample. Some Chironomini and Tanypodinae species are very tolerant to anoxic conditions prevailing in polluted environments. When other biota are eliminated, these surviving organisms may be suitable indicators of pollution changes. The relatively large body size of instar 4 larvae from certain species and their suitability to laboratory experiments is also to their advantage. Finally, their morphological deformity response can be quantified in a variety of ways, due to the indentation or segmentation of the larval structures involved. Other life stages of midges may become deformed as well, though data on this are almost nihil: malformed imagos (colour, shape of genitalia and wings, elongated legs) were obtained after lead exposure of larvae [88]. Egg masses with abnormal shape and texture were produced after exposure to an endocrine disruptor [89] (Table 3). A useful feature in the monitoring of historical changes in ecosystem contamination is the existence of chironomid subfossil head capsules in ancient sediment layers: the chitinous head capsules are preserved in the different sediment layers of lakes, so that shifting of species assemblages and the appearance of morphological deformities in certain sediment layers give useful palaeolimnological data on ecological and pollution events over time [90,91]. Furthermore, Chironomidae have proven their usefulness in environmental assessment at the community-level as well, with *e.g.* the lake trophic state classification system [*e.g.* 92,93]. Integration of the different functional levels applied to chironomids as bioindicators is strongly recommended [86] to achieve a more global view on structural and functional degradation of ecosystems. Chironomidae do not meet all requirements for an 'ideal' indicator, as put forward by Johnson *et al.*[2]: (1) the taxonomy is complex and species recognition by non-specialists is limited. (2) For the majority of species, ecological characteristics are not well known. Further, typical bioindicator species, used for deformity-monitoring, only occur in systems of medium to high saprobity.

An example of a limitation is the occurrence of *Chironomus riparius* in Belgian lowland rivers. From 61 net-samples taken in Flemish lowland rivers (Schelde and Meuse basins), 33% did not contain a single *Chironomus* larva, and another 16 % contained not enough larvae for statistical analysis. That means that only half the sites were appropriate for the deformity screening of *Chironomus* larvae for an ecotoxicological evaluation of sediment-bound xenobiotics. The preference of *Chironomus* gr. *thummi* for organically rich waters is well-documented [97,98,99]. *Chironomus* gr. *thummi* is classified as respectively polysaprobic and alpha-mesosaprobic [100,101]. This sets some limits to the use of *Chironomus* larvae as bioindicator.

Mechanical wear of the buccal structures
Chironomid larvae are subject to mechanical wear. Warwick [86] mentioned the structural differences between worn teeth of the mentum and mandibles and deformed teeth. Frank [102] also alluded to the problem and in his study he made sure to include deformities issued from teratogenic stress and not from mechanical damage, by using SEM-micrographs to pinpoint the differences. Lenat [103] included mechanically damaged structures in a deformity quantification scheme, hence probably pooling anomalies resulting from different stress factors. Scoring total perceived 'deformities' (better called 'anomalies', including pollution-induced 'deformities' and mechanically damaged structures), may mask the real effects of toxic compounds, as this method will often include anomalies caused by mechanical wear as well [104] . Mechanical damage may be indirectly caused by a weakening of the chitin by pollution stress [105]. In that case, a lack of pigmentation in the affected structures is observed and these anomalies were designated as 'integral deformities' [105]. Since it is often difficult to distinguish these deformities from mechanical damage, not related to pollution stress, the conservative approach would be to disregard specimens with broken structures [105, Warwick pers. comm.]. It is strongly adviced to separate both phenomena, since laboratory and field data showed that an inverse relationship between mentum wear and toxicant concentration was possible [104]. Mechanical wear of the mentum in *Chironomus* larvae decreases with increasing organic matter content in the substrate [15]. A direct relation between the occurrence of abrasion in third instars and the occurrence of deformities in fourth instars of *Chironomus riparius* could not be detected [106].

The Hamilton-Saether hypothesis
Hamilton and Seather [107] concluded from the distribution of deformed larvae that deformities might be induced by pollution originating from heavy metals and organic xenobiotics (especially pesticides) but not from domestic effluents. In latter publications, this statement is often being referred to as the 'Hamilton-Saether hypothesis'. In two case-studies the Belgian Biotic Index (BBI, [108]) was compared with some deformity metrics in the same *Chironomus* populations. The BBI is interpreted as a measure for organic pollution, but other pollution types might modulate the BBI as well [94, 95]. The first study was done on 36 sites in polluted lowland rivers in Northern Belgium, Flanders. Percentages of deformed larvae and mean deformity severities for *Chironomus* gr. *thummi* larvae were matched against the corresponding BBI-values, and this for 5 head capsule structures (mentum, mandibles, antennae, premandibles and pecten epipharyngis). Apart from only one negative correlation between the BBI and the percentage of larvae with deformed antennae (n=36, r=-0.40, $p<0.016$), no linear relations were detected. In a second study, a spatial and temporal comparison was carried out in the city of Ghent (Belgium, Flanders), which has a complex network of canals. The effectiveness of a water purification remediation was measured by means of the Belgian Biotic Index (BBI) and frequencies of head deformities in *Chironomus* larvae. From the 7 sampling points (20-30 4th instars per site), 6 sites had an increased BBI a few months after the remediation (average increase from BBI=2 to BBI=3), indicative of a slightly better water quality in terms of higher oxygen and lower toxicity of the water. In two sites, *Chironomus* larvae were found before and after remediation. The deformity responses of all structures together (40-100 %), mandibles (0-4 %) and menta (15-40 %), taken separately, were not different between the sites and between the years (Kruskal Wallis, p>0.05). This indicated that despite a slightly better BBI, sediment toxicity remained substantial. The lack of a direct relationship between deformities and the BBI, a measure of global river degradation and oxygen decline, is suggested in both studies. River saprobity or the trophic status of a lake may however have an indirect effect on the bioavailability of pollutants, and thus on the

deformity response of Chironomidae [90]. Lenat [103] concluded that the influence of the organic load in a river on the deformity response is rather negligible. Lien [145] did not find any concluent relationship between the saprobic state of a river (expressed by the Belgian Biotic Index, BBI) and the occurrence of morphological deformities in Flemish waters. The negative relation between antennal deformities and the BBI should be further investigated. Antennal morphological responses to toxic stress are rather independent from morphological responses of buccal structures and are probably expressing different stress factors as well [132,155]. Hakanson [168] demonstrated an increased PCB-toxicity and a decrease in heavy metal toxicity whith increasing bioproduction of Scandinavian lakes. This study demonstrated that the occurrence of buccal deformities in sediments of lowland rivers in Belgium is independent from the Belgian Biotic Index values, hence confirming the Hamilton-Saether hypothesis. Deformities thus could generate complementary information to the BBI about the pollution state of a water body.

The link to heavy metals and organic xenobiotics (see Table 3) makes the use of deformities in monitoring campaigns so promising because they could provide information about industrial and agro-chemical pollution in the sediments, which would be complementary to the existing biological methods which are rather oriented to eutrophication and saprobity problems. Three studies looked at the levels of single pollutants in deformed and normal larvae, and found higher PAH-levels in deformed *Chironomus anthracinus* larvae [143] , higher lead [144,162]- and copper [144]-levels in deformed *C.* gr. *thummi* larvae. The last study stressed that positive feedback mechanisms towards enhanced bioaccumulation in deformed larvae, due to longer exposure time and smaller size, might have provoked these results.

Since deformities appear during the molting process [106], which is strongly regulated by hormones, endocrine disruptors were put forward as possible deformity inductors. Two recent studies brought conflicting results: deformities in *Chironomus riparius* larvae could not be induced with Beta-sitosterol [167], while Meregalli and co-workers (pers. comm.) found a high morphological response of the same species to 4-N-Nonylphenol. The high number of field correlations between deformities in *Chironomus* larvae and phtalates found in an extensive field study [155], seem to support the hypothesis of a potential role played by endocrine disruptors.

Strategies in finding cause-effect relationships
An inverse relationship between quantified antennal deformities and the concentration of DDE was found [169]. The inverse relationship probably represented an isolated portion of the total response of antennae to DDE. Negative relationships between deformities and pollutants, or higher morphological responses at intermediate pollutant values were also reported [112,155,160]. Possible response models are represented in Fig. 1. Significant morphological responses of benthic midge larvae to pollutant concentrations may be overlooked when only linear correlation and regression methods between single deformity metrics and single pollutants are used. Synergistic or antagonistic interactions amongst pollutants, and deformity responses to several pollutants at the same time are possible. Ways of quantifying the toxicity of pollutant mixtures and relating it to deformities are provided by Vermeulen [169]. An example is given here of a more holistic approach based on multivariate analysis. In the following case-study (see Janssens de Bisthoven [96] for more details), deformity frequencies were calculated for each of 22 populations (min. 30 larvae per site) for a selection of 9 deformity types in fourth instar *Chironomus riparius* larvae, and related to sediment concentrations of 5 heavy metals. The deformity scores were first submitted to a cluster analysis (single linkage amalgamation, Euclidean distances).

Based on an arbitrarly defined cut-level, three site-clusters were defined. The site grouping was then used in a stepwise canonical discriminant function analysis on the metal concentrations (Fig. 2) to test whether the metals were able to separate the *a priori* defined site clusters (defined in the cluster analysis on deformity percentages) at a significant statistical level. All metals, except Cd and Zn, were retained by the analysis at significant p-levels (range $F(10, 30)=4.5-5.3$, $p<0.03$). Most sites fall in the three clusters inferred from the cluster analysis on the percentages of deformed larvae. The metals Cr and Pb score for the sites in the c-cluster, and Zn, Cu and Cd score for the sites in the b-cluster.

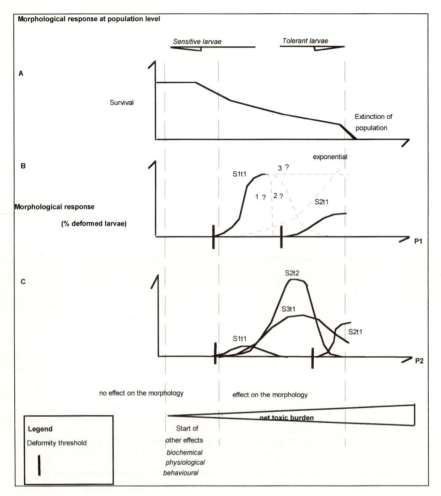

Fig. 1. Hypothetical representation (with the cooperation of Gerhardt and Goddeeris, pers. comm) of the deformity response of a population of benthic Chironomidae on different types and intensities of pollution stress. S=structure, t=deformity type, P=type of pollution stress. The term 'net toxic burden' was used in the scheme proposed by [169]. A, Survival of the population in function of pollution stress; B, theoretical possibilities of deformity response to pollution stress (quantal dose-response relationships hypothesis, [86,169]); C, example of a possible deformity response of a population to a hypothetical pollution situation. The responses are bell-shaped, sigmoid or linear (parts of the other curves)(from [96]).

Four positive linear regressions ($p<0.05$) between single metals and deformity metrics were found: for Pb and split medial mentum tooth, Cr and abnormal number of outer lateral mentum teeth, Cr and non-homologous antennal structures, and Cd and antennal vestigial last segment. Two negative correlations were found: for Cu and Zn and mandible assymmetry. In another multivariate approach, Giggins and Stewart [161] found a good linear correlation between deformities in *Chironomus* larvae and the factor scores generated by 9 heavy metals. Vermeulen *et al.* [166] made a similar approach too: 15 sites were grouped with a cluster analysis on the basis of a number of deformity variables. The sites fell neatly into 4 pollution categories. Like in [155], the sites (populations) were represented in function of the deformity variables in a factor analysis [166].

Site grouping (a,b,c) according to % deformed larvae for 9 types of deformities

Discriminant analysis on heavy metals in 22 sites

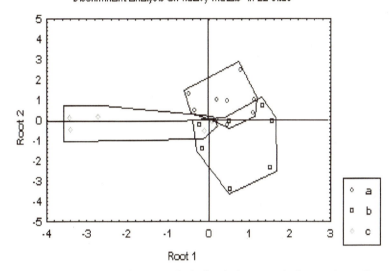

Fig. 2. Forward stepwise canonical discriminant analysis on the sediment concentrations of the heavy metals in 22 sites. The site grouping is derived from a cluster analysis on the deformity types, from [96].

Ontogeny of deformities
Until recently, very little information was available about the ontogeny of deformities, especially the question as to when and how deformities are formed. Are deformities really of teratogenic origin? Can deformities be induced in all larval instars and are they maintained throughout the different moults? Irradiation of *Chironomus riparius* egg ropes with cobalt 60 resulted in a weak chitinization in all larvae and a dose-dependent mentum deformity frequency in 4th instars [170]. This experiment showed (1) that deformities in Chironomidae larvae can be induced by damaging the genetic material of eggs and egg larvae and (2) that at least some of the resulting deformities are maintained through the different moults until instar four. Vermeulen and co-workers [106] have shed some new and interesting insights into the ontogeny questions. First, they could not find any deformed first instars in the field, while all other three instars exhibited deformities, with a cumulative effect over the subsequent instars.

This excluded a teratological cause in the analysed populations. In another study [16], however, deformed first instars were found after exposure of egg ropes to contaminated sediments. In the study by Vermeulen *et al.* [106], deformity percentages in winter 3rd and 4th instars were higher than in summer larvae, probably due to a longer exposure time. This seasonality in deformity frequencies has been found in other studies [16,96]. Most buccal deformities were passed to the next instar, while a few disappeared partly or completely ('repair') and a few new ones appeared, independent of the existing ones [106]. This explained the increasing deformity rate in subsequent instars.

It is important to know if the morphological deformities encountered in chironomid larvae can be passed from one generation to the next, because it could have consequences for biomonitoring.

Table 3. Chronological list of experimental and field studies with indication of reference, Chironomidae species, main prevalences of deformed larvae and associated pollution types. Only some minimal and maximal deformity frequencies are given, irrespective of the species mentioned in the list. (un)poll.=(un)polluted, (un)cont.=(un)contaminated, orgchl. pest.=organochlorine pesticides, organic=organic xenobiotics, agrochemicals, Chironomini=*Chironomus* spp. and other Chironomini, Pro.=*Procladius*, Other=other chironomid taxa, DEF= deformed, NOR=normal, ISAD= Index of Severity of Antennal Deformity, ISLD, Index of Severity of Ligula deformity, BBI=Belgian Biotic Index, FA=fluctuating asymmetry. Only the genera where deformed individuals were found are given. In the last column, reference values for less or not contaminated sites are underlined, and cases where known pollutants did **not** have an effect are in **bold**.

Taxon / Chironomini	Pro.	Other	Metal	Organic	Mixed, other	Deformity % and remarks	Reference
EXPERIMENTAL STUDIES							
C. decorus				coal liquid		2,7-4,6% DEF, **no clear** dose-response relationship	[109]
C. samoensis					UV-radiation	abnormal segment patterns	[110]
C. decorus			substrate Cu			pecten epipharyngis: >60%	[111]
C. gr. thummi						no diff. in emergence in NOR and DEF	[15]
Glyptotendipes barbipes			Pb			DEF larv., pupae, imag./DEF polytene chromosomes	[88]
Chironomus sp.				DDT/Dacthal		60%/15% DEF	[112]
C. tentans			**not** Pb/**not** 210Pb			10% median mentum tooth: inbreeding	[113]
Chironomus sp.				chlorpyrifos		**no clear effect on FA**	[114]
C. riparius			Cd/ Zn			12%/28% DEF	[96]
Chironomus etc...		X			metals/oil		[115]
C. gr. thummi				PAH/xylene/toluene		15% DEF	[116]
C. tentans				4-nonylphenol		up to 19% deformed egg masses	[89]
C. riparius					X	no diff. in pupation time in DEF and NOR	[16]
C. riparius			Cu/ Pb			ca. 10% DEF	[117]
C. riparius			Pb/ Hg	**not** sitosterol		up to 60% DEF	[167]
C. riparius			4-nonylphenol/ bisphenol-A			>20% DEF	Meregalli, 1999*
FIELD STUDIES							
C. tentans					radioactivity	inversion chromosomal polymorphism	[118]
Chironomus					X	first DEF reported	[119]
Chironomus					radioactivity	first DEF reported	[120]
C. semireductus	X	X		pesticides		21.4%-28.6% DEF	[121]

Table 3, continued

Taxon Chironomini	Pollution type					Deformity % and remarks	
	Pro.	Other	Metal	Organic	Mixed, other		
Chironomus, Stictochironomus	X	X		insecticides	X	**no** DEF next to domestic sewage Okanagan Lakes, Skaha L.	[107]
cf. Polypedilum						only 1 larva found	[122]
C. (s.s.)? cucini (salinarius group)					X	76-78% DEF	[123]
Micropsectra sp. (praecox group)/Procladius sp. ,					X	near the limit of 'industrial effluent tolerance range'	
C. (s.s.)? cucini (salinarius group)					X	<1% past industry/ 0% organic sewage, cottages	
C. anthracinus, C. plumosus				X		19.6% DEF	[124]
C. thummi/C. plumosus			X	X	X	**no** correl. with increased water temperature/ 27-38% DEF	[125]
Chironomus	X				X	living fauna (1972)1.49/0.50% DEF / fossil fauna (1951)0.35/0.71% DEF / fossil fauna (1087AD)0.09% DEF	[91]
Chironomus	X		X	orgchl.pest.		3.13% strongly eutrophic?/2.26% eutrophic	[126]
C. thummi					X	complex causes, high water temperature and metals	[163]
C.riparius, C. plumosus	X	X		coal?	X	up to 23% in poll./1-2% in unpolluted lab. culture	[102]
Chironomus						ca. 3% DEF	[127]
Chironomus sp.	X					cause? 43.9% DEF in poll./ 2.4% DEF in less poll.	[128]
C. salinarius	X				pulp/paper mill	anthropogenic cause? few larvae	[129]
Micropsectra, Tanytarsus					pulp	habitat impairment?	[130]
Chironomus			Hg, Zn,Cd			0-0.7% unpolluted/ 3.8% strongly polluted	[131]
Chironomus			Zn, Cu,Hg,Cr,Ni,Pb,Cd/ oil			0-0.8% unpolluted/up to 25% in polluted	[132]
Chironomus						0% unpolluted reference (subfossil 1830)	[133]
Chironomus				X	X (Rhine)	<2% in uncontaminated, up to 9.4% in poll.	[134]
C.thummi					radionuclides	2-12% less poll./17-50% poll./22% paper fabric	[135]
C.anthracinus			Ni		high temp.	83% DEF	
C.plumosus						14% DEF	
Chironomus			Cd,Hg,Pb,Cu		X	11-55% DEF	[136]
Chironomus					X	>40% DEF	[105]
Chironomus		X				0.9%/ ISAD **higher in unpolluted** than in polluted areas	

Table 3, continued

Taxon (Chironomini)	Pro.	Other	Metal	Organic	Mixed, other	Deformity % and remarks	Ref.
C. bernensis, C. plumosus	X		Zn, Pb, Cu, As, Hg/orgchl.pest.			24% DEF	[137]
C. plumosus			Cd, Zn, Cu	few toxic organics		59% DEF	[138]
	X	X	X	X		3-8% DEF unpolluted reference	[139]
	X	X	X	orgchl. pest.		0-9.6% DEF	[140]
	X	X			X	8% DEF	[141]
Chironomus	X		Hg,Pb	X		**less affected** by polluted waters, 0-5% in poll.	
Cryptochironomus	X	X				Chironomini 30%/ Procladius 16%	
Dicrotendipes	X	X		agricult.		up to 19% DEF	
						<5% DEF, control, only athmospheric pollutants	[142]
						% DEF, ISAD, ISLD and CCR	[143]
C. anthracinus				coal tar		14% DEF, PAHs higher in DEF/ 3.1% uncont.	
C.cucinii			X			2.8% external reference site	[144]
C. gr. thummi	X		X			48% DEF, Cu and Pb higher in DEF	[112]
Chironomus spp.		X		pesticides		42% DEF/ 0-1.3% DEF no pesticides	[145]
Chironomus, Glyptotendipes		X				5.6-20% DEF, BBI 2-6	[146]
C. cf. plumosus					X	55% DEF	[147]
C. salinarius				X		1% uncontaminated reference	[103]
Chironomus					X	FA 2.2/reference pond, FA 0.78, relatively unpolluted	[148]
C. gr. thummi	X	X				'toxic score', 3.6% ('excellent')-44.1% 'toxic'	[149]
C. plumosus		X	Cu, Ni, Zn	pesticides, atrazine		22-67% DEF	[150]
Chironomus, etc...	X	X				"abnormal mouth parts", eutrophic lake	[151]
Chironomus	X	X				pooled species	[152]
Polypedilum, etc...	X	X		X		0-18% DEF, **no** good correlation with heavy metals	[153]
Chironomus sp. A/Dicrotendipes sp. A		X		X		8.4-41.6% DEF	
Chironomus sp. B/cf. Kiefferulus sp.		X			X	0-26%/1-18% DEF	
C. cf. formosipennis		X		X		4-26%/0-29% DEF	
						4-65% DEF	
Chironomus spp.					X	16% DEF, **unknown poll, not cadmium, not anoxia**	[113]

Table 3, continued

Taxon			Pollution type			Deformity % and remarks	
Chironomini	Pro.	Other	Metal	Organic	Mixed, other		
C. gr. thummi					X	different behaviour in DEF and NOR	[154]
C. gr. thummi			Pb	X	X	linear relations, up to 80% DEF	[155]
species not specified				PCBs	X	DEF not specified	[156]
Chironomus		X		org.chl. pest.		0-20% DEF	[157]
C. gr. salinarius			Cd, Cr, Cu, Pb, Hg, Ni/PCB, PAH etc...			Rel. Nucleolus Diameter/buccal str. corr. with contam. sed.	[158]
Chironomus, Endochironomus		X		pulp mill		50% DEF	[159]
Chironomus, Stictochironomus, Prodiamesa				X	urban	5-75% DEF	[16]
C. spp.		X				not to 'moderately' contaminated. 0-13%	[160]
C. gr. thummi			X			15-45% DEF, correlations with factor scores from 9 metals	[161]
C. riparius			Cd, Zn			17% DEF, FA1.4-1.9	[21]
C. gr. thummi			Cd, Cu, Pb, Zn			linear and bell-shaped conc.-response curves/higher Pb in DEF	[162]
C. gr. thummi			X		X	lower fitness and diff. levels of metal-binding proteins in DEF	[14]
C. riparius					urban	10-37.5% DEF	[164]
	Prodiamesa olivacea				urban	2.3-9.1% DEF	
C. riparius					chem/phys	1.9-66.7% DEF	[165]
C. riparius			Hg	paper mill	X	up to 70% DEF	[166]
C. riparius					mixed stress	increased active nucleoli in DEF	Meregalli 1999,*

*unpubl.

The possibility that individuals with deformities inherited from populations of polluted sites might colonize an otherwise unpolluted site, through the adult terrestrial phase, could bias the possible conclusions drawn from an analysis of deformity frequencies at the latter site. A population of deformed *Chironomus riparius* larvae from the polluted Teltow canal (Germany) was reared in uncontaminated water [171]. After one generation, the deformity incidence dropped to near zero. In similar experiments, it was concluded that mentum deformities in *Chironomus riparius* are not passed from one generation to the other [96,106,138,172,143]. This suggests that the larvae with mentum gaps (which is the most obvious and most studied deformity) might be phenodeviants or the result of local mutations in somatic cells but not of mutations in the reproductive cells. In recent years, the link between morphological deformities and anomalies in the chromosomes and physiological parameters has been investigated, in a trial to understand the underlying physiological mechanisms of deformity induction and the whole 'stress reaction' of a midge larva. A difference in the amount of metal-binding proteins in normal and deformed larvae exposed to extreme concentrations of Cd and Zn was demonstrated [14]. Meregalli and co-workers (pers. comm.) demonstrated an increased activity in the nucleoli of the polytene chromosomes of deformed larvae, which could indicate an increased protein synthesizing activity. The inhibitory effect of lead on the replication activity of homologues, which could cause morphological and functional anomalies in the chromosomes of *Glyptotendipes barbipes* was pointed out [88]. A good correlation was found between concentrations in the sediments of the heavy metals Cu, Cd, Pb and Zn and structural changes in the polytene chromosomes of *Chironomus riparius*, before any appearance of deformities in the buccal structures [173].

Indicator species

Most deformity-pollution relations were inferred from field studies (ca. 60 publications), while experimental studies (ca. 16 papers) remain scarce. The majority of publications (Table 3) reports deformities in the genus *Chironomus*. Different genera show different deformity responses. Warwick [86,126,139,169] focussed especially on *Chironomus* spp. (Chironominae), a sessile detritus feeder and on *Procladius* (Tanypodinae), a more mobile predator, both very tolerant species. The generally higher deformity percentages found in *Chironomus* larvae are attributed to their direct uptake of polluted detritus [169]. *Procladius* is often the only surviving organism in toxic lake sediments and therefore may be a bioindicator when *Chironomus* is absent [148,169]. The deformity percentages reported by different authors are difficult to compare, because not always the same types and numbers of structures were screened [169]. Interpretation of the data is further hampered by badly documented spatial and temporal variations in pollution burden. In most cases, the pollution type mentioned is quite speculative or vague, based on information gathered from local authorities or "grey literature" from industries or research institutes. However, more recently, authors have become more precise in specifying the pollution burden in the environment where the deformed larvae have been found. Especially heavy metals, organochlorine pesticides, PCBs and PAHs are often mentioned as possible deformity inductors. Deformed mouth parts occur in several species of chironomids, including those that are not closely related or differ in their mode of living [131]. It seems that the susceptibility to malformation or perhaps the ability to survive malformation varies among species. Chironomids with deformed mouth parts may also occur in unpolluted bodies of water. Under such circumstances the incidence of deformation does not appear to exceed a few specimens per thousand (0.09% in several hundreds year old sediments form the Bay of Quinte, Lake Ontario, [91]). 'Unpolluted' sediments deposited since 1950 show 'background levels' of 1-2%. The question remains whether this figure, which is a factor 10 to 20 higher than the paleolimnological data, is not a reflection of pollution as well.

Larvae of *Prodiamesa olivacea* (Meigen, 1818) are often predominant in moderately polluted and well-oxygenated lowland rivers, while in the same rivers *Chironomus* larvae may be absent or in very low densities [174]. Therefore, it is of importance to assess the potential of *P. olivacea* for deformity screening. In a field comparison between *P. olivacea* and *C. riparius* [164], the deformity percentages in *P. olivacea* ranged between 2.3% and 9.1%, while the deformity percentages in *C. riparius* ranged between 10% and 37.5%. The morphological response of *P. olivacea* to pollution, in this case domestic and industrial, was one third of the response in *C. riparius*. Although in principle *P. olivacea* can be used in deformity screening, one should be cautious in evaluating the low deformity percentages of this and other morphologically less sensitive species. Because of the differences in sensitivity amongst species, the pooling of species to generate a higher number of larvae for statistical analysis (Table 3) is highly questionable.

The quantification of deformities and Rapid Assessment Methods
In order to be able to relate midge deformities to pollution data or to use them for ranking locations in function of their relative pollution degree, deformities have to be quantified. The most obvious way of representing deformity data is by calculating the fraction of the larval population bearing a particular <u>type</u> of deformity. This is referred to as a deformity percentage or a deformity <u>frequency</u>. Most authors use this method (see Table 3): percentages of certain deformity types are calculated for particular <u>structures</u>. A deformity percentage however does not tell anything about the <u>severity</u> (small or large gap in mentum?) of deformities in the individual larvae or on average in a certain population. Several attempts to compare the efficiency of severity and frequency metrics in predicting pollutant concentrations led to the conclusion that the use of both was redundant, while (the more easy) use of frequencies alone provided enough information [106,155]. The reason is a positive correlation between both.

Warwick [133] developed a detailed antennal indexing system (ISAD) for *Chironomus* spp. to illustrate the potential of using deformities in chironomid larvae as biological screening technique for the detection and assessment of contaminants in aquatic systems. To calculate ISAD, first, an "index of morphological response" (IMR) is calculated for each individual larva (see [133]). The values for IMR are summed and divided by the total number of larvae examined (n), to calculate an "index of severity of antennal deformation" (ISAD= ΣIMR/n) for the sample population. The indexing system, although based only on morphological deformities in the antennae of the genus *Chironomus*, has been set up in such a manner that deformities in other morphological structures could be incorporated. This leads to the "Total Morphological Response" (TMR= (Σ IMR+ k ΣOther)/n), where k is a sensitivity constant for other structures. Although ISAD proved to be very sensitive to toxic stress induced by organic compounds and/or heavy metals (field observations), it is not applicable to chironomid subfossil head capsules, as these lack the fragile antennae [133]. The practical application of ISAD should evaluate whether it makes any sense in terms of the ratio cost/information to add values of IMR for other structures. In cases of toxic stress, larvae of the genus *Chironomus* are often absent, while *Procladius* (subfamily Tanypodinae) larvae survive. Therefore, an analogous system was developed for the ligula (structure which is in function analogous to the mentum) and antennae of *Procladius* larvae [142]: an "index of severity of ligula deformation" (ISLD= ΣIMR/n). An ISAD for antennae of *Procladius* larvae was developed too [142]. Eventually, indices describing the response of other structures may be included in a comprehensive measure of "Population Morphological Response" (PMR) for *Procladius*. By extension, this may theoretically lead to a measure of "Chironomid Community Response" (CCR) [142].

The indexing system proposed by Warwick [133,142] has some intrinsic advantages: (1) It can be adapted for the number of points allocated in the calculation of the IMR, for the point scale or for the species- or structure-specific constants, when new information is available about the relationship between severity of morphological deformity response and contaminant stress for different structures and taxa. (2) It is a sensitive measure of deformity severity at the individual and the population level for a certain site. (3) The system is flexible for including other structures and taxa. It has proven to be a sensitive working instrument in assessing contaminant stress in the Canadian Great Lakes, as is reported in a number of publications by Warwick (Table 3). However, to our knowledge, few authors use it, because (1) the complexity and high resolution of the indexing scheme; one needs a very good light microscope and a thorough knowledge of midge morphology, (2) The lack of statistical calibration of the index, with which kind of contaminant stress does correspond a particular ISAD value? Since the index is open-ended, it also means that the index value range is not defined (not e.g. between 0 and 10), which makes it more difficult to visualize the real meaning of any ISAD value.

Lenat [103] classified mentum deformities in three groups, class I included slight deformities which were difficult to separate from "chipped" structures. Class II consisted of larvae with more conspicious deformities, including extra teeth, missing teeth, large gaps, and distinct asymmetry. Class III consisted of larvae with severe deformation, including at least two class II characters. A "Toxic Score" was computed for each site which gives greater weight to more severe deformities:

$$\frac{(\text{No. of class I} + 2(\text{No. of class II}) + 3(\text{No. of class III}))\times100}{\text{Total No. of larvae}}$$

The "Toxic Score" has been applied on *Chironomus riparius* and *Prodiamesa olivacea* [164]. The environmental impact of a fertilizer manufacturing facility, situated on the shore on the inlet of the Black Sea was assessed by means of the asymmetry in the pecten epipharyngis of *Chironomus salinarius* Kieffer [147]. Fluctuating asymmetry (FA) in normally bilaterally symmetrical organisms is a sign of developmental stress, which may be resulting from pollution. For each sample, asymmetry values were calculated as the squared signed differences between left (L) and right (R) sides divided by the number of individuals scored (N) ($FA=\Sigma(L_i-R_i)^2/N$). FA applied on chironomid larvae had a good predicting power for environmental stress [147].

The above mentioned methods quantify deformities, however without assigning deformity values to absolute or relative toxicity classes. In an attempt to enhance the predictive power of chironomid deformities in biomonitoring, two keys were recently proposed (one is adapted from [175], the other reported in [106]), which were independently developed on the basis of different sets of field data [155,162,166].

Assessment Method of Janssens de Bisthoven, adapted from [175]
1- Rapid Assessment (RA)
The present method represents the only attempt to use chironomid deformities in biomonitoring, without the need of microscopical preparation. This aims at reducing handling costs. Live or fixed fourth instars of *Chironomus* gr. *thummi* are put ventral side up in a Petri-dish with a little water, under a microscopical slide, under a binocular microscope. The mentum is observed for the occurrence of Köhn gaps, at high magnification (25 to 50x) under light pressure of the upper cover slip. A Köhn gap is an open gap with smooth edge (in contrast to mechanical damage) and lack of pigmentation or sclerotization at its basis. When handled gently, the screened larvae survive and may be reintroduced in the bioassay culture. A frequency of <2% puts the sediment in evaluation class 'A' (not to weakly contaminated), 3-12% in 'B' (weakly to moderately contaminated), and >13% in 'C' (moderately to very heavily contaminated). This deformity type rarely occurs in more than 20% of the larvae. A deformity percentage of 0% with this method does not completely exclude the absence of environmental contamination, as more subtle deformities may be indicative of pollution as well. Nevertheless, this method gives a fast and cheap first evaluation of a suspected sediment.

2-Detailed Assessment (DA)
A more detailed method is proposed, by looking at mentum, mandibles and antennae. The larvae need to be mounted for light microscopy. It is important not to include (1) anomalies resulting from mechanical damage, abrasion, mounting artifacts, (2) symmetrical deformities, (3) Köhn gaps (already included in the RA), and (4) split medial mentum teeth (too much related to inbreeding).

Table 4. Detailed method (DA) of Janssens de Bisthoven, adapted from [175]. In order to chose the right sediment toxicity class (A, B or C), take the highest score in the most right column.

Deformity %	Sediment contamination		
	A not to weak	B weak to moderate	C moderate to very heavy
Deformed mentum, excl. Köhn gap	0-10	11-30	>30
Deformed mandible(s)	0-5	6-30	>30
Deformed antenna(e)	0-10	11-30	>30
Deformed total, incl. other structures	0-40	41-80	>80

Sediments with a deformity percentage of <5% are considered of good quality. Class A sediments are acceptable. Class B sediments should be closely watched for further deterioration. Class C sediments should be chemically analysed and need remediation, when confirmed by other methods. Often, half of the visited sites in lowland rivers do not contain sufficient *Chironomus* larvae for deformity screening. In that case, two alternatives are possible:

1-Bioassay: cultured instar 1 or 2 larvae are exposed to the suspected sediments, eventually in a dilution series with sand and artificial fish food, until instar 4 is reached. These are then screened for deformities. These results can be combined with mortality and emergence data [14] and other analyses in a broader TRIAD-approach [176]. This approach, combining emergence, mortality and deformity data, is integrated in the ICT or 'Integrated Chironomid Toxicity Test' (Limco International).

2- Enclosure: Young instars are introduced in cages in the field, together with predried sediment of the locality, where they develop until instar 4. The advantage here is that the larvae are subject to the daily dynamism of their habitat. The disadvantages are that predation is difficult to avoid, mortalities are difficult to control, it is uneasy to accurately predict when the fourth instars should be collected for analysis (Meregalli, pers. comm.).

It is highly recommended to combine the deformity screening methods with a battery of other tests to cover all structural and functional levels of the ecosystem [151,160,177].

Key of Vermeulen [106]
Here, all deformity types of the mentum are taken into consideration. Six classes of sediment contamination are defined. A frequency (MDF) of <0.8% puts the sediment in evaluation class I (clean), 0.8-8% in II (very low), 8-18% in III (low), 18-38% in IV (moderate), 38-60% in V (high), and >60% in VI (very high). Caution should be taken when screening summer larvae, because the MDF is often lower than in the winter larvae. These two new methods (Janssens de Bisthoven RA and DA, Vermeulen) should be extensively tested and validated in other field studies.

When some differences in threshold percentages (e.g. 18 and 20 %, 30 and 38 %) are overlooked, roughly, the following classes from the Vermeulen key correspond with the Janssens de Bisthoven (detailed) classes: class I and II with class A, class III and IV with class B, and class V and VI with class C. The Janssens de Bisthoven rapid assessment is easy and truely 'rapid', but remains a first screening method, since it is based on a single deformity type, which does not necessarily appear with all types of contamination. It is expected that a population with 12% larvae with a Köhn gap (class B) will also contain larvae with a variety of other smaller mentum deformities, thus *de facto* reaching class B of the detailed Janssens de Bisthoven key or class III or IV of the Vermeulen key.

Conclusions

There are probably very few aquatic ecosystems left, where (in)vertebrate species have a natural deformity rate. Closed systems, or systems with reduced water exchange to the sea, like lenthic systems and the great American, African and Eurasian lakes, are especially at risk and have benefitted from closer scrutiny. Also at risk are shallow seas and bays, such as the Baltic Sea and more specially, the Bothnian Bay. But even the most remote areas in the world are not exempt from pollution, as is testified by the albatrosses of the Pacific. Paleolimnological data give testimony of pre-industrial conditions and serve as a distant reminder of how it could be. The most dynamic ecosystems, such as all lotic systems and the oceans, certainly have better chances of recovery, once a pollution source is stopped. Over the past two decades, deformities in most aquatic animal taxonomic groups have been documented. Apart from isolated studies, the bulk of marine research has focussed on commercial fish and other seafood. The assessment of deformities in these organisms is more an effort of quality control for human consumption, than a real undertaking to understand the deformity ontology and its biomonitoring value. Nevertheless, it appears from some studies that, despite their mobility, fish deformities can very well be associated with known contaminant sources and with particular ecological features or processes, such as a benthic lifestyle and migration patterns. The same applies for waterbirds, especially when one is considering colonial fish-eating birds in their breeding areas. In that case, bird deformities may be considered as the reflection of deleterious processes at the level of the fish fauna. It could be interesting to make parallels through the whole trophic chain. Amphibian deformities remain a bit of a mystery, as they are still subject to speculations about their origins. It seems appropriate to use amphibians for biomonitoring of very small wells and pools, where other vertebrates are missing. The screening of tadpoles is certainly recommended, given the partial terrestrial life of the adults, possibly subjected to terrestrial sources of pollution. Few studies deal with deformities in zooplankton, although I expect microcrustaceans and e.g. *Chaoborus* larvae to have potential for biomonitoring the water column. At the (epi)benthic level, Mysidaceae and Isopoda deserve more attention in marine, brackish and freshwater habitats, given their dominant role in the food chain. Larvae of Chironomidae have proven their usefullness in biomonitoring freshwater systems in general and sediments in particular. Although Chironomidae are often dismissed as 'a specialist's hobby', a lot can already be done with the easily identifiable genera *Chironomus* and *Procladius*. Meanwhile, a serious effort is undertaken worldwide to check for deformities in other genera and at species level, to link deformity rates with ecological and life history data, other ecotoxicological endpoints, and to construct concentration-response curves. The chironomid-deformity research and biomonitoring with Chironomidae in general, more and more use ecotoxicological theory and multivariate methods, and that can only be encouraged.

Acknowledgements

I would like to thank the following persons who contributed in one way or another to some of the results and ideas on the part concerning the Chironomidae, presented in this chapter: Frans Ollevier and Carla Huysmans (Catholic University Leuven, Belgium), Angelo Vermeulen and Boudewijn Goddeeris (Belgian Institute for Natural Sciences, Brussels, Belgium), Rudy Vannevel (Vlaamse Milieumaatschappij, Ghent, Belgium) and Almut Gerhardt (LimCo International, Germany).

References

[1] P.J. Sheenan and A.W. Knight: A multilevel approach to the assessment of ecotoxicological effects in a heavy metal polluted stream. Verh. Internat. Verein. Limnol., 22 (1985), 2364-2370.

[2] R.K. Johnson, T. Wiederholm and D.M. Rosenberg: Freshwater biomonitoring using individual organisms, populations and species assemblages of benthic macroinvertebrates. In: *Freshwater biomonitoring and benthic macroinvertebrates* (Eds. Rosenberg and Resh), Chapman and Hall, New York (1993), 40-158.

[3] J.A. Couch and J.C. Harshbarger: Effects of carcinogenic agents on aquatic animals: an environmental and experimental overview. Environ. Carcinogenesis Revs. 3, (1985), 63-101.

[4] L. Shugart and C. Theodorakis: Environmental genotoxicity: probing the underlying mechanisms. Env. Health Persp., 102 (1994), 13-17.

[5] J.E. Neigal and G.P. Schmahl: Phenotypic variation within bioto-compatibility-defined clones of marine sponges. Science 224, (1984), 413-415.

[6] Y. Hirano and N. Hiji: Effects of low pH and aluminium on root morphology of Japanese red cedar saplings. Environmental Pollution 101, (1998), 339-347.

[7] V. Yanko, M. Ahmad and M. Kaminski: Morphological deformities of benthic foraminiferal tests in response to pollution by heavy metals: implications for pollution monitoring. J. of Foraminiferal Res. 28, (1998), 177-200.

[8] B. Okamura: Genetic similarity, paraitism, and metapopulation structure in a freshwater bryozoan. in: Evolutionary Ecology of freshwater animals, ed. B. Streit, T. Städler, C.M. Lively, Birkhäuser Verlag Basel/Switzerland, (1997), p. 366.

[9] C. Pélabon and L. van Breukelen: Asymmetry in antler size in roe deer (*Capreolus capreolus*): an index of individual and population conditions. Oecologia 116, (1998), 1-8.

[10] J.A. Bantle and T.D. Sabourin: Standard guide for conducting the frog embryo teratogenesis assay. *Xenopus* (FETAX). E1439-91. In: *Annual book of ASTM Standards 11, (1991), 1-11.*

[11] J.R. Karr: Assessment of biotic integrity using fish communities. Fisheries 6, (1981), 21-27.

[12] G.J. Holloway, R.M. Sibly and S.R. Povey: Evolution in toxin-stresed environments. Functional Ecology 4, (1990), 289-294.

[13] E. Varkonyi, M. Bercsényi, C. Ozouf-Costaz and R. Billard: Chromosomal and morphological abnormalities caused by oocyte ageing in *Silurus glanis*. J. of Fish Biology 52, (1998), 899-906.

[14] L. Janssens de Bisthoven, P. Nuyts, B. Goddeeris and F. Ollevier: Sublethal parameters in morphologically deformed *Chironomus* larvae: clues to understanding their bioindicator value. Freshwater Biology 39, (1998), 179-191.

[15] L. Janssens de Bisthoven and F. Ollevier: Some experimental aspects of sediment stress in *Chironomus* gr. *thummi* larvae (Diptera: Chironomidae). Acta Biol. Debrec. Suppl. Oecol. Hung. 3, (1989), 147-155.

[16] W.A. Aston: Morphological deformities in chironomid larvae (Chironomidae: Diptera): biomarkers of urban polluted sediments. PhD-thesis, University of Leicester, UK, (1998), p.195.

[17] C.J. Sindermann: An examination of some relationships between pollution and disease. Rapp. P.-v. Réun. Cons. int. Explor. Mer 182, (1983), 37-43.

[18] J.F. Postma, P. van Nugteren, M.B. Buckert-de Jong: Increased cadmium excretion in metal-adapted populations of the midge *Chironomus riparius* (Diptera). Environ. Toxicol. Chem. 15, (1996), 332-339.

[19] B.E., Brown: The form and function of metal-containing granules in invertebrate tissues. Biol. Rev. 57, (1982), 621-667.

[20] D. Groenendijk: Dynamics of metal adaptation in riverine chironomids. PhD.-thesis, University of Amsterdam, (1999), p. 159.

[21] D. Groenendijk, L.W.M. Zeinstra and J.F. Postma: Fluctuating asymmetry and mentum gaps in populations of the midge *Chironomus riparius* (Diptera: Chironomidae) from a metal-contaminated river. Environ. Toxicol. and Chem. 17, 1998), 1999-2005.

[22] H.J. Auman, J.P. Ludwig, C.L. Summer, D.A. Verbrugge, K.L. Froese, T. Colborn, and J.P. Giesy: PCBs, DDE, DDT, and TCDD-EQ in two species of albatross on Sand Island, Midway Atoll, North Pacific Ocean. Environm. Toxicol. and Chem. 16, (1997), 498-504.

[23] M.A. Mora: Organochlorines and trace elements in four colonial waterbird species nesting in the lower Laguna Madre, Texas. Arch. of Environm. Contam. and Toxicol. 31, (1996), 533-537.

[24] D.J. Hoffman, G.J. Smith, and B.A. Rattner: Biomarkers of contaminant exposure in common terns and black-crowned night herons in the great lakes. Environ. Toxicol. Chem. 12, (1993), 1095-1103.

[25] D.E. Tillitt, T.J. Kubiak, G.T. Ankley, and J.P. Giesy: Dioxin-like toxic potency in Forster's tern eggs from Green Bay Lake Michigan North America. Chemosphere 26, (1993), 2079-2084.

[26] M. Gilbertson, T. Kubiak, J. Ludwig and G. Fox: Great Lakes embryo mortality edema and deformities syndrome GLEMEDS in colonial fish-eating birds similarity to chick-edema disease. J. Toxicol. Environ. Health 33, (1991), 455-520.

[27] D.S. Henshel, J.W. Martin, R. Norstrom, P. Whitehead, J.D. Steeves, and K.M. Cheng: Morphometric abnormalities in brains of great blue heron hatchlings exposed in the wild to PCDDs. Env. Health Persp. 103, (1995), 61-66.

[28] R.M. Kocan, H. Von Westernhagen M.L. Landolt and G. Furstenberg: Toxicity of sea-surface microlayer effects of hexane extract on baltic herring *Clupea harengus* and atlantic cod *Gadus morhua* embryos. Mar. Env. Res. 23, (1987) 291-306.

[29] R.M. Kocan, J.E. Hose, E.D. Brown, and T.T. Baker: Pacific herring (*Clupea pallasi*) embryo sensitivity to Prudhoe Bay petroleum hydrocarbons: laboratory evaluation and in situ exposure at oiled and unoiled sites in Prince William Sound. Can. J. of Fish. and Aq. Sci. 53, (1996), 2366-2375.

[30] N.T.H., Lien, D. Adriaens, and C.R. Janssen: Morphological abnormalities in African catfish (*Clarias gariepinus*) larvae exposed to malathion. Chemosphere 35, (1997), 1475-1486.

[31] I. Martinez, R. Alvarez, I. Herraez and P. Herraez: Skeletal malformations in hatchery reared *Rana perezi* tadpoles. The Anatomical record 233, (1992), 314-320.

[32] A. Bengtsson, B.E. Bengtsson, G. Lithner: Vertebral defects in fourhorn sculpin *Myoxocephalus quadricornis* L. exposed to heavy metal pollution in the Gulf of Bothnia Baltic Sea. J.Fish Biol. 33, (1988), 517-530.

[33] C. Kitajima, Y. Tsukashima, S. Fujita, T. Watanabe and Y. Yone: Relationship between uninflated swim bladders and lordotic deformity in hatchery-reared Red sea bream *Pagrus major*. Bull. Jpn. Soc. Fish. 47, (1981), 1289-1294.

[34] L.A. Norton, W.R. Proffit and R.R. Moore: Inhibition of bone growth in vitro by endotoxin histamine effects. Nature 221, (1969), 469-471.

[35] S. Corneillie, C. Agius and F. Ollevier: Comparison of the fatty acid profile of wild caught fingerlings and yolk sac sea bass (*Dicentrarchus labrax*) larvae with cultured healthy larvae and larvae suffering from whirling disease. Belg. J. Zool. 120, (1990), 157-164.

[36] A.W. Grady, D.L. Fabacher, G. Frame and B.L. Steadman: Morphological deformities in brown bullheads administered dietary beta naphtoflavone. J. Aquat. Anim. Health 4, (1992), 7-16.

[37] T. Matsusato: Studies on the skeletal anomalies of fishes. Bull. Natl. Res. Inst. Aquacult. 0, (1986), 57-179.

[38] J.A. Couch and L.A. Courtney: N-Nitrosodiethylamine-induced hepatocarcinogenesis in estuarine sheepshead minnow (*Ciprinodon variegatus*): neoplasms and related lesions compared with mammalian lesions. JNCI 79, (1987), 297-321.

[39] L.A. Vollestad, E. Fjeld, T. Haugen, and S.A. Osnevad: Developmental instability in grayling (*Thymallus thymallus*) exposed to methylmercury during embryogenesis. Environmental pollution 101, (1998), 349-354.

[40] D. Lemly: A teratogenic deformity index for evaluating impacts of selenium on fish populations. Ecotoxicology and Environmental Safety 37, (1997), 259-266.

[41] B.E. Bengtsson, A. Larsson, A. Bengtsson, and L. Renberg: Sublethal effects of tetrachloro-1,2-benzoquinone, a component in bleachery effluents from pulp mills, on vertebral quality and physiological parameters in fourhorn sculpin. Ecotox. and Environ. Saf. 15, (1988), 62-71.

[42] A. Bengtsson: Effects of bleached pulp mill effluents on vertebral defects in fourhorn sculpin *Myoxocephalus quadricornis* L. in the Gulf of Bothnia. Arch. Hydrobiol. 121, (1991), 373-384.

[43] R.O. Hermanutz: Malformations of the fathead minnow (*Pimephales promelas*) in an ecosystem with elevated selenium concentrations. Bull. Envir. Contam. Toxic. 49, (1992), 290-294.

[44] C.H. Jagoe, A. Faivre, and M.C. Newman: Morphological and morphometric changes in the gills of mosquitofish (*Gambusia holbrooki*) after exposure to mercury (II). Aquatic Toxicology 34, (1996), 163-183.

[45] J. Koyama: Vertebral deformity susceptibilities of marine fishes exposed to herbicide. Bull. Environ. Contam. Toxicol. 56, (1996), 655-662.

[46] E. Lindesjöö, J. Thulin, B.-E. Bengtsson, and U. Tjärnlund: Abnormalities of a gill cover bone, the operculum, in perch *Perca fluviatilis* from a pulp mill effluent area. Aquatic Toxicology 28, (1994), 189-207.

[47] A. Haaparanta, E.T. Valtonen, and R.W. Hoffmann: Gill anomalies of perch and roach from four lakes differing in water quality. J. of Fish Biology 50, (1997), 575-591.

[48] N.J. Gassman, L.B. Nye, and M.C. Schmale: Distribution of abnormal biota and sediment contaminants in Biscayne Bay, Florida. Bull. of Marine Sci. 54, (1994), 929-943.

[49] D.E. Barker, R.A. Khan, E.M. Lee, R.G. Hooper, and K. Ryan: Anomalies in sculpins (*Myoxocephalus* spp.) sampled near a pulp and papr mill. Arch. of Env. Cont. and Tox. 26, (1994), 491-496.

[50] A.D. Lemly: Teratogenic effects of selenium in natural populations of freshwater fish. Ecotox. and Envir. Safety 26, (1993), 181-204.

[51] J.A. Browder, D.B. MCClellan, D.E. Harper, M.G. Kandrashoff, and W. Kandrashoff: A major developmental defect observed in several Biscayne Bay Florida fish species. Environ. Biol. Fishes 37, (1993), 181-188.

[52] J.J. Pereira, E.J. jr. Lewis, R.L. Spallone and C. Sword: Bifurcation of gill filaments in winter flounder *Pleuronectes americanus* Walbaum from Long Island Sound. J. Fish Biology 41, (1992), 327-338.

[53] R. Wannemacher, A. Rebstock, E. Kulzer, D. Schrenk and K.W. Bock: Effects of 2378 Tetrachlorodibenzo-P-Dioxin on reproduction and oogenesis in zebrafish *Brachydanio rerio*. Chemosphere 24, (1992), 1361-1368.

[54] M.J. Van Den Avyle, S.J. Garvick, V.S. Blazer, S.J. Hamilton, and W.G. Brumbaugh: Skeletal deformities in smalmouth basss *Micropterus dolomieui* from Southern Appalachian Reservoirs. Arch. Environ. Contam. Toxicol. 18, (1989), 688-696.

[55] V. Pragatheeswaran, B. Loganathan, R. Natarajan, and V.K. Venugopalan: Cadmium induced vertebral deformities in an estuarine fish *Ambassis commersoni* Cuvier. Proc. Indian Acad. Sci. Anim. Sci. 96, (1987), 389-394.

[56] K. Baba, M. Nara, Y. Iwahashi, M. Matushima, and T. Sasaki: Toxicity of agricultural chemicals on marine fishes -II. On the vertebral deformation observed in young yellowtail during the determination of TL. Bull. Shizuoka Pref. Fish Exp. Stn 9, (1974), 43-52.

[57] P. Weis, and J.S. Weis: Abnormal locomotion association with skeletal malformations in the sheepshead minnow, *Cyprinodon variegatus*, exposed to malathion. Environ. Res. 12, (1976), 196-200.

[58] D.E. Wells and A.A. Cowan: Vertebral dysplasia in salmonids caused by the herbicide trifluralin. Environ. Poll. (A) 29, (1982), 249-260.

[59] Nath and Kumar: Nickel-induced histopathological alterations in the gill architecture of a troppical freshwater perch, *Colisa fasciatus*. Sci. Total Environ. 80, (1989), 293-296.

[60] R. Alvarez, M.P. Honrubia, and M.P. Herraez: Skeletal malformations induced by the insecticides ZZ-Aphox and Folidol during larval development of *Rana perezi*. Arch. of Env. Contam. and Toxicol. 28, (1995), 349-356.

[61] D.W. Kononen and R.A. Gorski: A method for evaluating the toxicity of industrial solvent mixtures. Env. Tox. Chem. 16, (1997), 968-976.

[62] C.L. Rowe, O.M. Kinney, A.P. Fiori and J.D. Congdon: Oral deformities in tadpoles (*Rana catesbeiana*) associated with coal ash deposition: effects on grazing ability and growth. Freshwater Biology 36, (1996), 723-730.

[63] G.S. Schuytema, A.V. Nebeker, W.L. Grifis and K.N. Wilson: Teratogenesis toxicity and bioconcentration in frogs exposed to dieldrin. Arch. Environ. Contam. Toxicol. 21, (1991), 332-350.

[64] J.W. Fournie, J.K. Summers and S.B. Weisberg: Prevalence of gross pathological abnormalities in estuarine fishes. Trans. of the Am. Fish. Soc. 125, (1996), 581-590.

[65] H. Moeller and K. Anders: Epidemiology of fish siseases in the Wadden Sea. J. Mar. Sci. 49, (1992), 199-208.

[66] C.W. Schmidt: Amphibian deformities continue to puzzle researchers. Env. Sci. and Technol./ News 31, (1997), 324-326.

[67] D.E. Kime: A strategy for assessing the effects of xenobiotics on fish reproduction. The Science of the Total Environment 225, (1999), 3-11.

[68] T.E. Tooby and D.J. Macey: Absence of pigmentation in corixid bugs (Hemiptera) after the use of the aquatic herbicide dichlobenil. Freshwater Biology 7, (1977), 519-525.

[69] D.B. Donald: Deformities in Capniidae (Plecoptera) from the Bow River, Alberta. Can. J. Zool. 58, (1980), 682-686.

[70] K.W. Simpson: Abnormalities in the tracheal gills of aquatic insects collected from streams receiving chlorinated crude oil wastes. Freshw. Biol. 10, (1980), 581-583.

[71] L.B.-M. Petersen and R.C. jr. Petersen: Effect of kraft pulp mill effluent and 4,5,6 trichloroguaiacol on the netspinning behaviour of *Hydropsyche angustipennis* (Trichoptera). Ecological Bull. 36, (1984), 68-74.

[72] J.A. Camargo: Toxic effects of residual chlorine on larvae of *Hydropsyche pellucidula* (Trichoptera, Hydropsychidae): a proposal of biological indicator. Bulletin of Environmental Cont. and Tox. 47, (1991), 261-265.

[73] K.Vuori: Rapid behavioural and morphological responses of Hydropsychid larvae (Trichoptera, Hydropsychidae) to sublethal cadmium exposure. Environmental pollution 84, (1994), 291-299.

[74] G. Milbrink: Characteristic deformities in tubificid oligocheates inhabiting polluted bays of Lake Vänern, southern Sweden. Hydrobiologia 106, (1983), 169-184.

[75] M. Lafont: Oligochaete communities as biological descriptors of pollution in the fine sediments of rivers. Hydrobiologia 115, (1984), 127-129.

[76] P.M. Chapman and R.O. Brinkhurst: Hair today, gone tomorrow: induced chaetal changes in tubificid oligochaetes. Hydrobiologia 155, (1987), 45-55.

[77] A.R. Fontaine, N. Olsen, R.A. Ring and C.L. Singla: Cuticular metal hardening of mouthparts and claws of some forest insects of British Columbia. J. Entomol. Soc. Brit. Columbia, 88, (1991), 45-55.

[78] D.E. Dockter: Developmental changes and wear of larval mandibles in *Heterocampa guttivitta* and *H. subrotata* (Notodontidae). Journal of the Lepidopterists' Society 47, (1993), 32-48.

[79] S.J. Blockwell, E.J. Taylor, D.R. Phillips, M. Turner and D. Pascoe: A scanning electron microscope investigation of the effects of pollutants on the hepatopancreatic ceca of *Gammarus pulex* (L.). Ecotoxicology and Environmental Safety 35, (1996), 209-221.

[80] T. Itow, R.E. Loveland and M.L. Botton: Developmental abnormalities in horseshoe crab embryos caused by exposure to heavy metals. Arch. Environ. Contam. Toxicol. 35, (1998), 33-40.

[81] P. Fioroni, J. Oehlmann and E. Stroben: The pseudohermaphroditism of Prosobranchs; morphological aspects. Zool. Anz., (1991), 1-26.

[82] M. Huet, P. Fioroni, J. Oehlmann and E. Stroben: Comparison of imposex response in three Prosobranch species. Hydrobiologia 309, (1995), 29-35.

[83] J. Oehlmann, P. Fioroni, E. Stroben and B. Markert: Tributylin (TBT) effects on *Ocinebrina aciculata* (Gastropoda: Muricidae): imposex development, sterilization, sex change and population decline. The Science of the Total Environment 188, (1996), 205-223.

[84] C. Alzieu, J. Sanjuan, J.P. Deltreil and M. Borel: Tin contamination in Arcachon Bay: effects on oyster shell anomalies. Marine Pollution Bulletin 17, (1986), 494-498.

[85] F. Gendron: Recherches sur la toxicité de peintures anti-salissures à base d'organostanniques et de l'oxyde de tributylétain vis-à-vis de l'huitre *Crassostrea gigas*. Thèse de Doctorat, Univ. Aix-Marseille, France, (1985).

[86] W.F. Warwick: Morphological deformities in Chironomidae (Diptera) larvae as biological indicators of toxic stress. In: *Toxic contaminants and ecosystem health; a great lakes focus* (M.S. Evans, Ed.). John Wiley and Sons, New York, (1988), p. 280-320.

[87] P.D. Armitage, P.S. Cranston and L.C.V. Pinder (eds.): Armitage, P.D.: Behaviour and ecology of adults. In: *The Chironomidae. Biology and ecology of non-biting midges.* (Chapman and Hall. London, (1995), p. 572.

[88] P. Michaelova and R. Belcheva: Different effect of lead on external morphology and polytene chromosomes of *Glyptotendipes barbipes* (Staeger) (Diptera, Chironomidae). Folia Biol. (Krakow) 38, 83-88.

[89] M.D. Kahl, E.A. Makynen, P.A. Kosian and G.T. Ankley: Toxicity of 4-Nonylphenol in a life cycle test with the midge *Chironomus tentans*. Ecotoxicol. and Environ. Safety 38, (1997), 155-160.

[90] W.F. Warwick: Chironomidae (Diptera) responses to 2800 years of cultural influence: a paleolimnological study with special reference to sedimentation, eutrophication, and contamination processes. Can. Ent. 112, (1980), 1193-1238.

[91] W.F. Warwick: Palaeolimnology of the Bay of Quinte, Lake Ontario: 2800 years of cultural influence. Can. Bull. Fish. Aquat. Sci. 206, (1980), p. 117.

[92] O.A. Saether: Nearctic chironomids as indicators of lake typology. Intern. Verein. Theor. Angew. Limn. Verh. 19, (1975), 3127-3133.

[93] O.A. Saether: Chironomid communities as water quality indicators. Holarctic Ecology. 2, (1979), 65-74.

[94] N. De Pauw, D. Roels, and P. Fontoura: Use of artificial substrates for standardized sampling of macroinvertebrates in the assessment of water quality by the Belgian Biotic Index. Hydrobiologia 133, (1986), 237-258.

[95] N. De Pauw and R. Vannevel: Macroinvertebraten en waterkwaliteit. In: *Dossier Stichting Leefmilieu 11*, Stichting Leefmilieu, Antwerpen, (1991), p. 316.

[96] L. Janssens de Bisthoven: Morphological deformities in *Chironomus* gr. *thummi* (Diptera, Chironomidae) as bioindicators for micropollutants in sediments of Belgian lowland rivers. PhD-thesis, KULeuven, Belgium, (1995), p. 253.

[97] M.A. Learner and R.W. Edwards: The distribution of the midge *Chironomus riparius* in a polluted river system and its environs. Air and Water Pollut. Int. J.10, (1966), 99-116.

[98] A.M. Gower and P.J. Buckland: Water quality and the occurrence of *Chironomus riparius* Meigen (Diptera: Chironomidae) in a stream receiving sewage effluent. Freshwater Biology 8, (1978), 153-164.

[99] L.C. Ferrington and N.H. Crisp: Water quality characteristics of receiving streams and the occurrence of *Chironomus riparius* and other Chironomidae in Kansas. Acta Biol. Debr. Oecol. Hung. 3, (1989), 115-126.

[100] J.M. Hellawell: *Biological indicators of freshwater pollution and environmental management.* Elsevier, London, New York (1986).

[101] G. Friedrich: Eine Revision des Saprobiensystems. Zeitschrift für Wasser und Abwasser Forschung 23, (1990), 141-152.

[102] C. Frank: Beeinflussung von Chironomidenlarven durch Umweltchemikalien ind ihre Eignung als Belastungs- und Trophieindikatoren. Verh. Dtsch.Zool. Ges., (1983), 143-147.

[103] D.R. Lenat: Using mentum deformities of *Chironomus* larvae to evaluate the effects of toxicity and organic loading in streams. J. N. Am Benth. Soc. 12, (1993), 265-269.

[104] C. Madden, A.D. Austin and P.J. Suter: Pollution monitoring using chironomid larvae: what designates a deformity? In: *Chironomids:from genes to ecosystems* (Cranston, P., Ed.) CSIRO (1995), p. 482.

[105] W.F. Warwick and N.A. Tisdale: Morphological deformities in *Chironomus, Cryptochironomus,* and *Procladius* larvae (Diptera: Chironomidae) from two differentially stressed sites in Tobin Lake, Saskatchewan. Can. J. Fish. Aquat. Sci. 45, (1988), 1123-1144.

[106] A.C. Vermeulen: Head capsule deformation in *Chironomus riparius* larvae (Diptera): causality, ontogenesis and its application in biomonitoring. PhD-thesis, KULeuven, Belgium, (1998), p. 207.

[107] A.L. Hamilton and O.A. Saether: The occurrence of characteristic deformities in the chironomid larvae of several Canadian lakes. Can. Ent. 103, (1971), 363-368.

[108] N. De Pauw and G. Vanhooren: Method for biological quality assessment of watercourses in Belgium. Hydrobiologia 100, (1983), 153-168.

[109] R.M. Cushman: Chironomid deformities as indicators of pollution from a synthetic, coal-derived oil. Freshwater Biology 14, (1984), 179-182.

[110] J. Percy, K.L. Kuhn and K. Kalthoff: Scanning electron microscopic analysis of spontaneous and UV-induced abnormal segment patterns in *Chironomus samoensis* (Diptera, Chironomidae). Roux's Archives of Developmental Biology 195, (1986), 92-102.

[111] P. Kosalwat and A.W. Knight: Chronic toxicity of copper to a partial life cycle of the midge, *Chironomus decorus*. Arch. Environ. Contam. Toxicol. 16, (1987), 283-290.

[112] C.P. Madden, P.J. Suter, B.C. Nicholson and A.D. and Austin: Deformities in chironomid larvae as indicators of pollution (pesticide) stress. Neth. J. of Aq. Ec. 26, (1992), 551-557.

[113] G.A. Bird, M.J. Rosentreter and W.J. Schwartz: Deformities in the menta of chironomid larvae from the Experimental Lakes Area, Ontario. Can. J. Fish. Aquat. Sci. 52, (1995), 2290-2295.

[114] G.M. Clarke, A.H. Arthington and B.J. Pusey: Fluctuating asymmetry of chironomid larvae a an indicator of pesticide contamination in freshwater environments. In: *Chironomids: from genes to ecosystems,* P. Cranston (ed.), CSIRO publication, Australia, (1995), 101-112.

[115] M. Dickman and G. Rygiel: Chironomid larval deformity frequencies, mortality, and diversity in heavy-metal contaminated sediments of a Canadian riverine wetland. Env. Int. 22, (1996), 693-703.

[116] L. Janssens de Bisthoven, C. Huysmans, R. Vannevel, G. Goemans and F. Ollevier: Field and experimental morphological response of *Chironomus* larvae (Diptera, Nematocera) to xylene and toluene. Neth. J. Zool. 47, (1997), 227-239.

[117] L. Janssens de Bisthoven, A. Vermeulen and F. Ollevier: Experimental induction of morphological deformities in *Chironomus riparius* larvae by chronic exposure to copper and lead. Arch. Environ. Contam. Toxicol. 35, (1998), 249-256..

[118] B.G. Blaylock: Chromosomal aberrations in a natural population of *Chironomus tentans* exposed to chronic low-level radiation. Evolution 19, (1965), 421-429.

[119] R.O. Brinkhurst, A.L. Hamilton and H.B. Herrington: Components of the bottom fauna of the St. Lawrence Great Lakes. (No. PR 33). Great Lakes Inst., Univ. Toronto, Toronto, (1968).

[120] W. Cook and D. Veal: A preliminary biological survey of Port Hope Harbour. Report to the Ministry of the Environment for Ontario (1968).

[121] O.A. Saether: A survey of the bottom fauna of the Okanagan valley, Brit. Col. Techn. Rep. 196. Fish. Mar. Serv. (Can.), (1970), 1-29.

[122] O.A. Saether and M.P. McLean: A survey of the bottom fauna in Wood, Kalamalka and Skaha lakes in the Okanagan Valley, British Columbia. Tech. Rep. 342, Fish. Res. Board Can., (1972), p. 27.

[123] L. Hare and J.C.H. Carter: The distribution of *Chironomus* (s.s.)? *cucini* (*salinarius* group) larvae (Dipera: Chironomidae) in Parry Sound, Georgian Bay, with particular reference to structural deformities. Can. J. Zool. 54, (1976),2129-2134.

[124] D.A. Birkholz, M.R. Samoiloff, W.F. Warwick, W.R.B. Webster, and J. Witteman: The Tobin Lake Project, Project proposal and progress report, November 1980. Off. Ind. Res., University of Manitoba, Winnipeg, (1980).

[125] T. Köhn and C. Frank: Effect of thermal pollution on the chironomid fauna in an urban channel. In: *Chironomidae. Ecology, systematics, cytology and physiology* (D.A. Murray, Ed.). Pergamon Press, Oxford, New York, Toronto, Sydney, Paris, Frankfurt, (1980), 187-194.

[126] W.F. Warwick: Pasqua Lake, Southeastern Saskatchewan: a preliminary assessment of trophic status and contamination based on the Chironomidae (Diptera). In: *Chironomidae Ecology, systematics, cytology and physiology* (D.A. Murray, Ed.). Pergamon Press, Oxford, New York, Toronto, Sydney, Paris, Frankfurt, (1980), 255-267.

[127] K.J. Tennessen and P.K. Gottfried: Variation in structure of ligula of Tanypodinae larvae (Diptera: Chironomidae). Ent. News 94, (1983), 109-116.

[128] A. Klink: Kopafwijkingen bij Chironomidae-larven onder invloed van chemische verontreiniging. Meded. Hydrobiol. Adviesburo K. Wageningen Rapp. 12, (1984), p. 31.

[129] K.A. Krieger: Benthic macroinvertebrates as indicators of environmental degradation in the southern nearshore zone of the central basin of Lake Erie. J. Great Lakes Res. 10, (1984) 197-209.

[130] R.A. Crowther and M.E. Luoma: Pattern recognition techniques to determine benthic invertebrate trophic and community responses to industrial input. Verh. Int. Ver. Limnol. 22, (1984), 3981-3989.

[131] T. Wiederholm: Incidence of deformed chironomid larvae (Diptera: Chironomidae) in Swedish lakes. Hydrobiologia 109, (1984), 243-249.

[132] A.G. Klink: Hydrobiologie van de Grensmaas. Hydrobiologisch Adviesburo Klink. Rapporten en Mededelingen 15, (1985), p.38+ bijlagen.

[133] W.F. Warwick: Morphological abnormalities in Chironomidae (Diptera) larvae as measures of toxic stress in freshwater ecosystems: indexing antennal deformities in *Chironomus* Meigen. Can. J. Fish. Aquat. Sci. 42, (1985), 1881-1914.

[134] G.Van Urk and F.C.M. Kerkum: Misvormingen bij muggelarven uit Nederlandse oppervlaktewateren. H$_2$O, (1986), 624-627.

[135] W.F. Warwick, J. Fitchko, P.M. McKee, D.R. Hart, and A.J. Burt: The incidence of deformities in *Chironomus* spp. from Port Hope Harbour, Lake Ontario. Internat. Assoc.Great Lakes Res. 13, (1987), 88-92.

[136] G. Van Urk and F.C.M. Kerkum: Bottom fauna of polluted Rhine sediments. Contaminated Soil, (1988), 1405-1407.

[137] V. Pettigrove: Larval mouthpart deformities in *Procladius paludicola* Skuse (Diptera: Chironomidae) from the Murray and Darling rivers, Australia. Hydrobiologia 179, (1989), 111-117.

[138] C. Van de Guchte and G. Van Urk: Discrepancies in the effects of field and artificially heavy metal contaminated aquatic sediments upon midge larvae. In: *Heavy metals in the environment*, 7th Int. Cont. 12-18 september, (1989), Genèva.

[139] W.F. Warwick: Morphological deformities in larvae of *Procladius* Skuse (Diptera: Chironomidae) and their biomonitoring potential. Can. J. Fish. Aquat. Sci. 46, (1989), 1255-1270.

[140] W.F. Warwick: Morphological deformities in Chironomidae (Diptera) larvae from the Lac St. Louis and Laprairie Basins of the St. Lawrence River. J. Great Lakes Res. 16, (1990), 185-208.

[141] R.M. Dermott: Deformities in larval *Procladius* spp. and dominant Chironomini from the St. Clair River. Hydrobiologia 219, (1991), 171-185.

[142] W.F. Warwick: Indexing deformities in ligulae and antennae of *Procladius* larvae (Diptera: Chironomidae): application to contaminant-stressed environments. Can. J. Fish. Aquat. Sci. 48, (1991), 1151-1166.

[143] M. Dickman, I. Brindle and M. Benson: Evidence of teratogens in sediments of the Niagara riverwatershed as reflected by chironomid (Diptera: Chironomidae) deformities. J. Great Lakes Res. 18, (1992), 467-480.

[144] L. Janssens de Bisthoven, K. Timmermans and F. Ollevier: The concentration of cadmium, lead, copper and zinc in *Chironomus* gr. *thummi* larvae (Diptera, Chironomidae) with deformed *versus* normal menta. Hydrobiologia 239, (1992), 141-149.

[145] N.T.H. Lien: Buccal deformities in chironomids as an indicator of sediment stress in running waters. MSc.-thesis, Ghent University, Belgium, (1992), p. 75.

[146] G. Van Urk, F.C.M. Kerkum and H. Smit: Life cycle patterns, density and frequency of deformities in *Chironomus* larvae (Diptera: Chironomidae) over a contaminated sediment gradient. Can. J. Fish. Aquat. Sci. 49, (1992), 2291-2299.

[147] G.M. Clarke: Fluctuating asymmetry of invertebrate populations as a biological indicator of environmental quality. Environmental pollution 82, (1993), 207-211.

[148] T.P. Diggins and K.M. Stewart: Deformities of aquatic larval midges (Chironomidae: Diptera) in the sediments of the Buffalo River, New York. J. Great Lakes Res. 19, (1993), 648-659.

[149] V.-P. Salonen, P. Alhonen, A. Itkonen and H. Olander: The trophic history of Enäjärvi, SW Finland, with special reference to its restoration problems. Hydrobiologia268, (1993), 147-162.

[150] G.A. Bird: Use of chironomid deformities to assess environmental degradation in the Yamaska river, Quebec). Environ. Monit. and Ass. 30, (1994), 163-175.

[151] T.J. Canfield, N.E. Kemble, W.G. Brumbaugh, F.J. Dwyer, C.G. Ingersoll and J.F. Fairchild: Use of benthic invertebrate community structure and the sediment quality triad to evaluate metal-contaminated sediment in the upper Clark Fork river, Montana. Env. Toxicol. and Chem. 13, (1994), 1999-2012.

[152] V. Pettigrove, W. Korth, M. Thomas and K.H. Bowmer: The impact of pesticides used in rice. agriculture on larval chironomid morphology: In: *Chironomids, from genes to ecosystems*, P. Cranston (ed.), CSIRO publication, Australia, (1994), 81-88.

[153] L. Janssens de Bisthoven and D. Van Speybroeck: Some observations of deformed midge larvae (Diptera, Chironomidae) in Kenya. Verh. Internat. Verein. Limnol. 25, (1994), 2485-2489.

[154] A. Gerhardt and L. Janssens de Bisthoven: Behavioural, developmental and morphological responses of *Chironomus* gr. *thummi* larvae (Diptera, Nematocera) to aquatic pollution. J. Aq. Ecosyst. Health 4, (1995), 205-214.

[155] L. Janssens de Bisthoven, C. Huysmans and F. Ollevier: The *in situ* relationships between sediment concentrations of micropollutants and morphological deformities in *Chironomus* gr. *thummi* larvae (Diptera, Chironomidae) from lowland rivers (Belgium): a spatial comparison. In: *Chironomids: from genes to ecosystems* (P. Cranston, Ed.), CSIRO-publication, Canberra, (1995), 63-80.

[156] V.G. Drabkova, V.A. Rumyantsev, L.V. Sergeeva and T.D. Slepukhina: Ecological problems of lake Ladoga: causes and solutions. Hydrobiologia 322, (1996), 1-7.

[157] L.A. Hudson and J.J.H. Ciborowski: Teratogenic and genotoxic responses of larval *Chironomus salinarius* group (Diptera: Chironomidae) to contaminated sediment. Env. Toxicol. and Chem. 15, (1996), 1375-1381.

[158] L.A. Hudson and J.J.H. Ciborowski: Spatial and taxonomic variation in incidence of mouthpart deformities in midge larvae (Diptera: Chironomidae: Chironomini). Can. J. Fish. Aquat. Sci. 53, (1996), 297-304.

[159] T.D. Slepukhina, I.V. Belyakova, Y.A. Chichikalyuk, N.N. Davydova, G.T. Frumin, E.M. Kruglov, E.V. Rubleva, L.V. Sergeeva and D. Subetto: Bottom sediments and biocoenoses of northern Ladoga and their changes under human impact. Hydrobiologia 322, (1996), 23-28.

[160] T.J. Canfield, E.L. Brunson, F.J. Dwyer, C.G. Ingersoll and N.E. Kemble: Assessing sediments from Upper Mississippi River navigational pools using a benthic invertebrate community evaluation and the sediment quality triad approach. Arch. Environ. Contam. Toxicol. 35, (1998), 202-212.

[161] T.P. Diggins and K.M. Stewart: Chironomid deformities, benthic commnity composition, and trace elements in the Buffalo River (New York) area of concern. J.N. Am. Benthol. Soc. 17, (1998), 311-323.

[162] L. Janssens de Bisthoven, J.F. Postma, P. Parren, K.R. Timmermans and F. Ollevier: Relations between heavy metals in aquatic sediments and in *Chironomus* larvae of Belgian lowland rivers and their morphological deformities. Can. J. Fish. Aquat. Sci. 55, (1998), 688-703.

[163] C. Frank: Glycolitic capacity of chironomid larvae from polluted and unpolluted waters. Verh. Int. Verein. Limnol. 21, (1981), 1627-1630.

[164] M.J. Servia, F. Cobo and M.A. Gonzalez: Deformities in larval *Prodiamesa olivacea* (Meigen, 1818) (Diptera, Chironomidae) and their use as bioindicators of toxic sediment stress. Hydrobiologia 385, (1998), 153-162.

[165] M.J. Servia, F. Cobo and M.A. Gonzales: Evaluacion del nivel de estres ambiental en diversos ecosistemas acuaticos de Galicia, mediante el estudio de deformidades en larvas de *Chironomus riparius* Meigen, 1804 (Diptera, Chironomidae). Nova Acta Compostelana (Bioloxia) 8, (1998), 271-280.

[166] A.C. Vermeulen, P.C. Dall, C. Lindegaard, F. Ollevier and B. Goddeeris: Improving the methodology of chironomid deformation analysis for sediment toxicity assessment: a case study in three Danish lowland streams. Archiv für Hydrobiologie, in press.

[167] A.C. Vermeulen, G. Liberloo, P. Dumont, F. Ollevier and B. Goddeeris: Exposure of *Chironomus riparius* larvae (Diptera) to lead, mercury and B-sitosterol: effects on mouthpart deformation and moulting. In: A.C. Vermeulen: Head capsule deformation in *Chironomus riparius* larvae (Diptera): causality, ontogenesis and its application in biomonitoring. PhD-thesis, KULeuven, Belgium, (1998), p.207.

[168] L. Hakanson: An ecological risk index for aquatic pollution control. A sedimentological approach. Water Research 14, (1980), 975-1001.

[169] W.F. Warwick: The use of morphological deformities in chironomid larvae for biological effects monitoring. National Hydrology Research Institute no. 43, (1990).

[170] R.N. Hooftman and I.M. Benschop: Onderzoek naar de mogelijke genetische achtergrond van het ontstaan van misvormingen bij chironomiden door middel van bestraling van embryo's en larven. abstract, "Chironomiden-studiedag", 23-12-1988, (1989).

[171] C. Frank and T. Köhn: The influence of waste water and thermal pollution on the benthos of an urban channel. In: *Urban ecology* (Bornkamm, R., Ed.), Blackwell Sc. Publ. Oxford, (1982), 345-346.

[172] E.M.M. Grootelaar, F.C.M. Kerkum and G. Van Urk: Effekten op chironomidelarven bij blootstelling aan waterbodems, verontreinigd met zware metalen. "Chironomiden-studiedag", 23-12-1988, (1989).

[173] G. Sella, L. Ramella, P. Michaelova, N. Petrova, F. Regoli and V. Zelano: Effets genotoxiques des sediments pollués chez deux populations de *Chironomus riparius* Meigen 1804. Actes du 4ème congrès du GRAPE, 'Sédiments et gestion des milieux aquatiques, Lyon 21-22 novembre, (1995), 98-101.

[174] L. Janssens de Bisthoven, E. Van Looy, R. Ceusters, F. Gullentops and F. Ollevier: Densities of *Prodiamesa olivacea* (Meigen) (Diptera: Chironomidae) in a second order stream, the Laan (Belgium): relation to river dynamics. Neth. J. Aq. Ec. 26, (1992), 485-490.

[175] L. Janssens de Bisthoven, F. Ollevier, B. Beyst and N. De Pauw: Misvormingen bij muggelarven als bioindicatoren voor microverontreiniging van waterbodems. In: *Macro-invertebraten en waterkwaliteit* (De Pauw and Vannevel, eds.) Dossiers Stichting Leefmilieu *(in press)*.

[176] P.M. Chapman: Sediment quality criteria from the sediment quality triad: an example. Env. Tox. Chem. 5, (1986), 957-964.

[177] G. Krantzberg: Ecosystem health as measured from the molecular to the community level of organization, with reference to sediment bioassessment. J. Aq. Ecosystem Health 1, (1992), 319-328.

[178] C.A. Bishop, P. Ng, K.E. Pettit, S.W. Kennedy, J.J. Stegeman, R.J. Norstrom, and R.J. Brooks: Environmental contamination and developmental abnormalities in eggs and hatchlings of the common snapping turtle (*Chelydra serpentina serpentina*) from the Great Lakes-St Lawrence River basin (1989-91). Environmental Pollution 101, (1998), 143-156.

Environmental Science Forum Vol.96 (1999) pp. 95-118
© 1999 Trans Tech Publications, Switzerland

Recent Trends in Online Biomonitoring for Water Quality Control

A. Gerhardt

'LimCo International', An der Aa 5, DE-49477 Ibbenbüren, Germany

Keywords: Automated Biomonitors, Continuous Biotests, Early Warning Systems, Water Quality Monitoring

Abstract

This review shortly describes the principles of various bacteria, algae, mussel, invertebrate and fish biomonitors, which are used for water quality monitoring or which are under recent development. Some new trends can be distinguished and should be encouraged: 1) simultaneous use of more than one species in a biomonitor, 2) use of indigenous test species in addition to reference organisms such as *Daphnia* sp., 3) use of a sequence of stress responses, which can be measured quantitatively and 4) relation of behavioral responses to instream ecologically relevant responses. As life is the ultimate ecologically relevant monitor of ecosystem/human health the development of online biomonitors for water quality control of toxic effluents, waste water purification plants and drinking water intakes is a necessary invaluable tool for environmental monitoring.

1) Introduction

As applied branch of ecotoxicology, biomonitoring is based on two pillars, trend biomonitoring to survey long-term changes in an ecosystem and biological early warning monitoring to detect exposure to spikes of toxic cocktails, providing data about possible permit violations. The development of online biotests follows the demands of Agenda 21, which highlights the protection of water as a restricted resource. The basic idea of the use of automated biological sensor systems for water quality management was first proposed by Cairns [1].

Automated biomonitors operate on a real-time basis and use living organisms as sensors, ideally providing a continuous flow of information about water quality [2].

Online biomonitors, continuous (dynamic) biotests, biotest automates or "Biological Early Warning Systems" (BEWS) are characterised by three components: a test organism, an automated detection system and an alarm system. Biomonitors based on subcellular biological responses connected to or integrated in a transducer, *e.g.* "Protoplastenbiotest", are better called biosensors [3,4]. However, other authors include organismic biomonitors in a broad definition of biosensors [5]. Suborganismal biosensors for environmental monitoring are excluded from this article, as they are described in several books [6]. Continuous biotests are based on the permanent renewal of test water in a flow-through system, which is realised in *Daphnia*-tests, fishtests, musseltests and some bacteria tests. In semi-continuous biotests the water flow is stopped during a defined contact time ($\leq$ 30 min.) where the biological parameters are measured [7].

The task of a biological early warning system is to detect the spike of a toxic substance cocktail in the environment as quickly as possible [3], by measuring the summation parameter, "biological effect" of the water to be tested, in contrast to chemical monitoring, which measures concentration levels of selected chemicals. Biological systems are a good complement to chemical monitoring, as they integrate potential toxic effects of different chemical compounds and their degradation products on the organism over time and they indicate the overall effects on the biocoenosis of the aquatic ecosystem [8]. A BEWS will never give quantitative nor qualitative information about a certain chemical pollutant. However, their strength lies in continuous operation, rapid response and broad spectrum detection of pollutants, sometimes even those which are not detected by routine chemical monitoring schemes [9].

Online biomonitors offer a fast and cheap method to evaluate potential negative effects of pollutants on organisms, especially when chemical monitoring is impossible (unknown degradation products of organic toxicants), unrealistic (synergistic or antagonistic effects of toxicants in a toxic cocktail) and expensive (the average cost for analysing the 126 U.S. EPA priority pollutants approaches 1000 USD per sample [2]). Polar substances increase in importance in aquatic pollution, as their use is preferred, due to low persistence and low bioaccumulation. Polar substances can, however, not be analysed with spectral methods, indicating hence the increased importance of biological detection methods [10].

Desired characteristics of automated biomonitors are: truely automated and real-time, online operation, sensitivity to pollution, rapid responses, easy operation, reliable alarm interpretation, minimal false alarms, low purchase costs, minimal installation, maintenance and training requirements, compatible with industrial environments, possibility of real-time remote data transmission and integration with online chemical monitors and automated water sample collection [2].

In Germany and the Netherlands the development and installation of online biomonitors has been encouraged after the Sandoz accident at the River Rhine in 1986. The following biomonitors are used in Germany at ca. 14 rivers since 1990: Dynamic *Daphnia*-test (at 27 sites), Fish-Rheotaxis-test (at 14 sites), Koblenz Verhaltensfischtest (at 1 site), *Dreissena*-Monitor (at 11 sites), Mussel-Monitor (at 1 site). Moreover, several algae- and bacteria-tests are in use since 1995 at about 8 sites [11]. The "Deutsche Kommission zur Reinhaltung des Rheins" (DK) recommends to replace fishtests with other, more sensitive biotests. One of the existing *Daphnia*-tests and one algae-test should be used in a biotest-battery, including the four different trophic levels, in order to cover as comprehensively as possible the spectrum of biocidal effects of toxicants [12]. For the different algae- and bacteria- tests, there is still a lack of information about online operation in the field [11].

2) Description of online biotest systems

In this chapter, existing (semi)continuous biomonitors will shortly be described and compared, firm names will be avoided. Detailed overviews and descriptions of different biotests are given elsewhere [3, 5, 13-15].

2.1) Bacteria tests

In general, there are two approaches for biomonitoring with bacteria, (1) using free living bacteria in a test cell, (2) using bacteria fixed to a substrate, or immobilised bacteria monocultures within electrodes acting as an intermediate for instrumental analysis such as potentiometry or amperometry [3].

The majority of continuous bacteria tests use the respiration activity of bacteria as test parameter (bacterial respirometers): test water rich in oxygen flows through a bioreactor and the oxygen concentration decreases as a function of the organisms' activity (Tab. 1). Oxygen consumption or carbon dioxide production is measured with electrodes. There are tests with free moving bacteria and with immobilised bacteria, either forming a biofilm on a substrate or immobilised in a substrate on a bioelectrode [5]. Activated sludge bacteria represent a group of mixed species and are much less sensitive than pure cultures of single bacteria species, such as *E. coli*, *P. aminovorans*, *P. putida*, *P. aeruginosa*), because their species composition changes during the test period, and they need up to two weeks to develop a functional biofilm before online operation can start [5, 16]. Some biotests use the growth of bacteria, measured photometrically as turbidity in the medium. This method is applied *e.g.* to monitor denitrification of waste and drinking water. Denitrification activity has been used to control fermentations and to detect 3,5-dichlorophenol [16].

Tab. 1: Biological Early Warning Systems (BEWS)

Organism	Biological response	Automatic detector	Development	Reference
Bacteria				
Pseudomonas sp.	respiration	O_2 or CO_2 electrodes	+	[5, 16]
E. coli	(O_2 or CO_2),			
mixed populations	growth or	turbidity	+	
Vibrio fischeri	bioluminescence	fluorimeter	++	"
Synechococcus sp.	electrontransport	amperometric	++	[5]
Algae				
Chlorella sp.	growth	chlorophyll content	+	"
	O_2-production	O_2 electrode	+	"
	fluorescence (spontaneous, delayed or modulated	fluorimeter	++	"
Euglena gracilis	motility	digital image processing	-	[22]
Higher Plants				
Protoplastenbiotest	O_2-production	amperometric	+	[5, 23]
Mussel				
D. polymorpha	shell movements	magnetic induction	+++	[33]
"	"	reed switch	+++	[38]
Corbicula fluminea	"	opto-electric switch	-	[44]
Anodonta cygnea	"	electric switch	-	[43]
Crustacea				
Daphnids	motility, phototaxis	IR-light	++	[47, 48, 54]
"	swimming velocity and direction	digital image processing	++	[55, 56]
Fish				
Leuciscus idus	rheotaxis	IR-light, pressure or	+++	[58]
		ultrasound sensors	+	[49]
diff. species	activity	ultrasonic	+	[60]
"	motility, direction	digital image processing	++	[62]
	schoaling, velocitiy	3d-image analysis	++	[63]
fish and *Daphnia* sp.	preference/avoidance		++	[61]
Salmonidae	ventilation, heart	bioelectrical responses	++	[65]
	and cough rate		n.m.	[3, 64]
	and locomotion			
Gnathonemus petersi	Electric Organ Dis-	potentiometric	+	[67]
Apteronotus albifrons	charge (EOD)			
Other invertebrates				
Tubificidae	motility	digital image processing	-	[71]
Insects, Crustaceans	motility, ventilation	2-polar impedance conversion	-	[75, 77]
all aquatic organisms (multispecies)	motility, ventilation	4-polar impedance conversion	++	[29,79, 82]

Development (three levels of development incl. degree of automation: + Min, +++ Max) [2], n.m. not mentioned, - not yet automated operation. (modified after [29]).

Photobacterium phosphoreum and *Vibrio fischeri* are used to register respiration and natural bioluminescence as function of inhibited energy metabolism, as they emit part of their metabolic energy as light (Microtox). During 30 min. measuring time bioluminescence in bacteria exposed to polluted water is compared to a reference and the result is plotted afterwards. As *P. phosphoreum* is a non-pathogen marine bacterium, it is necessary to add salt to the freshwater samples. Salt can, however, interfere with some toxicants and alter the real toxicity, which is intended to be measured with the biotest. Several types of this semi-continuous test have been developed [15]. However, the acceptance of these discrete bioluminescence biomonitors by the regulatory authorities in the U.S.A is low [2]. Bioluminescence can also be achieved in other bacteria by insertion of the luciferase gene "lux" into other bacteria species, *e.g. Klebsiella planticola*. Other genes, potentially interesting for biomonitoring, would be those for stress proteins (HSP) or metallothioneins. Such bioelectrodes could be substance-specific biosensors [5]. Also the yeast *Saccharomyces cerevisiae* immobilised in a gel of agar has been used to detect pollutants with help of measuring the NADP(H)-fluorescence at electron transport chains in the mitochondria [17]. A detailed description of different bacteria biotests can be found in [15]. A comparison shows, that not only the biological components (test species, test parameter, pure or mixed bacteria culture) but also the technical measurement details affect the sensitivity of a biomonitor [18].

Microbial biosensors (bioelectrodes, bioprobes) are a new and upcoming research field. They are invaluable tools for rapid, simple, low-cost online water quality monitoring at selected locations. The survival time of biolelectrodes varies between 3 to 30 days. For example, the measurement of the BOD (biological oxygen demand) relies on a bioelectrode with cyanobacteria or green algae. Their photosynthetic activity is measured via a mediator, which transfers the electron-flux from the organisms to the electrode. This bioelectrode has been used in waste water purification plants and gives indications about organic micropollutants which are biodegradable in the lower ppm range [5, 16]. The sensitivity of the bioprobes depends on the detector (O_2 or CO_2) and the characteristics of the biological membrane, *e.g.* species of bacterium, number of immobilised bacteria and method of fixation at the bioelectrode. The so called "EuCyanobacterienelektrode" is based on the measurement of inhibition of the photosynthetic electrontransport chain of *Synechococcus* sp. and has also been used for the measurement of the inhibition of the respiratory electrontransport chain of *E. coli* [19]. The cyanobacteria are immobilised on a graphit electrode, they produce photosynthetic intermediates which are re-oxidised at the electrode and amperometrically detected with a response time of 2 min. [19].

Problems with bacteria biotests have been as follows: Bacteria clogg the tubes, so that the daily maintenance effort is above that of other biotests. The mean exposure time of the bacteria to the test water is below 10 min, *i.e.* a fast reaction is necessary. Bacteria are generally less sensitive than algae. Bacteria tests are mostly useful in waste water purification plants, where the BSB is sufficiently high. For use in monitoring stations of surface water, nutrient addition is often needed, which might affect toxicity of the test water and sensitivity of the test organisms. Toximeters with separate bacteria cultures require more maintenance efforts than the biofilm reactors. Biofilm reactors are more ecologically relevant, but need adaptation times leading to decreased sensitivity [11, 15].

2.2) Algae biotests

Algae tests are invaluable for the detection of potentially toxic effects of herbicides in water.
The sensitivity of algae tests is governed by several factors, such as composition of the medium, type of limiting nutrients, cell density, chelator concentration, choice of response variables etc. Interlaboratory comparisons showed that EC_{50} values varied by several orders of magnitude [20]. Algae tests measure either cell growth, oxygen production, direct fluorescence or energy transfer in photosystems which can be registered as delayed fluorescence or fluorescence modulation (Tab.1). Algae tests use the following test species: *Scenedesmus* sp., *Chlorella* sp., *Chlamydomonas* sp. and *Microcystis* sp.. Photosynthesis tests based on algae, use the measurement principle of spontaneous or delayed fluorescence (luminescence) to measure photosynthetic activity. Photosynthesis rate and electrontransfer are negatively correlated with natural fluorescence, *i.e.* if a herbicide causes a

decrease in photosynthesis, the natural fluorescence increases because of electron accumulation. At the same time, oxygen production and the rate of fluorescence modulation diminish [16]. In biotests using direct natural fluorescence, no difference between alive and dead algae can be made. The measurement of post-illumination emission (delayed fluorescence) uses the ability of alive plant cells to transform light energy in electrons, store them and retransfer them after several minutes to hours after the light source has been switched off. Biotests based on fluorescence modulation combine both methods [16]. Distinction between signals from living chlorophyll-containing algae and other fluorescent material is achieved by using two modulated light sources resulting in a mean fluorescence rate. The measuring light induces changes in chlorophyll fluorescence with a frequency of 1 kHz and the actinic light modulates the redox state of the photosystem with a frequency of 1 Hz. This leads to a modulation of the yield which is detected by two phase sensitive rectifiers [16]. Whereas fluorimeters measure chlorophyll fluorescence and relate it to the chlorophyll content, toximeters add a defined amount of algae to a water sample and measure fluorescence or oxygen production. General problems with algae biotests have been 1) natural fluorescence of algae in the test water without addition of test algae suspension, 2) fluorescence emission is highly dependent on the physiological condition of the algae, 2) maintenance of a stable algae culture with avoidance of contamination by sterile working equipment and daily microscopical control of the culture, 3) high weekly maintenance efforts including sterilisation, cleaning, culture of algae, nutrient solutions etc. 4) only semi-continuous operation possible, measurement intervals ca. 30 min. [10]. Recently, fluorescence measuring bioelectrodes with immobilised algae have been developed, which in future might replace existing algae biomonitors [21].

The new biotest system ECOTOX measures motility of the flagellat *Euglena gracilis* with a digital image analysing system [22]. This test is semi-continuous, after an adaptation time of 2 min. of the cell suspension to the test water, the measurement of motility starts for 2-5 min. by registration of the following parameters: precision of gravitactic orientation, percentage of cells moving up in the flow-through cuvette, average swimming velocity and cell shape. Until now this system has only been tested in the laboratory with different toxins. The system needs a fast response of the organisms within 2-5 min., which might only happen during acute toxic pollution in the range of lethal concentrations.

2.3) Higher plant biotests

Isolated protoplasts of mesophyll cells of *Vicia faber* are immobilised in high polymer alginate and their oxygen production is measured with a polarographic method [23]. Isolated protoplasts keep for about one week. This biotest has been used to measure inhibition of light-dependent oxygen production during photosynthesis due to atrazine in France [5].

3) Animal biomonitors

Biological early warning systems using animals often measure behavioral parameters. Behavioral responses to toxins are amongst the most sensitive and the first to occur [24]. The behavior of an organism is the end point of the functional integration of the nervous system encompassing sensory, locomotory, and cognitive aspects [25]. Due to their sensitivity, non-destructivness, integrative nature and ecological relevance, behavioral responses allow for continuous long-term biomonitoring [26]. Behavioral plasticity originates from either differences in ontogenetic development or adjustments due to learning, both resulting in a "flexible reaction norm" of the species. In contrast to this "continuous" strategy, behavioral plasticity can also occur as "discontinuous" strategy, *e.g.* the innate ability to respond in different ways [27]. The latter case has been observed in several aquatic organisms and been described as dichotomy of the behavioral response [28]. Another possibility of a discontinuous strategy in behavioral plasticity is to switch from one type of behavior to another type, which has been described in the "stepwise stress model" [29]. With increasing environmental gradients, *i.e.* in more risky environments, the range of behaviors might increase [27]. These uncertainties might aggravate the evaluation of observed behavioral responses in any animal biomonitor.

3.1) Mussel biomonitors

As bivalves can circulate several tens of liters of water daily and thus enrich pollutants to tissue concentrations up to 10000 times of those in the surrounding water, they have been used as bioindicators for water pollution in trend biomonitoring programs [30]. Mussel biotests for online biomonitoring purposes measure valve movements and status of valve opening of *Dreissena polymorpha* (*Dreissena*-Monitor, Musselmonitor) (Tab. 1). *Dreissena polymorpha* is a neozoon from the pontocaspian area transported into the eurasian region via ships and channels. The use of such a "pest" species in online water quality monitoring is controversial due to their tolerance, and might only be accepted in streams where this species occurs. It is highly preferable to use local species than to import new populations or alien species [9]. The use of naturally occurring species such as *Unio* sp. and *Anodonta* sp. should be encouraged [31]. However, some species, *e.g. Corbicula fluminea*, have closing times of their valves of several days, which make them inadequate for online biomonitoring (Ortmann pers. communication).

The Musselmonitor measures the gape between the two valves of eight organisms with help of a high frequency electromagnetic induction system, in which miniature accelerator and detector coils are glued to opposite shell valves, producing a linear valve displacement response [32, 33]. The movement of the valves induces an electrical current that can be calibrated as a distance between both valves. The minimal and maximal reading of each mussel is recorded, and these values are set to 0 and 100%, respectively. All data are normalised and the average reading is recorded in a semi-continuous way each hour. Activity is defined as the number of valve movements per unit of time (usually per hour). Different states can be distinguished: 1) no alarm, when the mussels are open for respiration and filter feeding; 2) a sudden pollution peak reverses the behavior, the mussels are closed and only open their shells once in a while, 3) a proceeding chronic pollution causes the mussels to close more often than normal and to open the shells not as wide as normal, 4) in order to get rid of irritating substances the mussels frequently open and close the shells and 5) dead mussels have much wider open shells than living mussels [33]. This biotest has been used with several free living and sessile freshwater and marine species such as *D. polymorpha*, *M. edulis*, *O. edulis*, *U. pictorum*, *A. cygnea* and *C. fluminea* in laboratory and *in situ* experiments with organisms being collected from reference sites [33 - 35]. A comparative test of the Musselmonitor with the Fish rheotaxis biotest at the Rhine in the Netherlands showed good agreement in alarm generation due to toxicants [33]. Recently, a new data evaluation method for the Musselmonitor has been developed. This dynamic alarmsystem is based on Split-Moving-Window boundary analysis, a kind of moving average method for data matrices [36]. Changes in valve position and activity have been found in laboratory experiments at $\geq 80\ \mu g/l$ Cu and at $\geq 120\ \mu g/l$ Na-PCP. No diurnal and seasonal activity rhythms have been found during a 1-year test period in the field [36], which might be due to the low number of test organisms. The Musselmonitor is used at one site in Germany and at several sites in the Netherlands and Belgium, USA and Australia for the survey of water quality. The strength of the system is that it works with different mussel species of a size range between 1.5 and 12 cm and that it can be completely immersed in the water, which makes instream use possible, compared to the use in a bypass of most other biomonitors. The practical experiences reveal some drawbacks: 1) false alarms occurred, 2) eight organisms might not be enough to measure activity rhythms, 3) the quantitative measurement of the valve openings seems to be less important than the fact that the mussels are either open or closed (Penders, pers. communication) and 4) long closure times of the mussels, up to 24 h for *D. polymorpha* result in unreliable early warning alarm detection. The long-term use of a single individuum in this biotest (up to several months) can lead to less sensitivity to toxicants due to acclimation, however it allows for chronic long-term exposure measurements. The weekly maintenance effort is ca. 4,5 hours, including insertion of new mussels.

The *Dreissena*-Monitor is based on the measurement of valve movements of 2 x 42 specimens of the zebra-mussel with help of a reed switch. A small magnet is glued on the valve of a mussel. If the mussel opens, the magnet comes close to the reed switch, which closes the electrical circuit [37]. The dynamic limits of two of the parameters recorded by the system, 1) percentage of open mussels and 2) number of valve movements, are exceeded when the actual running average is less or greater than a long-term average, minus or plus its threefold standard deviation. Every second, the computer records whether a mussel is open or closed and after a 5 minute period of analysis, the percentage of open mussels and the number of valve movements are computed as running means for each channel. During toxicity tests, the number of valve movements normally increases before the percentage of open mussels decreases, which allows for the definition of a stepwise alarm system. A "water alarm" is defined if the stepwise alarms are verified simultaneously in both experimental channels [38]. Laboratory toxicity tests revealed sensitivity thresholds of 20 $\mu g/l$ pentachlorophenol, 150 $\mu g/l$ lindane, 200 $\mu g/l$ 2-nitrophenol and 3.3 mg/l atrazin [39]. A classic alarm has been measured in the field due to 5.5 $\mu g/l$ chloroform, resulting in an increase of valve movements from 0-1 to 3 mussels per hour and a decrease of the percentage of open mussels from 90 resp. 70 % (channel 1 and 2) to 40-50 % [39]. *In situ* toxicity tests revealed concentration limits and reaction times for the *Dreissena* Monitor of 0.9 mg/l ethylparation within 15 min., 33 mg/l chloroform within 15 min., 8.8 mg/l trichlorethylene within 20 min. and 0.1 mg/l $CuSO_4$ within 30 min and 120 $\mu g/l$ PCP. The sensitivities were comparable to those of the Musselmonitor [40, 41]. Diurnal and seasonal activity rhythms have been observed, which led to false alarms [42]. The *Dreissena*-Monitor seems to be more successful than the Musselmonitor, however false alarms due to diurnal activity rhythms have been observed at 60 % of the sites in Germany (Muscheltestanwendertreffen, pers. communication). Theoretically, every mussel remains ca. 10 months in use, however many mussels are excluded by the system from the alarm analysis, so that *e.g.* in the summer only 30 % of the mussels in the channel are considered for the calculations. The main reason for exclusion is bad adjustment of the reed switches. Exclusion criteria have to be revised and the software updated from a DOS to a WINDOWS system (Muscheltestanwendertreffen, pers. communication). The weekly maintenance effort is 2-3 hours [40].

Other trials to use mussels as biomonitors have been performed. Englund *et al.* [43] designed a system using two switches made of silver wire glued on the shells of *Anodonta cygnea*. A detection of open, half-open or closed valves can be done, whereas the mussels move freely in the freshwater environment to be tested. Ham and Peterson [44] used *Corbicula fluminea* in a flow-through system. Valve movements trigger an opto-electric switch and give qualitative information whether the shells are open or closed. In a biotest in the USA, *C. fluminea* is glued to a substrate and a magnet attached to one shell opposite of a proximity sensor [9].

3.2) Daphnia biomonitors

Daphnia sp. has often been utilized in toxicological studies and environmental monitoring of aquatic systems for a number of reasons, such as their sensitivity to toxins, their ease of culture and their importance in the zooplankton of many lakes [45]. There is a huge amount of toxicity studies about the effect of metals and other toxic chemicals on different *Daphnia* species, their survival, growth and reproduction. The most frequently used species have been *D. magna*, *D. pulex* and *Ceriodaphnia reticulata*. However, a comparison between the studies leading to generally accepted conclusions is almost impossible, because several abiotic and biotic factors varied in these tests, as well as the species-specific tolerances of the above mentioned species [45]. In order to standardise results, specific clones are produced to generate organisms of defined and equal conditions, *e.g.* clones selected on positive phototaxis for the dynamic waterflea biotest [46]. As daphnids are traditional toxicological test organisms, it is obvious that several automated biotests have been developed (Tab. 1).

The dynamic waterflea assay designed by Knie [47, 48] is based on the registration of the number of interruptions of infrared lightbeams by swimming waterfleas. Over a period of 10 min. the number of IR-lightbeam interruptions is recorded. It has a pair of 0.03 l chambers that contain each 20 2-d old waterfleas. The prefiltered water flows through at a rate of 0.5 l/h. If the flow rate through the biotest is much lower than *in situ,* toxic responses will be discovered much later, at expense of early warning functions. In order to reduce reproduction within one week, the temperature is kept low at 17 +/- 1 ^0C [49]. Within one week, *Daphnia* sp. grows and the probability of passing the lightbeams increases just due to increased body size. 7-d old *Daphnia* sp. showed 3 times higher impulse rates than 1-d old specimens [50]. Starved organisms are more active than fed daphnids. In this biotest, *Daphnia* sp. feeds on particles in the test water. The supply of particles with the water, however, varies throughout the year. In order to achieve a constant food supply, addition of food is provided in some tests [51]. Both, static (number of impulses during normal behavior) as well as dynamic (moving averages) thresholds are used in the dynamic waterflea biomonitor. The software is still based on a DOS-system, maintenance efforts are ca. 3-4 h per week [52]. During a 6 months long field test at the Weser, 10 failures of the biomonitor occurred due to reproduction, failure of pumps, software errors and external factors such as increased turbidity [53]. Moreover, increased maintenance efforts due to clogging of sieves, air bubble accumulation under the sieves and addition of food was necessary [53]. The application of the "adaptive Hinkley detector", an algorithm to detect abrupt sudden changes in noisy systems, as alarm evaluation method for the dynamic *Daphnia* test revealed delayed alarm detection [53].

Whereas the biotest of Knie relies on the motility of waterfleas, a further development of the test includes a light source and measures <u>phototaxis</u> of waterfleas [54]. Since then, the development of the dynamic waterflea phototaxis biotest evolved by including gentically homogeneous progeny with parthenogenetic reproduction, and by selecting clones for phototactic behavior [46, 51]. In the phototaxis biotest, two light sources, one on top and one on the bottom of the test chamber with waterfleas are alternatively switched on and off in order to stimulate the waterfleas to move and thus cross the IR-lightbeams. Already the concept of such an ecologically controversial approach cannot be recommended for surface water quality control: the use of specific genetic homogeneous clones is far from ecological reality. The alternative illumination of the test chambers rather represents a tremendous artificial stress factor for the daphnids, thus being no good basis for early detection of toxic stress in an ecological context.

During the last few years, <u>digital image processing</u> technology has become a rising application in online biomonitoring. Variables that determine the design of computer programs for digital image analysis include the number of objects to be analysed, the speed of the objects in relation to their size, whether the objects can be individually labelled (*e.g.* by applying coloured tags), whether the trajectory needs to be reconstructed in real-time, the types of behavior to be detected etc. [55]. There have been several trials to register the movements of *Daphnia* sp. with video cameras, some examples are following.

A system described by van Hoof *et al.* [51] presents a monitor for the continuous analysis of the activity of 20 swimming daphnids. The 2-dimensional image of the total test chamber is stored in a video recorder and off-line digitalised with a resolution of 256 x 256 pixels. The algorithm, witten in MS DOS Microsoft Pascal, defines an "activity index" as the number of pixels of moving organisms divided by the total number of pixels and then multiplied by 100. In order to obtain this value as a function of time, five sequential activity values from 14 sequential images are used to calculate a moving mean over all images. In a short-term toxicity test with Cu and Cr (0.1 and 0.2 mg Cu/l ; 0.5 and 3 mg Cr/l) van Hoof *et al.* [51] compared the image analysis system with the dynamic waterflea phototaxis biotest. In the image analysis system the behavioral changes due to the metals occurred much faster and responses to sublethal concentrations were recorded as well.

The image analysing system "BehavioQuant" has been used for recording the swimming behavior of daphnids [56]. A test unit contains a flow-through measurement chamber (7.5 x 6 x 1 cm) illuminated by an electroluminescence foil and a mirror in an angle of 45 ^{0} in front of the video camera. Every 15 minutes for a period of two minutes, movements of 2 times 10 organisms are taped from the image on the mirror, not directly from the test chamber. A reference background picture is created in the following way: the time interval between the records is chosen so that the position of the organisms has moved frequently. Thus the calculated mean values are low and are almost similar in brightness as the background itsself. This fact gives the reference background picture, where the objects do not appear due to their movements, even if they have been in the test chamber. After each record, the actual picture is compared with the background, where differences in brightness indicate an object. Its position coordinates are saved. Between the measurement cycles, velocity and homogeneity of the movements, position of the organisms (swimming height in 10 classes), number of changes in swimming direction and distance amongst the test organisms can be analysed. For *Daphnia* sp., the number of turnings is one of the most important parameters (Blübaum-Gronau, pers. comm.). With help of a Mann-Whitney-U test the behavior during a series of measurements is compared with the normal behavior which is based on 24 previous measurement cycles. This system has been successfully applied for the monitoring of fish and is now in development for *Daphnia* sp. There are indications that with increasing age or temperature (22 ^{0}C) the organisms swim faster and the number of changes in swimming direction increases compared to experiments at 11 ^{0}C [56]. In order to measure the realistic swimming behavior of *Daphnia* sp., the size of the test chamber is important, especially in the third dimension. Video systems often need flat test chambers as the accuracy of the measurements decreases with increasing depth. The size of the test chamber should also prevent the number of organisms to affect their swimming velocity, and allow for the measurement of the normal swimming behavior, instead of swimming artefacts.

Baillieul and Scheunders [56] present a digital image analysis system which allows the reconstruction of several trajectories simultaneously in real-time without limitations on the sequence length and without need to lower the resolution of the images. Using a simple geometrical model describing the displacement of the objects, the relative displacemnet of the objects is converted to an approximated value for the average velocity of the objects. The flow-through test chamber (10 x 7.5 x 1.5 cm) allows the 25 daphnids only to move in two dimensions, and is placed in a dark tunnel to exclude any visual stimuli other than the uniform background illumination. Every 1.5 sec. the program yields one result of the average velocity of 25 objects. The application of the geometrical model demands several assumptions, 1) identical shape of the objects and 2) effects of overlapping trajectories and possible disappearance from and/or reappearances into the active window are statistically cancelled out. To eliminate interference of small objects, such as offspring of daphnids or airbubbles, the object area threshold is defined, below which an object is excluded from the first sequence image. If too many objects are exluded, the average velocity is overestimated. Some general shortages of video systems are 1) for objects moving faster than their own diameter the trajectories are interrupted, which may lead to an underestimation of the results and 2) the movement of daphnids is only recorded in a two dimensional space of a small chamber, which might produce artifical stress to the organisms and 3) the constant background illumination might affect normal behavior of the organisms during the night.

3.3) Fish biomonitors

Amongst the freshwater animals, fish are the organisms with the most physiological similarity to human beings. Often, changes in behavior such as loss of orientation and damages at the gills or the skin can be seen. Fish might be used for monitoring drinking water intakes. However, ethical criticism in using fish for surface water quality control arise, due to often inadequate survival conditions. Differences in the species of fish used for biological early warning systems and different data analysis algorithms make it impossible to infer conclusions about fish biomonitors in general (Tab. 1) [57].

The first online biotests with fish were based on rheotactic swimming behavior [49, 58]. Four individuals of a gold-colored variety of the indigenous ide, *Leuciscus idus* (Cyprinidae) are exposed in a flow-through aquarium at a velocity rate of 30 l/min. The fish swim upstream against the current. When the fish are weakened or try to avoid toxicants with the incoming water, they touch pressure sensitive bars or pass an ultrasound barrier at the outflow of the aquarium, which produces an accoustic signal. Some other tests use an optical barrier, which has the disadvantage of being less sensitive in turbid water. Apart from loss of rheotaxis, surfacing can be recorded as well with this system by placing a kinetic screen above or below a horizontal axis [59]. Several short-term laboratoy tests with metals and organic compounds showed increasing inability of the fish to swim against the current ("manifestation"), usually accompanied or preceded by surfacing [59]. This biotest can also be equipped with a light barrier to count the passing fish, however both methods have the disadvantage that it is not known, whether one or many fish pass the light barrier or the kinetic screen. A variation of this biotest measures the surfacing behavior of fish in their resting phase with help of horizontal light barriers, and their rheotaxis during the current phase with help of vertical light barriers. During the current phase a shock generator produces electrical impulses between 50 - 400 V, in order to force the fish in the anterior measurement section of the test chamber [59]. In practice, this biotest has proven to be of low sensitivity and out of ethic reasons such a biotest device can no longer be recommended [11]. The fish rheotaxis biotests require ca. 3-6 h maintenance effort per week and have a stand alone time of 1 week, comparable to other biotests. An ultrasonic technique for monitoring the activity of different fish species has been developed by Morgan and Kühn [60]. Up to 10 fish, individually placed in one sensor chamber, can be monitored simultaneously. Each chamber is equipped with one ultrasonic transmitting element and one receiver. The generated signal comprises the sum of the various reflected sound waves. Whenever the fish moves, those reflections coming from the organism exhibit a Doppler shift in frequency which, after mixing in the receiver with the other reflections, appear as an amplitude modulation of the total signal. The system has been field tested in South Africa with *Poecilia reticulata* as test species.

Digital image processing has in the last years also been used to monitor fish behavior. Smith and Bailey [61] present a biomonitor system based on video tracking technique, in order to measure preference/avoidance of *Daphnia magna* and different fish species under toxic stress. The preference/avoidance chamber contains three compartments: control, test area and avoidance/preference area for 20 *Daphnia magna* or 30 fish. The software designs the paths of the movements of the organisms and generates other parameters such as linear and angular velocity, xy-coordinates, change of direction etc. Linear velocity and location in x and y coordinates proved to be the best parameters for automated early warning biomonitoring. An acclimation period of three (for fish), respective one (*Daphnia magna*) day, controlled by the measurement of linear velocity and location, is recommended before data registration. This first "multispecies" biomonitor system has been field tested and is still under development in the U.S.A [61].

The Koblenz behavioral fishtest measures behavioral changes of 2 x 6 juvenile golden orfs (*Leuciscus idus melanotus*) in a group with the image processing system "BehavioQuant" [56]. The principle of this system is the recognition of an object via a point by point comparison of sequential video pictures with a reference background picture as described above for the *Daphnia* sp. test. The behavior of the fish is registered by video observation for two minutes in 10 min. intervals. The video camera observes the fish through a mirror, which is set in an angle of 45 ^{0} in front of the test tank. The following parameters can be registered simultaneously: swimming height, mean horizontal position, number of turnings, motility, shoaling behavior and irregularity of swimming velocity. Ecological consequences of changes in motility, *i.e.* swimming distance per time, are alteration in the energy balance of the fish, which might lead to altered predation behavior. Toxic effects on the swimming hight of the fish and on their distance (shoaling) behavior may lead to altered predator-prey relationships [62]. The "BehavioQuant" biotest is more sensitive than the fishtests based on rheotaxis. Main criticism for all optical systems is the algal growth on the aquarium walls and turbidity, both factors which diminish the difference between object and background noise. The Koblenz behavioral fishtest based on "BehavioQuant" is being used and further developed at one monitoring station in Germany.

A recent progress in video image analysis has been the development of a 3-dimensional system based on 2 interactive underwater cameras. This technology is based on optical stereoscopy and computer image analysis for continuous monitoring of size, position, shape and spatial orientation of single fish in cages containing many fish [63]. The accuracy of stereo image analysis depends on the spatial sampling frequency of the sensors in the underwater TV cameras, the accuracy to which the image features can be found in the images and the spatial separation of the cameras, *i.e.* the stereo baseline. In the present system, the images cannot be processed in real-time, and the system is not completely automated. First experiments resulted in an error of less than 2 mm for determination of the size of fish. This method provides improved accuracy and detail of information at much higher data rates, allowing activities such as the swimming motion to be observed, however its use an a biological early warning system has to be proven [63].

A non-contact bioelectronic monitoring system was constructed to measure the muscular action potentials of one to four fish, each kept individually in one aquarium in which the positive and negative electrodes are placed above and below the fish and a neutral electrode between them. Electrical signals are obtained at a rate of 100 Hz, converted to digital mode, processed and filtered. Digital filters separate signals into ventilation (0.1 - 1.5 Hz), coughing (1.7 - 4 Hz) and ECG (4.5 - 50 Hz) [64]. Changes in heart rate, ventilation and total activity of brown trout (*Salmo trutta*) and swimming activity of roach (*Rutilus rutilus*) following stress due to exposure to parasites have been measured in the laboratory, and diurnal activity rhythms have been recorded for both species. This test system has not been developed as automated biomonitor. The WRc Mk III Fish Monitor has been available for about 15 years and is an established fish biomonitor in the UK [65]. Eight rainbow trout (*Oncorhynchus mykiss*) are placed in separate tanks with non-invasive electrodes. The electrodes immersed in the water detect small voltage oscillations from the fish muscular activity. From the part of the signals associated with gill activity, the ventilation frequency and the signal power (ratio gill ventilation/remaining muscular activity) for each fish are calculated each minute. This ratio also indicates swimming activity [66]. The stored ventilation data for each fish form a continuously updated historic data set, which is divided into a control period (1-2 h previous) and an observation period (the most recent 10 min.). Data in the observation period are compared against limits, set non-parametrically (5 and 95 percentiles) on the control period. Excursions outside the normal limits are summed up to provide the total number of excursions for the group of fish. The final fish monitor result is expressed as the number of standard deviations, by which the observed number of excursions differs from the expected mean value, derived from the control period. Eight non-fed trouts are kept under low level constant light conditions in their tanks from Monday to Friday, before they are replaced by new fish. After that, the fish need a time of at least 24 h to settle and acclimate, in which the monitor does not give reliable results. Ventilation frequency and signal power underlay natural variation,which could affect the detection of a sublethal effect of a toxin. Background noise levels between 80 and 800 Hz did not cause any reaction in the fish, however low frequency noise below 50 Hz was unfortunately not tested. During a 1 year test period at a borehole, the sensitivity of the WRc fishmonitor to several substances added to the testwater was investigated. Responses in ventilation behavior were seen at between 10 - 250 % of the LC_{50} within response times of 0 - 90 min., typically between 15 - 35 min., however depending on the toxin. The expected number of false alarms is estimated at about 1 in 3 months [57]. In the field, the weekly maintenance effort is 8 h and the fish often have to be changed twice a week. Above 20 ^{0}C, the monitor is unreliable, as the trout gradually die, and below 4 ^{0}C, problems due to condensation and heating of the system may occur.

A new biological early warning system is based on the analysis of the continuous electrical signals emitted by tropical fish, *e.g. Apteronotus albifrons* (Gymnotiformes) [67] or *Gnathonemus petersi* (Mormyridae) [68]. The biomonitor uses the electric organ discharge (EOD) frequency fluctuations as an indication of the presence of pollutants in the surrounding water. However, the fact that *Gnathonemus* EOD frequency time distribution is irregular by nature, even in absence of toxic substances, seriously weakens the attractiveness of this principle [67]. *Apteronotus albifrons* has never been used in toxicological studies, even though it presents two major advantages: (1), it possesses electric organs which generate weak electrical currents and (2), the electric information it emits is continuous and remarkably temporally stable [67]. The species is found in tropical surface waters of South America, is easy to keep and to raise, and adapts well to laboratory conditions. The biotest contains eight heat-proof test chambers, equipped with one fish in each and receiving thermoregulated test water. Two stainless steel electrodes, planted vertically in the median plane of each test chamber, tap the EODs of the fish. A third electrode is planted on the surface of the water in the centre of each test chamber and connected to the ground and to the symmetrical power sources. The signals are sent to a computer station which carries out frequency and temperature counts and digitizes the form of the EOD signal. Data processing contains 1) exploitation of the EOD frequency after signal smoothing in order to exclude random emissions of frequency peaks by the fish, 2) transformations and calculation of dynamic upper and lower bonds for accepted data variation and 3) Fast Fourier Transformation giving spectral analysis and amplitude ratios of different harmonics resulting in a waveform analysis. Alarm thresholds are based on the detection of points outside the defined dynamic bonds [67]. Laboratory experiments with potassium cyanide (44 μg/l) revealed changes in EOD frequency after 17 min., followed by changes in the waveform alaysis after 86 min., which is a much more sensitive and quick response than reported for *O. mykiss* (changes in positive rheotaxis at 400 μg/l after 30 min), *Dreissena polymorpha* (changes in valve movements at 400 μg/l) and *Daphnia magna* (changes in swimming at 1500 μg/l after < 240 min.) [67]. Unfortunately the recording quality of the signals is affected by movements of the fish. Moreover, the system has not yet been tested under *in situ* conditions.

Another principle of measuring spontaneous activity of fish is electromagnetic induction. A trout marked by a magnet moves freely in a test chamber made of electrical inert material, which on its outer side is covered with a magnetic coil. Movements of the fish induce voltage impulses which are processed. This system has been tested with rainbow trout (*O. mykiss*) exposed to detergents in the laboratory: at 19.75 mg/l the fish increased locomotion and ventilation became more irregular and frequent [69, 70]. This equipment measures only one single fish and is not developed as automated online biomonitor for use in the field.

3.4) Other aquatic invertebrates

A motility biotest with Tubificidae based on modern IR- video technique has been designed by Petry [71]. An amount of ca. 2 g wet weight of *Tubifex rivulorum* is kept in a flow-through measurement chamber and spontaneous activity is recorded, counted and presented on the PC as momentaneous value and as running sum. Normally Tubificidae remain in a compact mass, where normal activity is the motility of the whole mass. Due to toxins, the individuals first retract completely in the mass, thereafter become restless and at last leave the mass resulting in an increase in motility due to their undulative movements. In order to measure day- and night-activity simultaneously, two light sources, an IR light and a normal interval light are used. This biotest has only been used in the laboratory (Tab 1).

Another approach uses flatworms kept in a test chamber which consists of an inner and outer conductor. An electrical field is generated and the capacitance is determined as a function of the organisms' activity [72]. The behavior of aquatic invertebrates has been recorded with electrode chambers first developed by Spoor *et al.* [73]. Behavioral patterns, *i.e.* locomotion and ventilation of marine fish [74], chironomids and daphnids [75, 76], naids of Odonata [77], copepods [78] and diverse freshwater invertebrates have been recorded with impedance conversion in the laboratory [79]. Impedance is the apparent total opposition to electrical current flow in an alternating current circuit. Impedance is equal to the ratio of the root mean square electromotive force in the circuit to the root mean square current produced by it. Electrode grids in a bipolar arrangement were cemented on opposite chamber walls of the test chamber where an organism moved freely [77]. As big swimming movements often obscured small ventilation signals of *Gammarus* sp. [73], a system was developed where *Gammarus* sp. were glued with their cephalothorax to a stainless steel grid electrode. The second electrode was a needle, which was positioned as close as possible to the pleopods [80]. This test has only been used in the laboratory and seems to be unrealistic for long-term routine online biomonitoring, because no free movement of the organisms is allowed, which induces an additional stress and probably lowers the survival of the organisms.

4. Description of the Multispecies Freshwater Biomonitor (MFB)

4.1) Principle and measuring unit

The Multispecies Freshwater Biomonitor uses the measurement and analysis of different types of behavior from different aquatic organisms for online monitoring of water quality in fresh- and saltwater. Measuring unit is the test chamber, which can either be immersed directly in the water or placed in a flow-through aquarium, where the test water is pumped through.
The measurement principle is based on the analysis of changes in an electrical field of alternating current (impedance conversion). Theoretically this method is able to determine heart beat, but no practical online biomonitor has been developed using this measurement [81]. Contrary to electrode chambers based on bipolar impedance conversion technique, built by Spoor *et al.* [73] and Swain *et al.* [77], and used for the measurement of behavioral patterns of marine fish [74], chironomids and daphnids [76], copepods [78] and fish [66], the MFB is based on quadropole impedance conversion. In this technique, a high frequency alternating current of up to 100 kHz with adjustable amplitude (0 - 2 V) is generated over a test chamber filled with medium (*e.g.* water, sediment) by a pair of electrodes attached on opposite chamber walls. A second pair of non-current-carrying electrodes measures the changes in the impedance due to the organisms' movements in the chamber. The organism generates electrical signals, which mirror its' movement patterns [79]. As only relative

changes in the electrical field are evaluated instead of absolute signal amplitudes, this robust principle can be adapted to different materials, construction and position of the two electrode pairs in chambers of different sizes [29]. The chambers can be constructed with more than one level of sensing electrodes, *e.g.* to measure the vertical migration of plankton organisms, and to determine the swimming speed of organisms. Moreover, the chambers can be placed in the water column, on the sediment or digged in the sediment. Chambers completely filled with sandy sendiment still register movements of sediment dwelling organisms. For the control and collection of emerging adults, the chambers can also be equipped with emergence cages. The size, form and construction of the test chamber is variable, which makes it possible to adapt to the ecological and behavioral demands of all kinds of aquatic organisms.

4.2) Description of the MFB

The MFB is composed of a Pentium-PC, the impedance measurement instrument, the test chambers and a flow-through aquarium. The measurement instrument has a power supply, a signal-generation card for the generation of the sinus signal over up to 96 chambers with adjustable amplitude, and 8 signal-processing cards, each for 8 channels. The latter collect data from the test chambers with a sampling frequency of 20 Hz [29]. The data analysis algorithm is based on the spectrogram for impedance variations which is calculated by splitting the signals successively in intervals of 64 samples each and calculation of the discrete Fast Fourier Transform (FFT) with the Haming function [82]. The result of the organisms' movement signals is a histogram of different frequencies in bands of 0.5 Hz width between 0.5 and 10 Hz, typical for different behaviors, *e.g.* slow movements at low frequencies (crawling) and fast movements at high frequencies (gill ventilation) (Fig. 1).

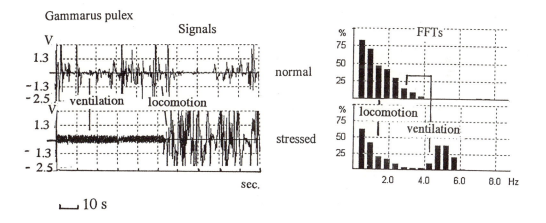

Fig.1: Behavioral pattern of *Gammarus pulex* consisting of locomotion (high amplitudes and low frequencies (< 2 Hz) and ventilation (low amplitudes and high frequencies (> 2 Hz). Above: normal animal. Below: stressed animal (acid, salt, metals etc.). Left: Original signals produced by the animal in the chamber. Right: calculated frequency histogram in frequency bands of 0.5 Hz in width.

The operation of the MFB-program in Windows-NT 4.0 is easy and user friendly. Up to 96 channels can be defined and named individually. Up to 12 virtual alarm channels can be defined, either for single or a sum of different measurement channels. The following alarm criteria can be defined: 1) The percentage of deviation or the number of standard deviations, seen in the last measurement of a fequency band from the running mean out of the x last measurements, 2) mortality after a defined number of measurements with no activity and 3) a weightening of selected frequency bands. For example, a small change in ventilation may be defined as a higher alarm value than a big change in locomotion. General adjustments of the system include: 1) the background noise level, which depends mainly on the conductivity of the water, the flow-through conditions etc., 2) the amplification of the signals (1, 2 and 4x) and 3) the calculation of the FFT can be adjusted according to different noise and threshold values. The threshold value determines the limit of how often and how long a fequency band has to occur in the signal, in order to be taken into account for the FFT calculation. Also the mathematical method of the FFT can be chosen (Normal or Haming) [29].

After these adjustments, the program starts to measure all selected channels online for a preset measurement interval. In case 96 channels are active, the data transfer and the calculations need ca. 5 min. after each measurement. The desktop can be built up with different types of windows, such as 1) the actual behavioral signals of selected test chambers, 2) the FFT-histograms and 3) longterm diagrams of selected frequencybands (mean and standard deviation) over a selected time period. All open windows are actualised after each measurement. An alarm bar gives the status of the "water quality alarm" according to the previously defined alarm criteria in 3 steps (ok, warning, alarm). For each species, such alarms are generated. In order to define alarm criteria for a new test species, simulations of an exemplary data set can be performed by changing the alarm settings of the program.

4.3) Practical aspects

The MFB is a continuous biotest for aquatic (in)vertebrates and can therefore only be compared to other continuous biomonitors, such as *Daphnia*-tests, mussel- and fishtests. Following the criteria of the "Wirkungstest Rhein" (Umweltbundesamt1994) and the LAWA (1996), the MFB has a reaction time of ca. 10 min depending on the test species, type of behavior recorded and type of pollution. The stand alone time is more than one week, depending on the test species, thus comparable to the *Daphnia*-tests, musseltests and fish-tests. The starting up of the MFB and training of staff takes 1 day. The maintenance requirements including cleaning of the system, collecting and replacing test organisms, are estimated to 1 - 3 h per week depending on the survival of the test species in the MFB, its abundance and ease of collection in the field. Similar times of weekly work efforts are mentioned for the other biotests. The refractory time, *i.e.* the time needed to reinstall the whole system after damage of the organisms is ≤ 1 day. The alarm analysis of the MFB relies on dynamic thresholds like in other above mentioned biotests. The whole data analysis is transparent and alarms can be followed up backwards stepwise through the calculation procedure until the original data, thus allowing for high plausibility control. Background data for stress responses to pollution are available, longterm operation data mostly for freshwater amphipods. The costs for a MFB-system with 96 channels are comparable to those of the *Daphnia*-tests and fish biotests with the advantage of simultaneously monitoring several species in a higher number of specimens than other biotests.

4.4) Experiments with the MFB

The MFB based on quadropole impedance conversion technique has been built and tested with different aquatic organisms [79]. Changes of the behavioral pattern due to increasing amounts of toxicants have been proven in several acute toxicity tests (LC_{50}-120 h), and early warning responses have been measured in static bioassays (summarised in [29]). Changes in locomotion (escape, avoidance) followed by changes in ventilation at metal concentration levels of ≥ 1/100 of the respective LC_{50} values were observed. Early warning responses were found within 15 to 90 min. depending on the types of toxicants, their site and mode of action and the sensitivity and reaction time of the test organism [29]. Sensitivity also depended on the choice of population and origin of the test species. For example, a population of *G. pulex* from a clean mountain stream showed lower activity levels in the stream than a local population, adapted to the water quality parameters of that stream [82]. The MFB has been used *in situ* for about three week long field tests, one along a small stream polluted by copper and agricultural pesticides with *G. pulex* as test species [82], and one along the river Rhine in winter 1997 with *Chaetogammarus* sp., *Dinocras cephalotes* and *Aphelocheirus* sp.(Gerhardt unpubl.). Other field tests along the rivers Rhine and Ruhr are ongoing and will be published elsewhere.

For *Gammarus pulex*, a general behavioral response pattern to toxicants such as dissolved metal ions, salinity, acidity and oil has been observed, consisting first of increased swimming behavior ("avoidance", trial to "escape" from the pollution source as 1. threshold of stress) followed by increased gill ventilation, both in time spent on ventilation and in ventilation frequency as 2. and 3. threshold of stress (Fig. 2). This series of stress responses has been summarised in a graphic model called "stepwise stress model" [29]. The main essence of this model is, that in reaction to a toxin an organism shows a sequence of behavioral stress responses above their respective threshold of resistance. Within a tolerance range, the organism can regulate the behavior in order to keep its body functions unaffected, *e.g.* ventilation behavior. *Gammarus* sp. has been reported to be highly regulative in its oxygen consumption in contrast to many other rheophilous macroinvertebrates [83]. Above the organisms resilience range, toxic effects appear, such as hyperactivity, resulting in exhaustion, or decreased activity and morbidity, finally resulting in lethality. However, in order to avoid these toxic effects, the organism can switch to a second stress behavior. Such a discontinuous strategy as one type of behavioral plasticity has been discussed earlier [27]. If the respective thresholds are known, the appearance of the different behaviors allow for a quantification of the toxins. The stepwise stress model might also be linked to the Dynamic Energy Budget (DEB) theory, which gives qualitative links between key-processes such as survival and reproduction [84].

Stepwise Stress Model

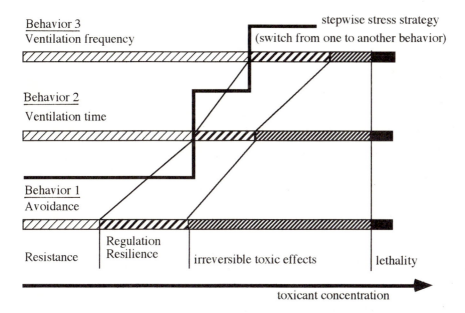

Fig. 2: Stepwise Stress Model (modified from [29]).
There are three different behaviors with inherent ranges for resistance, regulation and irreversible toxic effects. With increasing toxicant concentration, the organism switches from one behavior to the next behavior according to their respective ranges of regulation.

5) Concluding evaluation of the biomonitors

Compared to the types of biomonitors used ten years ago [3], new methods *e.g.* image processing and quadropole impedance conversion have been developed and more aquatic species, *e.g.* insect larvae, benthic crustacea and tubificids have been incorporated. Which systems are to be advised for a specific situation strongly depends on the application and the monitoring purpose [3]. In the following, existing biotests are compared and evaluated under the criterium of being an ecologically relevant method for online biomonitoring of surface water quality, in order to protect the natural environment. The aim is to protect the local ecosystem from pollution waves, which originate sporadically from point pollution sources. To cover the four different trophic levels, bacteria biotests, algae biotests and (in)vertebrate biotests are all necessary components of an integrated ecological monitoring. A comparison between the different biotest groups would be a comparison between apples and pears. The comparative evaluation below concentrates on the animal biotests, as there exist long-term *in situ* experiences and there have been immense development and innovation efforts in recent times.

5.1) Test species

Biotests using standard laboratory cultures of organisms, which even do not occur in the respective ecosystem, *e.g.* selected clones of *Daphnia* sp., a tolerant neozoon such as *Dreissena polymorpha* or tolerant fish species like *Leuciscus idus* are of less ecological relevance than biotests which can be adapted to the local situation by use of locally important species, *e.g.* keystone species of the ecosystem. The MFB principally allows for monitoring all kinds of aquatic (in)vertebrate species. For the first time in Europe, the MFB uses benthic insects and crustaceans. Benthic macroinvertebrates have been attractive targets for pollutant risk assessment and trend biomonitoring, because they are a diverse group, reacting strongly and predictably to aquatic pollution [85]. They are important links in the aquatic foodweb [86]. With the use of the MFB, it is now possible to relate trend biomonitoring data to online monitoring data, which strengthens the conclusions and evaluations for risk and hazard assessment, thus closing the gap between these two different approaches [29].

5.2) Behavioral variability

A common problem in online biotests based on behavioral responses, is the behavioral variability of the test species due to other factors than pollution waves. The behavior of a single individual in its test chamber can change during prolonged exposure, due to lack of competition and predation pressures, lack of food, lack of appropriate habitat structures, flow conditions, size of chamber etc. Circadian and seasonal activity rhythms, activity decreases just before moulting, different inherent activities, habitat and food choices of different instars (*e.g.* Hydropsychidae) and availability of organisms throughout the year are further biological variables to be considered for the evaluation of an ecologically relevant early warning system [29]. Many of these problems can be avoided by adapting the early warning system to the test site and the season by choosing the respective dominant test species in its respective stage of life cycle, by taking a high number of replicates and by exposing the system directly in the stream, where unfiltered stream water together with organismic drift can flow through the test chambers, thus delivering food for the organisms and optical and olfactorial presence of predators and competitors. *In situ* exposure furthermore allows for realistic flow-through conditions and light regime. The more artificial the conditions in a biomonitor, the more aberrant the behavior of the test organisms, the more stressed the organisms and thus the more unrealistic a "water quality alarm" (more false alarms, alarms not interpretable) [29].

5.3) Multispecies measurement

Most of the existing biotests use one test species. Therefore, water authorities have been equipped with biotest batteries, in order to monitor different species from different trophic levels and habitats. This is necessay as each test species has different thresholds of toxic effects and reaction times to pollution waves. The MFB measures simultaneously different test species, *e.g.* 3 species with each 32 individuals with the same method. The different reactions of the 3 test species can be better compared than when each species is measured in its own biomonitor. In the latter case, an alarm can be caused either by the response of the species to the toxin or the alarm can be due to method-specific technical characteristics and problems in the biomonitor, *e.g.* flow-through conditions changed, clogging of sieves etc. In the MFB, all data from different species are simultaneously available and can be analysed directly with the same software in an intelligent, stepwise and ecologically relevant "water quality alarm". In the biotest batteries, the alarm responses of the different biomonitors have to be transferred to a data centre and evaluated by the personnel.

5.4) Parameters

All animal biotests rely on behavioral and/or physiological response parameters, as they are amongst the first reactions to acute pollution at the whole organism level [87]. Physiological parameters, *e.g.* respiration, are highly resistant to toxic stress with low regulative abilities, as they are essential body functions. The same holds for basic essential reflexes, *e.g.* photo- or geo-taxis. Such stress parameters usually exhibit high toxic thresholds and seem to be less important for early warning purposes. Locomotion and especially inter-individual behaviors are more complex and therefore more regulative than resistant [88]. In complex behaviors, several sense organs can be involved with each containing different information content, directionality and persistence [89]. Within the range of regulative behaviors, *e.g.* avoidance, the first sensitive and reversible responses to toxic stress can be measured, which is important for environmental protection. Avoidance can be measured at the organism level, but also indicates changes at the population level, as it is directly related to habitat selection and abundances [88]. Such stress parameters should be favoured in biological early warning systems aiming at the protection of ecosystem health.

In the recent development of biotests, a shift from the qualitative registration of a certain behavior, *e.g.* mussel shells closed or open, towards quantitative measurement of one, or even better, different types of behavior can be noticed. Not only different species but also different types of behavior have different toxicant sensitivities, which allows for stepwise definition of an alarm-gradient. Together with the biotests based on digital image analysis for fish and *Daphnia* sp., the MFB offers quantitative multiparameter measurement as described above in the stepwise stress model.

In order to reach an ecologically relevant integrated online biomonitoring, parameters measured at the whole organism level should be backed up with response parameters at the population and community levels. Here an integrated approach of a combination of biological early warning biomonitoring with trend biomonitoring by use of local bioindicator species or keystone species in the online biotest is a first step in that direction. The simultaneous measurement of responses of different species from different habitats, trophic niveaus, sensitivity and importance in the ecosystem, *e.g.* keystone species, in one biomonitor is another step in that direction.

5.5) Field operation and remote sensing

Up to now, many biomonitors are set up in test batteries in field stations, where the river water is pumped through. However, the ultimate operation of a biomonitor is instream. In this case, the organisms are exposed to the realistic stream flow and particle load of the water, which is crucial in detecting realistic alarms. To that aim, portable biomonitors with instream exposure are being developed. Online biomonitors generate biological early warning data in case of pollution events. Recent developments concentrate on real-time data transfer from remote monitor stations via satellite communication to a data centre [90]. Remote biosensing technologies combined with biotelemetry for data retrieval are promising tools for integrated water quality monitoring [91].

5.6) Propositions for future developments in online biomonitoring

To reach the aim of ecologically relevant online biomonitoring, several improvements in measurement technologies and data analysis methods, but mainly biological optimisations have to be made. Survival of the organisms in online biotests should be increased, because biotests which require the organisms to be changed once to twice a week only record acute toxic effects, not sublethal long-term exposure to chemical substances. Biomonitors, which simultaneously monitor different test species should be supported in their development, as they have the potential to replace the costly and space demanding biotest batteries, which lack comparability due to different technical solutions and different stress parameters measured. Biomonitors which relate whole organism responses to higher ecological organisation levels should be encouraged in order to reach an ecologically relevant water quality monitoring. Biomonitors which are flexible and use locally relevant test species as well as standard test organisms should be favoured, as they allow for an integrated biomonitoring approach by linking trend biomonitoring to biological early warning biomonitoring. The combination of monitoring local and standard test species in one biomonitor allows for locally adapted ecologically relevant biomonitoring as well as comparability between different streams and sites. The latter is an important standardisation for setting up surface water quality alarm criteria in the legislation.

6) Online biomonitoring in an ecologically relevant biomonitoring program

Integrated ecological online biomonitoring might be performed in future according to the following program (Fig. 3). Online biomonitoring is supposed to become an important part of ecological risk assessment in aquatic environments. The use of indigenous and standard test organisms of different habitats and trophic levels in biotest batteries or multispecies biomonitors, *e.g.* Multispecies Freshwater Biomonitor, creates a link to seasonally generated *in situ* trend biomonitoring and ecotoxicological data with the species used in the biomonitors, performed as whole effluent toxicity tests. These data can be related to results from data bases, mainly generated for standard toxicity test species. The three approaches complement and back up each other, which enhances the reliability of their results (Fig. 3). In case of water quality alarm in the online biomonitors, WET-toxicity assessment of automatically collected water samples is recommended. Simultaneously, several measurements in the field might be performed, depending on which biomonitor taxon reacted to the event, *e.g.* in case of alarm responses in *Gammarus* sp., drift measurements in the field are a good complement and verification of the obtained biomonitor result. They also indicate responses at the population level, *e.g.* the whole local population of *Gammarus* sp. might have drifted further downstream, whereas in the biomonitor increased locomotory activity might have been found during the event. The WET-tests might show the dilution level, where *Gammarus* sp. starts to react with increased activity, which might result in increased drift in the field. The biological data are to be compared with online chemical data. If the toxic cocktail has to be specified, further chemical analysis is necessary. SPMD (semipermeable membrane devices) are a good tool to concentrate toxicants in order to allow for better chemical detection. Toxicity and bioaccumulation studies for the biomonitor species and some field species are recommended to relate the measured toxicant concentration in the water to that in the organisms.

7) Acknowledgments

Herewith I would like to thank Beckmann & Farwerk Ingenieurgesellschaft for cooperation in the recent development of the MFB soft- and hardware as well the water authorities and institutions which tested the MFB in the laboratory and in the field.

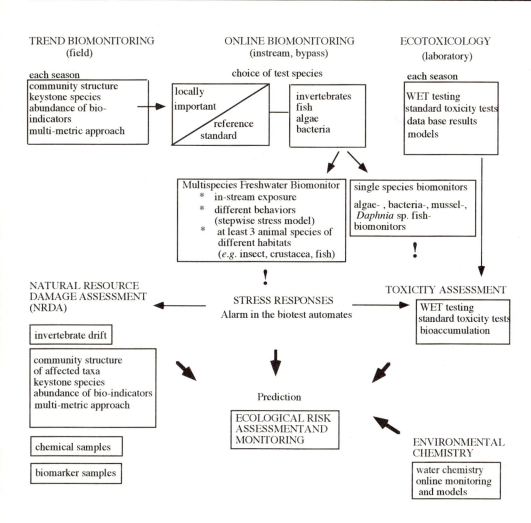

Fig. 3: Online biomonitoring in the framework of an integrated ecological risk assessment and monitoring. Routine biomonitoring methods for each season are proposed as well as toxic event oriented methods in case online biomonitors give an alarm (!).

8) References

[1] J. Cairns Jr., K. L. Dickson, R. E. Sparks and W.T. Waller: A preliminary report on rapid biological information systems for water pollution control. J. Water Pollut. Control. Fed. 42(5) (1970): 685-703.

[2] D. Gruber, C. H. Frago and W. J. Rasnake: Automated biomonitors- first line of defense. J. Aquat. Ecosystem Health 3 (1994): 87 - 92.

[3] K. J. M. Kramer and J. Botterweg: Aquatic, biological early warning systems: An overview. In: D. W. Jeffrey and B. Madden (eds.): Bioindicators and Environmental Management (1991), 95 - 126, Academic Press, London, UK.

[4] P. D. Hansen, P. Stein and H. J. Löbbel: Entwicklung, Erprobung und Implementation von Biotestverfahren zur Überwachung des Rheins. Teil 3 (1994). Institut für Wasser, Boden und Lufthygiene des Bundesgesundheitsamtes Berlin.

[5] D. Osbild, M. Babut and P. Vasseur: Revue- État de l'Art. Les biocapteurs appliqués au controle des eaux. Revue des Sciences de l'Eau 8 (1995), 505-538.

[6] D. P. Nikolelis, U. J. Krull, J. Wang and M. Mascini (eds.): Biosensors for direct monitoring of environmental pollutants in the field. Proceedings of the NATO advanced research workshop, Smolenice, Slovakia, May 1997.

[7] P. Schmitz, U. Irmer and F. Krebs: Automatische Biotestverfahren in der Gewässerüberwachung. In: H.-J. Pluta, J. Knie and R. Leschber (eds.): Biomonitore in der Gewässerüberwachung. Schriftenreihe des Vereins für Wasser-, Boden- und Lufthygiene 93, (1994), 1 - 19, Gustav Fischer Verlag, Stuttgart.

[8] A. Gerhardt: Monitoring behavioral responses to metals in *Gammarus pulex* (L.) (Crustacea) with impedance conversion. Environ. Sci. & Pollut. Res. 2(1) (1995), 15-23.

[9] K. J. M. Kramer and E. M. Foekema: The "Musselmonitor" as biological early warning system: The first decade. In: F. M. Butterworth, M. E. Gonsebatt Bonaparte and A. Gunatilaka (eds.): Biomonitors and biomarkers as indicators of environmental change: A Handbook, Vol. 2; Plenum New York, 1999 (in press).

[10] R. Schlink: Vergleichsuntersuchungen zum praktischen Einsatz von kontinuierlichen Algentestgeräten in der Gewässerüberwachung. Diplomarbeit der privaten Fachhochschule Fresenius Idstein, Juni 1998, 178 pp.

[11] LAWA (Länderarbeitsgemeinschaft Wasser) (ed): Empfehlungen zum Einsatz von kontinuierlichen Biotestverfahren für die Gewässerüberwachung (1996), 37 pp, Kulturbuchverlag, Berlin.

[12] UBA (Umweltbundesamt) (ed): Continuous Biotests for Water Monitoring of the River Rhine. Summary, Recommendations, Description of Test Methods. Umweltbundesamt Texte 58 (1994), Berlin.

[13] VDI (Verein Deutscher Ingenieure) (ed): Grundlagen zur Kennzeichnung vollständiger Meßverfahren.- Begriffsbestimmungen VDI-Richtlinien 2449. VDI Handbuch Reinhaltung der Luft, Band 5 (1987).

[14] D. Gruber and J. Diamond (eds.): Automated Biomonitoring: Living Sensors as Environmental Monitors (1988). Ellis Horwood Limited Publishers, Chichester, 206 pp.

[15] H.-J. Pluta, J. Knie and R. Leschber (eds.): Biomonitore in der Gewässerüberwachung. Schriftenreihe des Vereins für Wasser-, Boden- und Lufthygiene 93, (1994), 315 pp., Gustav Fischer Verlag, Stuttgart.

[16] C. Plieth and C. Moldaenke: Biomonitoring: Methoden zur kontinuierlichen Gewässerüberwachung. Teil 2. Gewässerbiomonitoring mit Algen und Mikroorganismen. WAP 2 (1995), 22-28.

[17] L. Campanella, G. Favero and M. Tomassetti: Un nuovo biosensore a lieviti immobilizzati per misure di tossicita integrale. Inquinamento 7 (1994), 46-59.

[18] H. Fritz-Langen, B. Blessing and F. Krebs: Bakterientoximeter mit Reinkulturen (*Photobacterium phosphoreum* und *Pseudomonas putida*) in der Gewässerüberwachung. In: H.-J. Pluta, J. Knie and R. Leschber (eds.): Biomonitore in der Gewässerüberwachung. Schriftenreihe des Vereins für Wasser-, Boden- und Lufthygiene 93, (1994), 249 - 275, Gustav Fischer Verlag, Stuttgart.

[19] P. D. Hansen and P. Stein: Biosensoren im Umweltmonitoring am Beispiel der EuCyano-Bakterienelektrode. In: H.-J. Pluta, J. Knie and R. Leschber (eds.): Biomonitore in der

Gewässerüberwachung. Schriftenreihe des Vereins für Wasser-, Boden- und Lufthygiene 93, (1994), 237 - 249, Gustav Fischer Verlag, Stuttgart.

[20] C. Y. Chen and K. C. Lin: Optimization and performance evaluation of the continuous algal toxicity test. Environmental Toxicology and Chemistry 16 (7) (1997), 1337-1344.

[21] D. Merchant, P. J. Scully, R. Edwards and J. Grabowski: Optical fibre fluorescence and toxicity sensor. Sensors and actuators chemical, 1999 (in press).

[22] E. Stallwitz and D-P. Häder: Effects of heavy metals on motility and gravitactic orientation of the flagellate *Euglena gracilis*. Europ. J. Protistol. 30 (1994), 18-24.

[23] S. Overmeyer, E. Hostert, S. E. Lindner, H. Schnabl, L. Peichl: Der Protoplastenbiotest- ein Wirkungstest zur summarischen Schadstofferfassung in der Umwelt. In: H.-J. Pluta, J. Knie and R. Leschber (eds): Biomonitore in der Gewässerüberwachung. Schriftenreihe des Vereins für Wasser-, Boden- und Lufthygiene 93 (1994), 291 - 299, Gustav Fischer Verlag, Stuttgart.

[24] R. E. Warner: Bioassays for microchemical environmental contaminants with special reference to water supplies. Bull. W. H. O. 36, (1967), 181 - 207.

[25] N. K. Mello: Behavioral toxicology: A developing discipline. FASEB 34 (1975), 1832 - 1834.

[26] E. Scherer: Behavioural responses as indicators of environmental alterations: approaches, results developments. J. Appl. Ichthyol. 8 (1992), 122 - 131.

[27] P. E. Komers: Behavioural plasticity in variable environments. Can. J. Zool. 75 (1997), 161 - 169.

[28] R. E. McNichol and E. Scherer: Behavioural responses of lake whitefish (*Coregonus clupeaformis*) to cadmium during preference-avoidance testing. Environm. Toxicol. and Chem. 10 (1991), 225 - 234.

[29] A. Gerhardt: A new Multispecies Freshwater Biomonitor for ecologically relevant control of surface waters. In: F. M. Butterworth, M. E. Gonsebatt Bonaparte and A. Gunatilaka (eds.).: Biomonitors and biomarkers as indicators of environmental change: A Handbook, Vol. 2; Plenum New York, 1999 (in press).

[30] V. P. M. Englund and M. P. Heino: Valve movement of *Anodonta anatina* and *Unio tumidus* (Bivalvia, Unionidae) in a eutrophic laek. Ann. Zool. Fennici 31 (1994), 257-262.

[31] V. P. M. Englund and M. P. Heino: The freshwater mussel (*Anodonta anatina*) in monitoring 2,4,6-trichlorophenol: Behaviour and environmental variation considered. Chemosphere 32 (2) (1996), 391-403.

[32] M. Hoffmann, E. Blübaum-Gronau and F. Krebs: Die Schalenbewegung von Muscheln als Indikator von Schadstoffen in der Gewässerüberwachung. In: H.-J. Pluta, J. Knie and R. Leschber (eds): Biomonitore in der Gewässerüberwachung. Schriftenreihe des Vereins für Wasser-, Boden- und Lufthygiene 93 (1994), 87 - 119, Gustav Fischer Verlag, Stuttgart.

[33] D. de Zwart, K. J. M. Kramer and H. A. Jenner: Practical experiences with the biological early warning system "Musselmonitor". Environmental Toxicology and Water Quality 10 (1995), 237-247.

[34] H. A. Jenner, F. Noppert and T. Sikking: A new system for the detection of valve-movement response of bivalves. Kema Scientific & Technical Reports 7 (1989), 91-98.

[35] J. Borcherding and M. Volpers 1994: Der '*Dreissena*-Monitor' - Erfahrungen mit dem neuen biologischen Frühwarnsystem. In: H.-J. Pluta, J. Knie and R. Leschber (eds): Biomonitore in der Gewässerüberwachung. Schriftenreihe des Vereins für Wasser-, Boden- und Lufthygiene 93 (1994), 119 - 123, Gustav Fischer Verlag, Stuttgart.

[36] H. Sluyts, F. van Hoof, A. Cornet and J. Paulussen: A dynamic new alarm system for use in biological early warning systems. Environmental Toxicology and Chemistry 15 (8) (1996), 1317-1323.

[37] C. Plieth: Biomonitoring: Methoden zur kontinuierlichen Gewässerüberwachung. Teil 1: Gewässerbiomonitoring mit Fischen, Kleinkrebsen und Muscheln. WAP 1 (1995), 20-24.

[38] J. Borcherding: Another early warning system for the detection of toxic discharges in the aquatic environment based on the valve movements of the freshwater mussel *Dreissena polymorpha*. In. D. Neumann and H. A. Jenner (eds.): Limnologie aktuell - Vol 4- the zebra mussel *Dreissena polymorpha*- ecology, biological monitoring and first applications in water quality management (1992), pp. 127-146. Gustav Fischer Verlag, Stuttgart.

[39] J. Borcherding and M. Volpers: The 'Dreissena-Monitor' improved evaluation of dynamic limits for the establishment of alarm-thresholds during toxicity tests and for continuous water control. In: I. A. Hill, F. Heimbach, P. Leeuwangh and P. Matthiessen (eds.): Freshwater field test for hazard assessment of chemicals (1994), 477-484, Boca Raton, Fl, Lewis Publishers.

[40] U. Matthias and S. Römpp: Erprobung des 'Dreissena-Monitors', eines neuen Biotestsystems mit der Zebramuschel (*Driessena polymorpha*) in der Rhein-Gütemesstation Karlsruhe. Acta hydrochim. hydrobiol. 22 (1994), 161 - 165.

[41] J. Borcherding and B. Jantz: Valve movement response of the mussel *Dreissena polymorpha*- the influence of pH and turbidity on the acute toxicity of pentachlorophenol under laboratory and field conditions. Ecotoxicology 6 (1997), 153-165.

[42] M. Marten: Erfahrungen mit dem Routinebetrieb kontinuierlicher Biotest-Verfahren in der Gütemesstation in Karlsruhe am Rhein. Erweitertes Abstract. DGL-Jahrestagung 1996 in Schwedt/Oder.

[43] V. P. M. Englund, M. P. Heino and G. Melas: Field method for monitoring valve movements of bivalved molluscs. Water Research 28 (1994), 2219-2221.

[44] K. D. Ham and M. J. Peterson: Effect of fluctuating low-level chlorine concentrations on valve-movement behavior of the asiatic clam (*Corbicula fluminea*). Environmental Toxicology and Chemistry 13 (1994), 493-498.

[45] P. Tomasik and D. M. Warren: The use of *Daphnia* in studies of metal pollution of aquatic systems. Environ. Rev. 4 (1996), 25-64.

[46] L. de Meester: The vertical distribution of *Daphnia magna* genotypes selected for different phototactic behaviour: outdoor experiments. Arch. Hydrobiol. Beih. Ergeb. Limnol. 39 (1993), 137 - 155.

[47] J. Knie: Der Dynamische Daphnientest- ein automatischer Biomonitor zur Überwachung von Gewässern und Abwässern. Wasser und Boden 12 (1978), 310 - 312.

[48] J. Knie: Der Dynamische Daphnientest. Decheniana 26 (1982), 82 - 86.

[49] A. J. Hendricks and M. D. A. Stouten: Monitoring the response of microcontaminants by dynamic *Daphnia magna* and *Leuciscus idus* assays in the Rhine delta: Biological early warning as a useful supplement. Ecotoxicology and Environmental Safety 26 (1993), 265-279.

[50] U. Matthias and H. Puzicha: Erfahrungen mit dem Dynamischen Daphnientest- Einfluss von Pesticiden auf das Schwimmverhalten von *Daphnia magna* unter Labor- und Praxisbedingungen. Z. Wasser- und Abwasserforschung 23 (1990), 193 - 198.

[51] F. van Hoof, H. Sluyts, J. Paulussen, D. Berckmans and H. Bloemen: Evaluation of a biomonitor based on the phototactic behaviour of *Daphnia magna* using infrared detection and digital image processing. Water, Science & Technology (1994), 79-86.

[52] BEO (Projektträger Biologie, Energie, Ökologie des BMFT): Biomonitore zur kontinuierlichen Überwachung von Wasser und Abwasser (1993), 48 pp., Berlin.

[53] J. de Haas: Untersuchung zur Bewertung des Dynamischen Daphnientests als Instrument der Abwasserüberwachung. Diplomarbeit der Universität Oldenburg (1996), 83 pp.

[54] K. Kerren: Aqua-Tox-Control-*Daphnia* (1991), 1 - 4,. Kerren Umwelttechnik, Viersen.

[55] M. Baillieul and P. Scheunders: On-line determination of the velocity of simultaneously moving organisms by image analysis for the detection of sublethal toxicity. Water Research 32 (4) (1998), 1027-1034.

[56] E. Blübaum-Gronau and M. Hoffmann: Steigerung der Sensitivität eines kontinuierlichen Daphnientestes durch die Berücksichtigung einer Vielzahl von Verhaltensweisen. Vom Wasser 89 (1997), 163-173.

[57] I. G. Baldwin, M. M. I. Harman and D. A. Neville: Performance characteristics of a fish monitor for detection of toxic substances. 1. Laboratory trials. Water Research 28 (10) (1994), 2191-2199.

[58] I. Juhnke and W. K. Besch: Eine neue Testmethode zur Früherkennung akut toxischer Inhaltsstoffe im Wasser. Gewässer und Abwasser 51/52 (1971), 107 - 114.

[59] W. K. Besch, A. Kemball, K. Meyer-Waarden and B. Scharf: A biological monitoring system employing rheotaxis of fish. In: J. Cairns Jr, K. L. Dickson and G. F. Westlake (eds.) Biological Monitoring of Water and Effluent Quality, ASTM STP 607, American Society for Testing and Materials (1977), 56-74.

[60] W. S. G. Morgan and P. C. Kühn: Effluent discharge control at a South African industrial site utilizing continuous automatic biological surveillance techniques. In: D. Gruber and J. Diamond (eds): Automated Biomonitoring: Living Sensors as Environmental Monitors (1988), 91 - 104, Ellis Horwood LtD, Chichester, UK.

[61] E. H. Smith and H. C. Bailey: Development of a system for continuous biomonitoring of a domestic water source for early warning of contaminants. In: D. Gruber and J. Diamond (eds):

Automated Biomonitoring: Living Sensors as Environmental Monitors (1988), 182 - 206, Ellis Horwood LtD, Chichester, UK.

[62] O. H. Spieser, W. Scholz, E. Blübaum-Gronau, M. Hoffmann, B. Grillitsch and C. Vogel: Das System Behavio-Quant zur Bioindikation anhand des Verhaltens von Fischen und von anderen aquatischen Organismen. in: K. Alef, H. Fiedler, O. Hutzinger (eds.): Umweltmonitoring und Bioindikation. Eco-Informa-'94, Band 5 (1994), 429 - 448, Umweltbundesamt Wien.

[63] B. P. Ruff, J. A. Marchant and A. R. Frost: Fish size and monitoring using a stereo image analysis sytem applied to fish farming. Aquacultural Engeneering 14 (1995), 155-173.

[64] M. Laitinen, R. Siddall and E. T. Valtonen: Bioelectronic monitoring of parasite-induced stress in brown trout and roach. J. Fish Biology 48 (1996), 228-241.

[65] J. P. Evans and J. P. Wallwork: The WRc fish monitor and other biomonitoring methods. In: D. Gruber and J. Diamond (eds.): Automated Biomonitoring: Living Sensors as Environmental Monitors (1988), 75 - 91, Ellis Horwood Limited Publishers, Chichester.

[66] P. Stein, P. D. Hansen and H. J. Löbbel: Erprobung des WRC-Fischmonitors zur Störfalluberwachung. In: H. J. Pluta, J. Knie and R. Leschber (eds): Biomonitore in der Gewässerüberwachung (1994), 75 - 87, Gustav Fischer Verlag, Stuttgart.

[67] M. Thomas, A. Florion, D. Chretien and D. Terver: Real-time biomonitoring of water contamination by cyanide based on analysis of the continuous electrical signal emitted by the tropical fish, *Apteronotus albifrons*. Water Research 30, (1996), 3083-3091.

[68] P. R. Campbell, J. W. Lewis and I. P. Toms: Microcomputer-based detection and analysis of the electric organ discharges in the Mormyrid fish *Gnathonemus petersi*. Environ. Technol. 11 (1) (1990), 41 - 50.

[69] H. Petry: Versuche zur messtechnischen Erfassung von Fischgiften. Fisch und Umwelt 2 (1976), 175-181.

[70] H. Petry: Der Motilitätstest, ein Frühwarnsystem zur biologischen Gewässerkontrolle. Zbl. Bakt. Hyg. 1. Abt. Orig. B 176 (1982), 391-412.

[71] H. Petry: Automatisiertes Frühwarnsystem zur kontinuierlichen Gewässerkontrolle mit Tubificiden als Schadstoffindikatoren. Z. Wasser- Abwasser- Forsch. 22 (1989), 120-124.

[72] J. Kostelecky: Verfahren und Vorrichtung zur Bestimmung von Schadstoffen in Gewässern und Abwässern. Patent application, publication nr. and date: DE 37.08.753, (1988).

[73] W. A. Spoor, I. W. Neiheisel and R. A. Drummond: An electrode chamber for recording respiratory and other movements of free-swimming animals. Trans. Am. Fish. Soc. 100 (1971), 22-28.

[74] C. J. Wingard and C. J. Swanson: Ventilatory responses of four marine teleosts to acute rotenone exposure. J. Appl. Ichthyol. 8 (1992), 132-142.

[75] F. Heinis, K. R. Timmermans and W. R. Swain: Short-term sublethal effects of Cd on the filter-feeding chironomid larva *Glyptotendipes pallens* (Meigen) (Diptera). Aquatic Toxicology 16 (1990), 73 - 86.

[76] F. W. Heinis and R. Swain: Impedance conversion as a method of research for assessing behavioral responses of aquatic invertebrates. Hydrobiol. Bull. 19 (1986), 183-192.

[77] W. R. Swain, R. M. Wilson, R. P. Neri and G. S. Porter: A new technique for remote monitoring of activity of freshwater invertebrates with special reference to oxygen consumption by naids of *Anax* sp. and *Somatochlora* sp. (Odonata). Can. Entomol. 109 (1977), 1-8.

[78] C. W. Gill and S. A. Poulet: Utilization of a computerized micro-impedance system for studying the activity of copepod appendages. J. Exp. Mar. Biol. Ecol. 101 (1986), 193-198.

[79] A. Gerhardt, M. Clostermann, B. Fridlund and E. Svensson: Monitoring of behavioral pattern of aquatic organisms with an impedance conversion technique. Environment International 20(2) (1994), 209-219.

[80] J. Kahlert and D. Neumann: Die automatische Erfassung der Ventilationsrhythmik von Gammariden. Erweiterte Zusammenfassung, DGL- Jahrestagung (1994), 544-546, Hamburg.

[81] I. G. Baldwin and K. J. M. Kramer: Biological early warning systems (BEWS). In: K. J. M. Kramer (ed.): Biomonitoring of Coastal Waters and Estuaries. CRC Press Inc Boca Raton, Chapt. 1 (1994), 1- 25.

[82] A. Gerhardt, A. Carlsson, C. Ressemann and K. P. Stich: New online biomonitoring system for *Gammarus pulex* (L.) (Crustacea): *In situ* test below a copper effluent in South Sweden. Environm. Sci. & Technol. 32(1) (1998), 150-156.

[83] M. J. Toman and P. C. Dall: Respiratory levels and adaptations in four freshwater species of *Gammarus* (Crustacea: Amphipoda). Internat. Rev. Hydrobiol. 83 (1998), 251-263.

[84] B. Kooijman: Process-based analysis of toxic effects of chemicals on organisms. Abstract. 9th Annual Meeting of SETAC-Europe, 25 - 29 May 1999, Leipzig, Germany.

[85] J. Cairns, Jr. and V.H. Pratt, 1993. A history of biological monitoring using benthic macroinvertebrates. In: D. M. Rosenberg and W. A. Resh (eds): Freshwater Biomonitoring and Benthic Macroinvertebrates (1993), 10 - 28, Chapman & Hall, New York.

[86] A. Garmendia-Tolosa and B. Axelsson:*Gammarus*, their biology, sensitivity and significance as test organisms. Swedish Environmental Research Institute, IVL-report B 1095 (1993), 88 pp, Stockholm.

[87] P. J. Sheehan: Effects on Individuals and Populations. In: P. J. Sheehan, D. R. Miller, G. C. Butler and P. Bourdeau (eds.): Effects of pollutants at the ecosystem level. Scope (1984). John Wiley & Sons LtD.

[88] T. L. Beitinger and R. W. McCauley: Whole-animal and physiological processes for the assessment of stress in fishes. J. Great Lakes Research 16 (4) (1990), 542-575.

[89] J. H. S. Blaxter and C. C. Ten Hallers-Tjabbes: The effect of pollutants on sensory systems and behviour of aquatic animals. Neth. J. Aquat. Ecol. 26(1) (1992), 43 - 58.

[90] E. L. Morgan, R. C. Young and J. R. Wright, Jr.: 1988. Developing portable computer-aided biomonitoring for a regional water quality surveillance network. In: D. Gruber and J. Diamond (eds): Automated Biomonitoring: Living Sensors as Environmental Monitors (1988), 127 - 145, Ellis Horwood LtD, Chichester, UK.

[91] E. L. Morgan, K. D. Roberts and T. E. Pride: Remote Biosensing Applications in Environmental Assessment. Environm. Auditor Vol. 2 (4) (1991), 213 - 227.

Environmental Science Forum Vol.96 (1999) pp. 119-140
© 1999 Trans Tech Publications, Switzerland

Groundwater Biomonitoring

F. Mösslacher[1] and J. Notenboom[2]

[1] Institute of Limnology, Austrian Academy of Sciences, Gaisberg 116, AU-5310 Mondsee, Austria

[2] Laboratory for Ecotoxicology, National Institute of Public Health and the Environment,
P.O. Box 1, NL-3720 BA Bilthoven, The Netherlands

Keywords: Groundwater Contamination, Surface/Subsurface Water Interactions, Metazoa, Microbial Communities, Bioaccumulation Monitoring, Toxicity Monitoring, Ecosystem Monitoring

ABSTRACT

Groundwater systems display highly complex and heterogeneous environmental characteristics and are generally difficult to access and to explore. A combined approach of physico-chemical monitoring and biomonitoring provides therefore the most appropriate information on the state of the system investigated. This information should be collected in a way to discern natural fluctuations from anthropogenic induced changes. The studies available on groundwater metazoans indicate that the structure of metazoan communities is strongly related to hydrodynamics and the supply of organic carbon sources. No studies are available showing clear relationships between community structure and exposure to toxic chemicals. We also give an overview of available studies on bioaccumulation and toxicity monitoring conducted with groundwater metazoans. The problems that arise when applying groundwater metazoans for routine biomonitoring are shown and biological and methodological explanations are discussed. In general it is concluded that analyses of microbial communities appear an effective tool to collect information on the ecological status of groundwater systems. They are the most abundant group of biota, occur almost in every type of groundwater system, and play, because of their large physiological diversity, a predominant role in biodegradation processes. Novel approaches for future research in groundwater biomonitoring are introduced.

INTRODUCTION

Ecological research of subterranean waters revealed that these systems are dynamic, displaying heterogeneous habitat characteristics, and harbour a very specialised biota [*inter alia* 1-4]. Due to exchange processes between the subterranean water, the substrate, and organisms, groundwater systems are considered as true ecosystems [5,6]. If information on the impact of human activities on the properties of these ecosystems is required, physico-chemical monitoring alone provides not sufficient information and biological investigations (biomonitoring) are additionally required.

From a theoretical point of view water saturated subsurface systems can be divided into three phases: (1) the aqueous phase, i.e. the groundwater itself which is maintained in the system and transports substances; (2) the solid phase, i.e. the surroundings to which the water has contact (dead organic and inorganic material); and (3) the living phase, i.e. the biota occurring in the ecosystem, free in the water and attached to solids. The groundwater reservoir is, with the exception of the ice masses, the largest freshwater reserve on earth [7] and is mainly used by humans as potable water, for agricultural and industrial purposes. Therefore, most groundwater monitoring activities are dedicated to the quantitative and qualitative properties of the groundwater (aqueous phase) in respect to human demands. The entire subsurface ecosystem itself is rarely subject to study.

Monitoring of physical, chemical, and biological aspects is usually conducted with the aim to assess the natural state of the system and its dynamics in space and time. Changes can be due to natural influences, due to hydrology and climate, or due to anthropogenic impact such as contamination. The natural state of a particular investigated ecosystem should be well understood in order to recognise deviations induced by human activities. Such unnatural deviations can strongly influence the structure and function of a groundwater ecosystem: the water quality could debase, and indigenous organisms and ecological processes could be negatively influenced.

For the overview monitoring approaches are divided in three main categories following De Zwart [8]: (1) bioaccumulation monitoring, (2) toxicity monitoring, and (3) ecosystem monitoring. We will not only discuss activities within the strict definition of monitoring (standardised, long-term biological measurements and evaluation in order to define environmental status and trends) but also surveys (intensive measurement programmes with a finite duration to evaluate and report the quality of an environment) will be considered. The experiences obtained from these different activities, including those related to basic science (to understand the ecosystem structure and function), and routine biomonitoring (applied science), and novel approaches for groundwaters will be discussed.

STRESS FACTORS FOR GROUNDWATER ECOSYSTEMS

Most shallow groundwater systems are recharged by and/or are in close contact with any kind of water originating from the surface. Infiltrating surface water and percolating soil water can impose a direct negative influence caused by human activities and/or can uptake disturbing factors (stressors) from the surroundings (atmosphere or soil) and carry them into the subsurface environment (see Table 1). Deep groundwaters hardly ever have direct hydrological connections with the surface and if pollutants occur there they display much lower concentrations than in the shallower compartments.

Table 1: The origin and source of main stressors for shallow groundwater ecosystems (see also [9-11])

ORIGIN	SOURCE	STRESSOR
natural	leaching	salts and metals (e.g. chloride, sulphate, nitrate, fluoride, iron, etc.)
	radiation	radioisotopes (e.g. uranium)
human activities	atmospheric deposition	natural organics (e.g. nitrogen and phosphorous compounds), xenobiotica (e.g. pesticides), acids, heavy metals
	agriculture	pesticides, fertilizers and manure (including heavy metals, minerals, salts), sludge
	industry	natural organics, xenobiotica, salts, heavy metals, heat
	line source (motorway, railway, sewerage systems)	heavy metals, salts (e.g. chloride), natural organics
	mining	groundwater depression, salts, heavy metals
	overdrafting	groundwater depression, saltwater intrusion (in coastal areas)
	urban areas	natural organics, xenobiotica
	waste water and waste disposal	natural organics, xenobiotica, fertilizers and manure (including heavy metals, minerals, salts), sludge, microorganisms (i.e. pathogens)

The main anthropogenic stressors can be grouped using their chemical properties and overall effects as criteria (Table 2). In the following discussion we will mainly address these groups rather than single stressors, because pollution effects are generally due to a combination of single stressors belonging to one or more of these groups.

Table 2: Division of main disturbing factors into categories.

GROUP	DISTURBING FACTORS
toxic chemicals	1. organic chemicals (e.g. pesticides) 2. heavy metals
organic loads	1. organic matter often associated with macro-nutrients (nitrogenous and phosphorous compounds) 2. sewage water and sludge (complex mixture of organic matter, macro-nutrients, toxic chemicals, minerals, salts)
inorganic substances	salts, metals
thermal influence	heat

GROUNDWATER SYSTEM FEATURES

Accessibility

The inaccessibility of most groundwater environments is the main reason why our knowledge on groundwater ecosystems is minute compared to surface water systems. The exact and complete investigation of the properties of a groundwater system is often limited by technical and financial problems mainly caused by the difficult access to the system. Different methods for exploring groundwater systems and sampling metazoans and microbial communities are described in the literature [12-18].

Abiotic parameters

In comparison to surface water systems groundwaters display some characteristic features such as the lack of light, generally restricted habitat variety (absence of plants) and lower environmental fluctuations, suppressed day/night cycles and seasonallity, and dependency on allochthonous energy resources i.e. organic matter and oxygen (see also [19]). Nevertheless, subterranean environments are known to be highly complex systems whose inter- and intrasystematic heterogeneity is mainly based on regional and small scale geologic (unconsolidated and consolidated rock, different aquifer material and distinct sorption capacity of particles) and hydrologic characteristics (water regime in connection with the climate: rainfall, snow, dry periods). It is on these factors that all other parameters such as electric conductivity, alkalinity, acidity, pH, temperature, oxygen content, content of organic matter, and dissolved solids are dependent, and they can vary strongly (near to the surface) or be relatively constant (deep in the earth). Karst groundwater fills the cavities in consolidated rock. The heterogeneity of these systems is mainly due to the different sizes of fissures, voids, cracks, and even larger spaces like caves. Porous groundwater flows through unconsolidated sediments. The structure of porous groundwater environments is dominated by the granulometry of the sediment: layers and/or patchy accumulations of different grain size classes (gravel, sand, clay) influence the porosity and permeability of the aquifer [20]. The surface/subsurface water interface zone (ecotone zone, e.g. hyporheic zone) can be considered as a type of groundwater ecosystem too [21]. The environmental conditions of these habitats are strongly dependent on the properties of the connected surface water system. Compared to karst and porous groundwater systems the stability of abiotic and biotic parameters in the ecotone zone is low. A more detailed description of the abiotic properties of groundwater ecosystems would be out of the scope of this chapter and can be found in the literature [inter alia 20-25].

The influence of surface waters alters the habitat conditions in karst, porous, and ecotones zones temporally and spatially. The larger the openings of the system and the higher the intensity of hydrologic exchange, the more frequently the habitats are supplied with surface water, energy resources, surface water organisms, and also stressors. Shallow groundwater habitats have higher impact of energy and nutrient resources and provide therefore generally better living conditions for the groundwater biota [20,22,26]. However, these kind of habitats may suffer from anthropogenic stressors and are therefore the most at risk.

Biota in groundwater systems

Numerous classifications for the fauna found in subterranean habitats are proposed in the literature (for a review see [19,27]). In this chapter we will use the classification into stygobites, stygophiles, and stygoxenes, which refers to the ecological requirements and habitat type preferences of a certain species and/or population. Stygobites are considered true groundwater organisms which fulfil their whole life cycle in groundwater environments. Stygoxenes are surface water organisms which were passively transported to the groundwater environment and are not able to reproduce there. Stygophiles are surface water organisms too, but they can actively and successfully colonise groundwater habitats if they are not too constraining, for instance too low energetic resources. True groundwater organisms (stygobites) display characteristic features such as a long life span (e.g. up to 10 years for Amphipoda [28]), low reproductive rates, and a low metabolism. These adaptations make the culture of groundwater metazoans difficult and satisfying results were not obtained up to now (see also [29]). Moreover, their abundance is generally low and their distribution in the groundwater environment is very heterogeneous. Due to this, it is often impossible to obtain stygobites in a reasonable amount to investigate them in physiological and ecological studies. General information on the heterogeneous distribution and abundance of stygobite, stygophile, and stygoxene species is given elsewhere [30-32].

The groundwater biota consists of microorganisms, invertebrates and vertebrates. Higher plants and organisms using sun light as an energy source are generally absent in groundwaters, except if they were passively transported into it. Vertebrates build the smallest group within the groundwater biota. Few stygobite fish and urodel species have been found only in a restricted number of habitat types (karst and macro-porous systems) [30]. In contrast to this, a manifold invertebrate fauna, including an enormous portion of endemic species, develop in many groundwater habitats. The most common groundwater invertebrate taxa are: Turbellaria, Oligochaeta, Nematoda, Mollusca, Rotifera, Crustacea (e.g. Copepoda, Ostracoda, Amphipoda, and Isopoda), and Acari. Their distribution in groundwater seems restricted only by the size of the pore space and the availability of energetic resources. For a proper interpretation of most biomonitoring results, information on the physiology and ecology of the species is required. Since laboratory studies with stygobite species are difficult to conduct, knowledge of their autecology is limited. This limitation restricts the interpretability of groundwater metazoan data obtained in biomonitoring programmes.

Microorganisms i.e. bacteria and protists (mainly flagellates and amoeba), are the most abundant and diverse groups of organisms in the subterranean environment. Their extension reaches "to the depth of temperature limited biosphere" [33] and they are widespread throughout all geological formations [34]. Furthermore, they transfer dissolved compounds into biomass and degrade organic compounds, therefore they form the basic link in the groundwater food chain. Due to the diverse physiological functions of microorganisms they are able to degrade numerous pollutants and play therefore an important role in bioremediation of groundwater systems [34-39]. It is expected that microbes are relevant organisms for all kinds of biomonitoring activities due to their wide distribution, high abundance, and diverse physiological functions. Furthermore, it is suggested that changes in the structure and function of microbial communities are effecting more highly organised organisms and the whole function of the groundwater ecosystem.

Subsurface microbiology has become an increasingly important research discipline since the contamination of aquifers requires an understanding of microbial processes in the subterranean surrounding. Unfortunately, groundwater microbiology, from an ecological and/or ecotoxicological point of view, is still poorly developed [33,36], despite progress in recent years (see [35]).

BIOMONITORING IN GROUNDWATERS

Basic considerations

Biomonitoring methods have the advantage that they reflect bioavailable fractions and the intrinsic toxicity of chemicals, whereas the overall environmental concentration is measured with chemical analyses. Bioindicators integrate effects over space and time. Physico-chemical analyses only provide information on the momentary state of the environment, consequently, temporary pollution could remain undetected. Further, biomonitoring activities can evaluate the influence of stressors to the fitness of populations. These advantages show that biomonitoring in addition to physico-chemical monitoring enhances the probability of detecting pollution in groundwaters, enables evaluation of the ecological consequences of pollution, and allows possible remediation and protection activities to be developed.

The selection of possible groundwater organisms for biomonitoring has to be based (1) on general requirements for indicator species i.e. clear taxonomy, wide distribution, high abundance, well investigated syn- and autecology, easy to handle and to breed in the laboratory, etc. [see 40,41] and (2) on practical requirements i.e. feasible time and costs. Microorganisms appear to better fulfil these requirements compared to groundwater metazoans.

Bioaccumulation monitoring

This approach gives an indication of the bioavailable pollutant concentration in the environment, and of the pollutant transfer and biomagnification in the food chain. Applying this method BioConcentration Factors can be calculated whereby the amount of chemicals in parts and/or the whole body of the organism is measured and set in relation to the combined exposure of the organism to its environment and food [8].

Heavy metals are the only pollutants which have been investigated in groundwater bioaccumulation monitoring up to now. Plénet et al. [42], Plénet [43,44] and Dickson et al. [45] carried out studies on the metal accumulation in groundwater Amphipoda and crayfish, respectively, in comparison to surface water relatives. Plénet et al. [42] determined metal (Cd, Cr, and Zn) concentrations in two polluted sediments and in the stygobite amphipod *Niphargus* cf. *rhenorhodanensis* inhabiting those sediments. The results show that metal concentrations in the organisms positively correlate to sediment concentrations. *N.* cf. *rhenorhodanensis* was here considered as "potential integrator" of heavy metal pollution in subterranean environments [42]. The authors suggested exposing caged animals to the environment (wells) and analysing the metal concentrations in the organisms after a defined exposure time. With this method the pollutant concentration available for indigenous groundwater metazoans could be measured *in situ*. However, for such assays organisms of a defined age and physiological condition (cultured animals) are required, but breeding groundwater metazoans in the laboratory has not yet been possible (see above). In other studies Plénet [43,44] correlated heavy metal (Cu, Zn) concentrations in the Amphipoda *N.* cf. *rhenorhodanensis*, *N. casparyi* (both stygobite), and *Gammarus fossarum* (stygophile population) with concentrations in their interstitial environment (water and sediment). The stygobite organisms showed higher body metal concentrations than the stygophile but, in contrast to the previous study [42], a positive relation between environmental concentrations and body concentration could only be determined for the stygophile individuals. Plénet [43,44] supposed therefore that stygobite Amphipoda are not useful for biomonitoring heavy metal pollution in the hyporheic zone, but the use of stygophile Amphipoda populations for this purpose was supported [see also 46].

Dickson et al. [45] investigated metal concentrations in different tissues of the stygobite (*Orconectes australis australis*) and the stygophile crayfish (*Cambarus tenebrosus*) both sampled in an unpolluted cave stream. In the stygobite species significantly higher tissue concentrations could be found for non essential metals (Cd and Pb), but not for essential metals (Cr, Cu, Fe, Mg, Mn, Zn, Ca, and K). The different results obtained can be explained by active control processes in the organism for uptake and elimination of essential metals [47]. The authors [42-45] explained the relatively higher metal concentrations in the stygobite species by their longer life spans, their close contact to the sediment (for Amphipoda), and their feeding behaviour (see below).

In the following an example of the impact of heavy metals on cave biocoenoses is introduced. Stygobite crayfish from an unpolluted habitat contain relatively high concentrations of non essential metals [45]. This is explained by their longevity [48] and feeding behaviour. Longevity leads to an enormous time period in which they are feeding on invertebrates (80% of their food consists of stygobite Amphipoda and Isopoda [49]), which can accumulate heavy metals from their surrounding [42-44]. This illustrates the relatively high accumulation potential of heavy metals in cave food chains. Moreover, the detoxification mechanisms could use up energy which would be needed for reproduction [45]. The statement that groundwater environments are vulnerable ecosystems where low chronic pollution could accumulate to severe toxic levels in the organisms is supported by metal accumulation and ecological studies of cave crayfish.

Findings of the few available accumulation studies combined with general knowledge on groundwater organisms assume that groundwater macro-crustaceans, e.g. Amphipoda and crayfish, cannot be efficiently used for bioaccumulation monitoring: they can not be cultured for laboratory and *in situ* investigations and their body metal concentrations do not show clear relations to the metal concentrations of their surrounding. Nevertheless, the use of micro-crustaceans such as Copepoda and Ostracoda, cannot be completely discounted, although problems in obtaining the required information on their physiology and ecology are likely to arise. The presence of bioavailable pollutant concentrations could be indicated by using well investigated surface water organisms for groundwater bioaccumulation monitoring. However, when applying this approach the consequences of pollutants on groundwater biota and further groundwater ecosystem damage would remain unknown.

Toxicity monitoring

Toxicity (or effect) monitoring is conducted to measure the physiological response of organisms to pollutants. This approach includes mainly toxicity tests and biological early warning systems [8]. Toxicity tests using cultured organisms and following a standard protocol are generally conducted to prove the properties of single chemical compounds. Many of these tests are also adapted, as *in vitro* bioassays, to check the toxicity of environmental water samples [see 38-40,50]. For surface water ecosystems, organisms from different taxonomic and functional groups are applied for ecologically significant test results [40]. Since groundwater ecosystems harbour different biota than surface water systems and display distinct environmental conditions, the application of bioassays using surface water organisms e.g. *Daphnia spp.* or fish [see 40], appears not to be fully applicable for groundwater systems. However, such assays are in use to check the toxicity of groundwaters after bioremediation [39,40] or to calculate risk quotients for potential pollutants e.g. pesticides in groundwaters [51]. The calculation of risk quotients is based on the combination of results of current aquatic (mainly surface water) toxicity data and ecological information on groundwater systems. Such approaches are acceptable as indication for potential risks but give little information on the specific effects that may occur. To apply routine laboratory bioassays with groundwater organisms, microorganisms are thought to be the most effective and ecologically significant target group.

Another approach in ecotoxicology is not to test the toxicity of the environmental water sample, but to investigate the sensitivity of indigenous organisms to chemical stressors. For this purpose, organisms are brought from their natural habitat to the laboratory and are exposed to a concentration series of specific substances under controlled conditions. Toxicity data collected in this way can be used to establish ecologically based environmental quality objectives for groundwater. Moreover, they can be used to check the sensitivity of groundwater organisms in comparison to closely related surface water species [52-57]. The hypothesis [11,52] that groundwater metazoans differ systematically in sensitivity from closely related surface water species was not clearly verified or rejected up to now (see below and [52,57]). The lack of information on a broad range of toxicity data of different surface water and groundwater species under similar experimental conditions make a true verification of this hypothesis impossible.

In some studies [11,52,57] the above mentioned hypothesis was checked by comparing the very few available EC 50 values (pollutant concentration which effects 50 % of the test organisms, most of them deriving from tests conducted under different laboratory conditions) of groundwater organisms with those of closely related surface water species. It was found that acute toxicity values were not significantly different between organisms originating from surface water and groundwater habitats. An overview of the few existing toxicity data for groundwater organisms is given in Table 3. It shows that crustaceans are the most frequently used organisms and metals the most often applied chemicals.

Table 3: Overview on acute toxicity data for stygobites and stygophile populations.

Species		Test Substance: T = [°C], H = mg CaCO₃/l LC 50 (96h) mg/l	Ref.
Amphipoda	*Metacrangonyx spinicaudatus* (stygobite)	Cadmium: LC 100 (48h) 4 T = 21 Copper: 0.2 Zinc: 0.5 Lead: 1.5 Ammonium: 1.5	54
	Niphargus aquilex (stygobite)	Cadmium: T = 12, H = 103 6.25 (pH 5), 5 (pH 6), 4.5 (pH 7), 2.45 (pH 8) Zinc: 163 (pH 5), 118 (pH 6), 180 (pH 7)	56 55
Isopoda	*Caecidatea bicrenata* (stygobite)	Cadmium: T = 13 2.2 (pH 7, H = 220), 1.2 (pH 7.2, H = 83) Copper: 2.2 (pH 6.4, H = 82) Zinc: 20 (pH 7, H = 220) Total Residual Chlorine: 0.11 (pH 7, H = 220)	52,53 53 52 52
	Caecidatea stygia (stygobite)	Cadmium: T = 13 0.29 (pH 7.5) H = 70 Copper: 2.3 (pH 6.9) H = 70 Chromium: 2.4 (pH 7.1) H = 86	53
	Proasellus cavaticus (stygobite)	Cadmium: T = 13, H = 103 0.56 (pH 5), 0.56 (pH 6), 0.5 (pH 7), 0.5 (pH 8) Zinc: 124 (pH 5), 90 (pH 6), 127 (pH 7)	56 55
	Proasellus coxalis africanus (stygophile population)	Cadmium: LC 100 (48h) 1 T = 21 Copper: 4.1 Zinc: 10 Lead: 15 Ammonium: 190	54
	Proasellus slavus vindobonenis (stygobite)	Cadmium: T = 10 < 4.7 Potassium Chloride: 205 (pH 8) Potassium Nitrate: 347 (pH 8)	*
	Proasellus strouhali strouhali (stygobite)	Permethrin: T = 11 10 µg/l (EC 100, 0.5 h) Sodiumhypochloride: 5 mg/l Chlor (LC 90, 0.5 h)	58
	Typhlocirolana haouzensis (stygobite)	Cadmium: LC 100 (48h) 180 T = 21 Copper: 18 Zinc: 20 Lead: 110 Ammonium: 100	54

Table 3 continued.

Species		Test Substance: T = [°C], H = mg CaCO₃/l LC 50 (96h) mg/l	Ref.
Decapoda	*Orconectes a. australis* (stygobite)	Total Residual Chlorine: T = 10.5, H = 242, pH 8.9 3.39, acclimated individuals 2.7, unacclimated Free Chlorine: 3, acclimated individuals 2.25, unacclimated	 59
Ostracoda	*Fabaeformiscandona wegelini* (stygobite)	Potassium Chloride: T = 10, pH 8 LC 50 (24h) 1930, (48h) 1577	*

Copepoda	*Parastenocaris germanica*
	(stygobite)
		Cadmium: T = 10.5
		2.2 (pH 6.8)
		Zinc:
		Adults: 1.7 (pH 6.8)
		Copepodites: (Zinc 130491): 1.27 (T = 13, pH 6.5-7.5)
		                   (Zinc 030292): 0.46
		Pentachlorophenol:
		Adults: 0.036 (T = 10.5, pH 6.8)
		      (PCP 150692) > 0.09 (T = 13, pH 6.5-7.5)
		      (PCP 200192) > 0.16 (T = 13, pH 6.5-7.5)
		Copepodites: 0.11 (T = 13, pH 6.5-7..5)
		3,4-Dichlorophenol:
		4.6 (pH 6.8)
		Aldicarb:
		Adults: 2.9 (T = 13, pH 6.5-7.5)
		Copepodites: 0.036 EC 50 (96h)
		Thiram:
		Adults: 0.003 (T = 13, pH 6.5-7.5)
			
			60
			
			60
			61
			
			60
			61
			
			60
			
			61

	Acanthocyclops vernalis (stygophile population)	Cadmium: T = 10 5.09 Potassium Chloride: 1609 (pH 8) Potassium Nitrate: 1986 (pH 8)	*
	Diacyclops bicuspidatus (stygophile population)	Cadmium: T = 10 >4.9; LC 50 (48h) 6.7 Potassium Chloride: 1050 (pH 8) Potassium Nitrate: 1460 (pH 8)	*
	Diacyclops aff. *disjunctus* (stygophile population)	Potassium Chloride: T = 10, pH 8 590	*
	Diacyclops n. sp[1] (stygophile population)	Potassium Chloride: T = 10, pH 8 564	*
	Megacyclops viridis (stygophile population)	Cadmium: T = 10 4.07 Potassium Chloride: 460 (pH 8)	*
	Paracyclops fimbriatus (stygophile population)	Potassium Chloride: LC (48h) T = 10, pH 8 1200	*
Oligochaeta	*Trichodrilus tenuis* (stygobite)	Cadmium: T = 12, H = 103 1.47 (pH 5), 1.15 (pH 6), 1.05 (pH 7), 0.8 (pH 8) Zinc: 9.3 (pH 5), 8.5 (pH 6), 8.3 (pH 7)	 56 55

* [Mösslacher unpubl. results]

[1] [*Diacyclops felix* n.sp, Stoch and Pospisil unpubl.]

Biological early warning systems (on-line systems) are already established at sites where surface water is introduced into groundwater ecosystems [40]. In these on-line systems, such as the Dynamic Daphnia Test [62], surface water organisms are used to check the quality of infiltrating surface water. This approach is suited to providing information on the overall toxicity of infiltrating surface water, but it does not provide data on the possible damages to the groundwater ecosystem itself.

Application of indigenous microbial communities seems to be a reasonable approach for detecting the impact of contaminants on the groundwater community. A modification e.g. of the toxiguard system [63] which is based on the physiological response of a bioreactor, and is used as a biological early warning system for running waters, could serve for groundwaters too. The toxiguard system is built up by a biofilter supplied with river water to allow surface associated bacteria to grow (bioreactor). After a water specific (indigenous) microbial community has settled on the filter, the oxygen demand of the biofilm can be measured at the outlet of the system. Changes in the oxygen consumption of the test system indicate changes in the water composition e.g. loads of pollutants. The considerations of the modification of this system are very preliminary but should introduce possibilities for new methods in groundwater biomonitoring. A sterile biofilter could be exposed to groundwater (*in* or *ex situ*) to grow an indigenous microbial community whose oxygen demand can be measured as in the toxiguard system. Studies on the colonisation of sterile biofilters in groundwaters have already been successfully conducted [64,65], and it is known that the attached microbial community provides a good representation of the indigenous groundwater microbial biocoenoses [14,66]. Laboratory and *in situ* studies on the response of groundwater microbial communities to different water properties and chemicals are basic requirements for a proper evaluation of the results i.e. changes in the oxygen consumption of the microbial community. Although the type of stressor can not be clearly identified with this early warning system, it would have the advantage of detecting continuously, changes in the groundwater quality using organisms which are quantitatively and qualitatively important for the function of groundwater ecosystems.

The development of a micro-biological early warning system for groundwaters can either follow the strategy of using indigenous microbes (bioreactor, see above), or changing the bacterial genom to produce whole-cell biosensors [67,68]. The latter method is based on genom engineering. Specific (i.e. for single substances) or less specific (i.e. for substance groups) promotor sequences and/or promotorless reporter genes (e.g. bacterial luciferase) are fused into the genom of active bacteria [69]. The bioavailable fraction of the specific pollutant and/or general stressors induce gene expressions such as the formation of degrading and light emitting enzymes. The emitted light (bioluminescence) can be easily measured using standard techniques. The microbiological research group of the EAWAG (Switzerland) [67,68] introduced a hypothetical model of a method to investigate bioavailable pollutant concentrations in groundwaters by whole-cell biosensors. They suggested a system whereby whole-cell biosensors on top of lightning glass fibres are immobilised and luminometers are installed to measure the light emission of the active bacterial cells. To prevent genetically engineered bacteria leaking into the environment, a membrane unpermeable for bacteria but permeable for pollutants could be placed on top of the glass fibre. The assay conditions of these methods have to be standardised to prevent interpretation failures due to unexpected instability of the reporter gene, and eventually non optimal physiological state of the strain used as donor for the whole-cell biosensors [69-71]. Another source leading to misinterpretations of the light emission is the occurrence of possible inhibitors and other stimulators beside the one for which the specific biosensor has been developed. Promising results using whole-cell biosensors for the detection of environmental pollution are already published [67-75], and should encourage microbiologists and ecological engineers to further develop such approaches.

Ecosystem monitoring

Ecosystem monitoring is the term for the coordinated collection of physico-chemical and biological information and it measures the integrity of ecosystems in relation to natural and human induced perturbations. This approach provides the most complete information on the state of an ecosystem and is required for water pollution control and risk management [8]. Ecosystem biomonitoring aims are: (1) to monitor the natural state of the ecosystem (in space and time) and (2) to monitor stressor induced effects on the ecosystem.

(1) Monitoring the natural state of the groundwater ecosystem

The relationship between the occurrence of invertebrates and the natural environmental conditions of the investigated groundwater system is the main topic of this approach. In many studies [1-4,21,22,26,76-97,etc.], the subterranean community has been explored with regard to the abundance, the size of organisms, and the ratio of stygobite to stygophile to stygoxene species. A high individual abundance in a groundwater aquifer suggests a good nutrient and oxygen supply, and the size of individuals is directly related to the sediment structure. Habitats with small interstitial pore volume or narrow fissures contain only few meiofaunal organisms (30μm-1mm). The larger the pore space of the habitats, the more macrofaunal individuals (>1mm) can be found. Indirectly the pore space influences the water current velocity and the supply of oxygen and organic carbon sources. Further, relatively high numbers of stygoxene and stygophile species indicate strong hydrological exchange between the investigated aquifer with surface waters. Although, several types of groundwater environments can be discerned (karst, porous and ecotone zone) and the heterogeneity within these categories is enormous, the relationships mentioned above are basically valid for all groundwater systems regardless of scale.

The search for single (stygobite) indicator species for specific environmental conditions in groundwater has not been successful up to now [83,84,88,90,91]. Due to the general low abundance, heterogeneous distribution of groundwater organisms, and limited quantitative sampling a complete interpretation of field data is often difficult. Furthermore, the high portion of endemic groundwater species [30,31] and the ability of the organisms to adapt very closely to their specific habitat, what can lead to differences in the ecology and physiology of populations, show that single (stygobite) invertebrate species are not a useful tool for routine biomonitoring. Nevertheless, analysis of the entire invertebrate community structure in particularly in the more productive parts of the groundwater system is a well investigated and promising tool for biomonitoring natural environmental conditions (see also [26,76,77,98]).

(2) Monitoring stressor induced effects

Monitoring anthropogenic impacts using metazoans is highly appropriate in surface waters and is believed to be applicable also to groundwater systems. Poulson [99,100] created the idea of an Index of Biological Integrity (IBI) for karst waters combining habitat characteristics with population and community data. In these studies the Mammoth Cave System (USA) was intensively investigated over 30 years and non-lethal effects on populations due to several stressors e.g. toxic chemicals and siltation, were found. The Index of Biological Integrity, based on the entire groundwater metazoan fauna of the investigated system, is a useful but limited approach: the natural state and function of the system have to be known, meaningful statistical evaluation is difficult due to the high variety of qualitative and quantitative data [99], and it is only applicable to easily accessible groundwater systems. The use of an IBI is proposed to investigate and evaluate the state of well known cave ecosystems, such as the Mammoth Cave (USA), the Postojna-Planina Cave (Slovenia), and the Movile Cave (Romania). However, wider application of the IBI is not possible especially in the case of less accessible karst systems or porous groundwaters owing to enormous difficulties in quantifying the structure of the entire community.

Table 4 gives a general view of studies concerning the influence of pollution on groundwater metazoan and microbial communities. Most of these studies deal with the influence of organic loads and toxic chemicals, whilst studies on the impact of thermal stress are lacking.

Table 4: List of studies on the *in situ* influence of stressors to the groundwater biota.

Ecosystem	Stressor	Investigated Organisms	Reference
Karst	organic loads	vertebrates	101,102
		invertebrates	85,87,94,101-111
		microorganisms	94
	toxic chemicals	vertebrates	-
		invertebrates	112
		microorganisms	112
Porous	organic loads	invertebrates	84,87,89,91,93,113-118
		microorganisms	35,119-124
	toxic chemicals	invertebrates	125
		microorganisms	35,119,120,126-131
Ecotone (hyporheic zone)	organic loads	invertebrates	88,132-136
		microorganisms	132
	toxic chemicals	invertebrates	42-44,121-124,132
		microorganisms	106

From the studies listed in Table 4 some general deductions can be made concerning the impact of pollution on groundwater organisms (see also [107,108]).

The entire metazoan stygobite community is strongly constrained by natural factors such as oligotrophy, oxygen dynamics, and space. Anthropogenically derived impacts are difficult to establish probably due to the dominance of natural factors.
The indicative power of single stygobite species for human induced perturbations appears to be limited.
Surface water species i.e. stygoxenes and stygophiles, appear to react most sensitively to pollution. Insect larvae, for example, occur in high abundance in ecotone zones. Low abundance or even absence of insect larvae in these areas was found to be related to elevated heavy metal concentrations.
Low organic pollution can induce increasing abundance and activity of the groundwater microbial community. When sufficient oxygen is available, this fact also holds for groundwater metazoans.
Moderate pollution with organic loads can decrease the abundance of stygobites and enhance the penetration of stygoxenes and stygophiles into the groundwater. When there is sufficient supply of resources, surface water organisms can settle in parts of the groundwater system and outcompete the indigenous biocoenoses.
Severe pollution can strongly diminish or even exterminate metazoan populations in groundwaters.
Recovery of the stygobite metazoan community after pollution can be time consuming due to the general long reproduction cycles of groundwater organisms. Recolonisation after severe

pollution is often slow and/or hindered by little or no hydrological connections to adjacent systems.

An interdisciplinary approach (combining hydrogeology, physics, chemistry, microbiology, zoology, and ecology) is necessary to evaluate the state and functions of a certain groundwater ecosystem, and to detect deviations from the natural state due to anthropogenic influence.

The fact that no clear relationship between stygobite metazoan species and environmental pollution was found up to now, raises the need to search for other ecologically relevant approaches. It has been shown that microbes from a polluted site can degrade the specific pollutant better than microbes from an unpolluted site [126,130,137] and that pollution decreases the overall degradative capacities of microbial communities [138-141]. Therefore, the capacity of a microbial community to degrade substrates and the diversity of its physiological functions could be used to indicate pollution effects. For these purposes, the investigation of the functional properties and community parameters of microbial biocoenoses from different groundwater systems influenced by human activities and/or undisturbed systems, could be conducted. Such community analyses result in specific community patterns (fingerprints). Fingerprints provided in this way could be compared with those of the study location and deviations from the natural state could be detected. The development of such fingerprints would undoubtedly be an enormous task, but could provide promising results, since this approach has already proved successful with aquatic microbial communities and soils [138-141, M. Rutgers unpubl. results].

CONCLUSION

Stressors for groundwater ecosystems (toxic chemicals, organic and inorganic loads, and heat) are mainly introduced by infiltrating surface water and percolating soil water. The effects of these stressors on the groundwater biota and/or on the entire ecosystem are as yet poorly understood. However, it is supposed that due to the rather extreme life history adaptations of many groundwater species, pollution consequences for population viability could indeed be more severe than for surface water communities. The knowledge obtained from bioaccumulation studies and chronic toxicity tests with metazoans, however, and general information on the response of microbial communities to pollution, could be useful for extrapolating the effects of pollutants to the entire groundwater ecosystem, and to derive safety levels based on ecotoxicological risks.

Investigations on the groundwater metazoan community show that the analyses of the community structure can indicate natural environmental conditions such as water flow patterns, supply with energy resources, and sediment structure. Groundwater metazoans do not clearly reflect anthropogenic influence, but the results of structure analyses of groundwater communities can provide information on the amount of infiltrating surface water, which is a potential indicator of pollution risks. The potential use of microorganisms in ecotoxicological risk assessment is described by Cairns et al. [142] for aquatic systems and by van Beelen and Doelman [143] for soils and sediments. It seems that the microbial communities will also be the most effective tool for biomonitoring pollution in groundwater ecosystems.

Prospective for future research:

The survival, reproduction, and competitiveness of groundwater organisms under stress conditions induced by chronic and/or acute pollution, should be investigated to explore the extent to which toxicity data on surface water organisms are applicable to groundwaters.

Further research on biomonitoring groundwaters should concentrate on the development of novel microbiological approaches, e.g. bioreactors, biosensors, the measurement of microbial *in situ*

activity and the investigation of the physiological diversity of microorganisms.

\# Existing methods have to be adapted and/or new techniques e.g. sampling, test procedures, developed, to conduct rigorous biomonitoring for the different groundwater environments (karst, porous, ecotone).

\# An interdisciplinary approach (hydrology, physics, chemistry, biology, etc.) and research on the self purification capacity and recovery potential of groundwater ecosystems is needed, to develop models for assessing ecological risks as a basis for sustainable groundwater management.

ACKNOWLEDGEMENTS

We thank D.L. Danielopol, C. Griebler and R. Rutgers for their critical comments on an early manuscript. J. Gibert and A. Gerhardt reviewed the first draft of the manuscript and contributed to its improvement. H. Bennion corrected the English style. Financial support for the preparation of this chapter was given by the Austrian Academy of Sciences, by the Büro für Internationale Beziehungen (Vienna), and the Fonds zur Förderung der wissenschaftlichen Forschung (FWF-Project 11149Bio attributed to D.L. Danielopol). Further we have to thank R. Dallinger (Univ. Innsbruck) for providing the facilities to conduct toxixicty tests with heavy metals.

REFERENCES

[1] J. Gibert, D.L. Danielopol, and J.A. Stanford (1994): Groundwater Ecology, 1st edition, Academic Press, San Diego.

[2] A.I. Camacho (1992): The Natural History of Biospeleology, 1st edition, Museo Nacional de Ciencias Naturales, C.S.I.C., Madrid.

[3] J.A. Stanford and J. Simons (1992): Proceedings of the First International Conference on Groundwater Ecology, American Water Resource Association, Bethesda, MD.

[4] J.A. Stanford and H.M. Valett (1994): Proceedings of the Second International Conference on Groundwater Ecology, American Water Resource Association, Atlanta.

[5] R. Rouch (1977): Considérations sur l'écosystème karstique. C. R. Acad. Sci. Paris, Sciences de la vie / Life sciences,284 (D),1101-1103.

[6] D.L. Danielopol (1982): Phreatobiology reconsidered. Polskie Archiwum Hydrobiologii,29,375-386.

[7] A. Baumgartner and H.J. Liebscher (1990): Lehrbuch der Hydrologie, 1st edition, Gebrüder Borntraeger, Berlin, Stuttgart.

[8] D. De Zwart (1995): Monitoring water quality in the future. Volume 3 Biomonitoring, RIVM, Bilthoven, The Netherlands.

[9] J.J. Fried (1975): Groundwater Pollution, 1st edition, Elsevier, Amsterdam.

[10] E.A. Laws (1993): Aquatic Pollution, 2nd edition, John Wiley and Sons Ltd, New York.

[11] J. Notenboom, S. Plénet, and M.-J. Turquin (1994): Groundwater Contamination and Its Impact on Groundwater Animals and Ecosystems, 477-504. In: J. Gibert, D.L. Danielopol, and J.A. Stanford (1994): *Groundwater Ecology*, 1stedition, Academic Press, San Diego.

[12] A.I. Camacho (1992): Sampling the subterranean biota cave (aquatic environment), 135-168. In: A.I. Camacho (1992): *The Natural History of Biospeleology*, 1st edition, Museo Nacional de Ciencias Naturales, C.S.I.C., Madrid.

[13] D.R. Cullimore (1993): Practical manual of groundwater microbiology, 1st edition, Lewis Publishers, Chelsea, MI.

[14] E.L. Madsen and W.C. Ghiorse (1993): Groundwater microbiology: subsurface ecosystem processes, 167-213. In: T.E. Ford (1993): *Aquatic microbiology*, 1st edition, Blackwell, Boston.

[15] F. Malard, J. Gibert, R. Laurent, and J.-L. Reygrobellet (1994): A new method for sampling the

fauna of deep karstic aquifers. C. R. Acad. Sci. Paris, Sciences de la vie / Life sciences,317,955-966.

[16] F. Malard, J.-L. Reygrobellet, R. Laurent, and J. Mathieu (1997): Developements in sampling the fauna of deep water-table aquifers. Archiv für Hydrobiologie,138,401-432.

[17] J. Mathieu, P. Marmonier, R. Laurent, and D. Martin (1991): Collection of subsurface biological material and sampling techniques. Hydrogeologie,3,187-200.

[18] P. Pospisil (1992): Sampling methods for groundwater animals of unconsolidated sediments, 109-134. In: A.I. Camacho (1992): *The Natural History of Biospeleology*, 1st edition, Museo Nacional de Ciencias Naturales, C.S.I.C., Madrid.

[19] J. Gibert, J.A. Stanford, M.-J. Dole-Olivier, and J.V. Ward (1994): Basic Attributes of groundwater ecosystems and prospects for research, 7-40. In: J. Gibert, D.L. Danielopol, and J.A. Stanford (1994): *Groundwater Ecology*, 1st edition, Academic Press, San Diego.

[20] M. Creuzé des Châtelliers (1994): Geomorphology of alluvial groundwater ecosystems, 157-185. In: J. Gibert, D.L. Danielopol, and J.A. Stanford (1994): *Groundwater Ecology*, 1st edition, Academic Press, San Diego.

[21] J. Gibert, J. Mathieu, and F. Fournier (1997): Groundwater/Surface water Ecotones: Biological and Hdyrological Interactions and Management Options, 1st edition, Cambridge University Press, Cambridge.

[22] P. Vervier, J. Gibert, P. Marmonier, and M.-J. Dole-Olivier (1992): A perspective on the permeability of the surface freshwater-groundwater ecotone. Journal of North American Benthological Society,11,93-102.

[23] L. Zilliox (1994): Porous media and aquifer systems, 69-96. In: J. Gibert, D.L. Danielopol, and J.A. Stanford (1994): *Groundwater Ecology*, 1st edition, Academic Press, San Diego.

[24] M. Bakalowicz (1994): Water geochemistry: water quality and dynamics, 97-127. In: J. Gibert, D.L. Danielopol, and J.A. Stanford (1994): *Groundwater Ecology*, 1st edition, Academic Press, San Diego.

[25] R. Maire and S. Pomel (1994): Karst geomorphology and environment, 129-155. In: J. Gibert, D.L. Danielopol, and J.A. Stanford (1994): *Groundwater Ecology*, 1st edition, Academic Press, San Diego.

[26] F. Mösslacher (1998): Subsurface dwelling crustaceans as indicators of hydrological conditions, oxygen concentrations, and sediment structure in an alluvial aquifer. Internationale Revue der gesamten Hydrobiologie,83,349-364.

[27] A.I. Camacho (1992): A classification of the aquatic and terrestrial subterranean environment and their associated fauna, 57-103. In: A.I. Camacho (1992): *The Natural History of Biospeleology*, 1st edition, Museo Nacional de Ciencias Naturales, C.S.I.C., Madrid.

[28] R. Ginet (1960): Ecologie, éthologie et biologie de Niphargus (Amphipodes Gammaridés hypogés). Annales de Spéléologie,15,1-254.

[29] F. Lescher-Moutoué (1973): Cyclopides des eaux souterraines de l'Ain et de l'Isère (France). Annales de Spéléologie,28,429-502.

[30] L. Botosaneanu (1986): Stygofauna mundi, 1st edition, E.J. Brill/Backhuys, Leiden.

[31] F. Stoch (1995): The ecological and historical determinants of crustacean diversity in groundwaters, or: why are there so many species? Mémoires de Biospéologie,22,139-160.

[32] D.L. Strayer (1994): Limits to biological distributions in groundwater, 287-310. In: J. Gibert, D.L. Danielopol, and J.A. Stanford (1994): *Groundwater Ecology*, 1st edition, Academic Press, San Diego.

[33] W.C. Ghiorse (1997): Subterranean Life. Science,275,789.

[34] F.H. Chapelle (1993): Groundwater microbiology and geochemistry, 1st edition, John Wiley and Sons Ltd, New York.

[35] Symposium of Subsurface Microbiology (1996): Abstracts, Swiss Society of Microbiology,

Davos, Switzerland.

[36] A.M. Gounot (1994): Microbiology of groundwaters, 189-215. In: J. Gibert, D.L. Danielopol, and J.A. Stanford (1994): *Groundwater Ecology*, 1st edition, Academic Press, San Diego.

[37] R.M. Gersberg, R.M. Dawsey, and M.D. Bradley (1993): Nitrate enhancement of *in situ* bioremediation of monoaromatic compounds in groundwater. Remediation,3,233-245.

[38] R.M. Gersberg, M.J. Carroquino, D.E. Fischer, and J. Dawsey (1995): Biomonitoring of toxicity reduction during *in situ* bioremediation of monoaromatic compounds in groundwater. Water Research,29,545-550.

[39] G.D. Marty, S. Wetzlich, L.M. Nunez, A. Craigmill, and D.E. Hinton (1991): Fish-based biomonitoring to determine toxic characteristics of complex chemical mixtures documentation of bioremediation at a pestizide disposal site. Aquatic Toxicology,19,329-340.

[40] G. Gunkel (1994): Bioindikation in aquatischen Ökosystemen, 1st edition, Gustav Fischer Verlag, Stuttgart.

[41] J.M. Hellawell (1986): Biological indicators of freshwater pollution and environmental management, 1 edition, Elsevier Applied Science Publishers, London and New York.

[42] S. Plénet, P. Marmonier, J. Gibert, J.A. Stanford, A.-M. Bodergat, and C.M. Schmidt (1992): Groundwater Hazard Evaluation: A Perspective for the Use of Interstitial and Benthic Invertebrates as Sentinels of Aquifer Metallic Contamination, 319-329. In: J.A. Stanford and J.J. Simons (1992): *Proceedings of the First International Conference on Ground Water Ecology*, 1st edition, American Water Resource Association, Bethesda, MD.

[43] S. Plénet (1993): Sensibilité et role des invertebrates vis à vis d'un stress metallique à l'interface eau superficielle/eau souterraine. Ph.D. Thesis, Univ. Lyon.

[44] S. Plénet (1995): Freshwater amphipods as biomonitors of metal pollution in surface and interstitial aquatic systems. Freshwater Biology,33,127-137.

[45] G.W. Dickson, L.A. Briese, and J.P. Giesy Jr. (1979): Tissue metal concentrations in two crayfish species cohabiting a Tennessee cave stream. Oecologia,44,8-12.

[46] G.-P. Zauke (1982): Monitoring aquatic pollution using Gammaridae (Amphipoda: Crustacea) with emphasis on Cadmium. Polskie Archiwum Hydrobiologii,29,289-298.

[47] K. Liebscher and H. Smith (1968): Essential and non-essential trace elements. Archiv of Environmental Health,17,881-890.

[48] J.E. Cooper (1975): Ecological and behavioural studies in Shelta Cave, Alabama, with emphasis on decapod crustaceans. Ph.D. Thesis, Univ. Kentucky.

[49] D.L. Weingartner (1964): Production and trophic ecology of two crayfish species cohabiting and Indiana cave. Ph.D. Thesis, Michigan State Univ.

[50] G.M. Rand (1995): Fundamentals of aquatic toxicology, 2nd edition, Taylor & Francis, London.

[51] J. Notenboom and K. van Gestel (1992): Assessment of Toxicological Effects of Pesticides on Groundwater Organisms, 311-317. In: J.A. Stanford and J. Simons (1992): *Proceedings of the First International Conference on Ground Water Ecology*, American Water Resources Association, Bethesda, MD.

[52] A.D. Bosnak and E.L. Morgan (1981): Acute toxicity of cadmium, zinc, and total residual chlorine to epigean and hypogean isopods (Asellidae). NSS Bulletin,43,12-18.

[53] A.D. Bosnak and E.L. Morgan (1981): Comparison of acute toxicity for Cd, Cr, and Cu between two distinct populations of aquatic hypogean isopods (Caecidatea sp.). 8th International Congress of Speleology,1,72-74.

[54] C. Boutin, M. Boulanouar, and M. Yacoubi-Khebiza (1995): Un test biologique simple pour apprecier la toxicite de l'eau et des sediments d'un puits. Toxicite comparee, *in vitro*, de quelques metaux lourds et de l'ammonium, vis-a-vis de trois genres de crustaces de la zoocenose des puits. Hydroécologie Appliquée,7,91-109.

[55] W. Meinel and R. Krause (1988): Zur Korrelation zwischen Zink und verschiedenen pH-Werten in ihrer toxischen Wirkung auf einige Grundwasser-Organismen. Zeitschrift für angewandte Zoologie,75,159-182.

[56] W. Meinel, R. Krause, and J. Musko (1989): Experimente zur pH-Wert-abhängigen Toxizität von Kadmium bei einigen Grundwasserorganismen. Zeitschrift für angewandte Zoologie,76,101-125.

[57] F. Mösslacher (1997): Effects of inorganis pollutants on the survival and respiration of groundwater crustaceans. Do groundwater organisms reflect water quality?, 13th international Symposium of Biospeleology, Marroko,57.

[58] D.L. Danielopol (1983): Bekämpfungsmöglichkeiten der Grundwasserassel *Proasellus strouhali* Karamann im Trinkwasserversorgungsnetz der Stadt Linz. Internal Report, Limnological Institute Mondsee, ÖAW,unpubl.,1-26.

[59] R.C. Mathews, A.D. Bosnak, D.S. Tennant, and E.L. Morgan (1977): Mortality curves of blind caryfish (*Orconectes australis australis*) exposed to chlorinated stream water. Hydrobiologia,53,107-111.

[60] J. Notenboom, K. Cruys, J. Hoekstra, and P. van Beelen (1992): Effect of Ambient Oxygen Concentration upon the Acute Toxicity of Chlorophenols and Heavy Metals to the Groundwater Copepod *Parastenocaris germanica* (Crustacea). Ecotoxicology and Environmental Safety,24,131-143.

[61] J. Notenboom and J.-J. Boessenkool (1992): Acute toxicity testing with the groundwater Copepod *Parastenocaris germanica* (Crustacea), 301-307. In: J.A. Stanford and J.J. Simons (1992): *Proceeding of the First Conference on Ground Water Ecology*, American Water Resource Association, Bethesda, MD.

[62] J. Knie (1978): Der Dynamische Daphnientest - ein automatischer Biomonitor zur Überwachung von Gewässern und Abwässern. Wasser und Boden,12,310-312.

[63] B. Blessing, H. Fritz-Langen, and F. Krebs (1992): Einsatz von Bioreaktoren zur kontinuierlichen Gewässerüberwachung. Schriften Reihe Verein WaBoLu,89,247-254.

[64] P. Hirsch and E. Rades-Rohkohl (1990): Microbial colonization of aquifer sediment exposed in a grounwater well in Northern Germany. Applied and Environmental Microbiology,56,2963-2966.

[65] J. Marxsen (1982): Ein neues Verfahren zur Untersuchung der bakteriellen Besiedelung grundwasserführender sandiger Sedimente. Archiv für Hydrobiologie,95,221-233.

[66] R.W. Harvey, R.L. Smith, and L. George (1984): Effect of organic contamination upon microbial distributions and heterotrophic uptake in a Cape Cod, Mass., Aquifer. Applied and Environmental Microbiology,48,1197-1202.

[67] H. Harms, M. Jaspers, P. Sticher, and J. Roelof van der Meer (1996): Entwicklung von Biosensoren für die bestimmung bioverfügbarer Schadstoffkonzentrationen. Jahresbericht 1996, EAWAG, Zürich,20-25.

[68] J. Roelof van der Meer, M. Jaspers, P. Sticher, R. Tchelet, and H. Harms (1997): Bakterien können die bioverfügbaren Konzentrationen verschiedener Umweltschadstoffe anzeigen. EAWAG News,43D,25-27.

[69] P. Sticher, M. Jaspers, K. Stemmler, H. Harms, A.J.B. Zehnder, and J. Roelof van der Meer (1997): Developement and characterization of a whole-cell bioluminescent sensor for bioavailable middle-chain alkanes in contaminated groundwater samples. Applied and Environmental Microbiology,63,4053-4060.

[70] A. Heitzer, O.F. Webb, J.E. Thonnard, and G.S. Sayler (1992): Specific and quantitative assessment of naphtalene and salicylate bioavailability by using a bioluminescent catabolic reporter bacterium. Applied and Environmental Microbiology,58,1839-1846.

[71] M. Korpela, P. Mäntsälä, E.-M. Lilius, and M. Karp (1989): Stable-light-emitting *Escherichia coli* as a biosensor. Journal of Bioluminescence and Chemiluminescence,4,551-554.

[72] A. Heitzer, K. Malachowsky, J.E. Thonnard, P.R. Bienkowski, D.C. White, and G.S. Sayler (1994): Optical biosensor for environmental on-line monitoring of naphtalene and salicylate bioavailability with an immobilized bioluminescent catabolic reporter bacterium. Applied and Environmental Microbiology,60,1487-1494.

[73] O. Selifonova, R. Burlage, and T. Barkay (1993): Bioluminescent sensors for detection of bioavailable Hg(II) in the environment. Applied and Environmental Microbiology,59,3083-3090.

[74] O. Selifonova and R.W. Eaton (1996): Use of an *ipb-lux* fusion to study regulation of the isopropylbenzene catabolism operon of *Pseudomanoas putida* RE204 and to detect hydrophobic pollutants in the environment. Applied and Environmental Microbiology,62,778-783.

[75] T.K. Van Dyk, W.R. Majarian, K.B. Konstantinov, R.M. Young, P.S. Dhurjati, and R.A. Larossa (1994): Rapid and sensitive pollutant detection by induction of heat shock gene-bioluminescence gene fusions. Applied and Environmental Microbiology,60,1414-1420.

[76] D.L. Danielopol (1991): Spatial distribution and dispersal of interstitial Crustacea in alluvial sediments of a backwater of the Danube at Vienna. Stygologia,6,97-110.

[77] M.-J. Dole-Olivier and P. Marmonier (1992): Patch distribution of interstitial communities: prevailing factors. Freshwater Biology,27,177-191.

[78] R. Rouch (1991): Structure du peuplement des Harpacticides dans le milieu hyporheique d'un ruisseau des Pyrenees. Annales de Limnologie,27,227-241.

[79] M. Creuzé des Châtelliers, P. Marmonier, M.-J. Dole-Olivier, and E. Castella (1992): Structure of interstitial assemblages in a regulated channel of the river Rhine (France). Regulated Rivers: Research and Management,7,23-30.

[80] M. Creuzé des Châtelliers and J.-L. Reygrobellet (1990): Interactions between geomorphological processes, benthic and hyporheic communities: first results on a by-passed canal of the french Upper Rhone river. Regulated Rivers: Research and Management,5,139-158.

[81] D.L. Danielopol (1976): The distribution of the fauna in the interstitial habitats of riverine sediments of the Danube and Piesting (Austria). International Journal of Speleology,8,23-51.

[82] D.L. Danielopol (1989): Groundwater fauna associated with riverine aquifers. Journal of North American Benthology Society,8,18-35.

[83] C.C. Hakenkamp, M.A. Palmer, and B.R. James (1994): Metazoans from a sandy aquifer: dynamics across a physically and chemically heterogeneous groundwater system. Hydrobiologia,287,195-206.

[84] S. Husmann (1975): The boreoalpine distribution of groundwater organisms in Europe. Verhandlungen der Internationalen Vereinigung Limnologie,19,2983-2988.

[85] F. Malard, J.-L. Reygrobellet, J. Mathieu, and M. Lafont (1994): The use of invertebrate communities to describe groundwater flow and contaminant transport in a fractured rock aquifer. Archiv für Hydrobiologie,131,93-110.

[86] J. Mathieu and C. Amoros (1982): Structure et fonctionnement des écosystèmes du Haut-Rhône francais. XX. Evolution des populations de Copépodes Cyclopoides de deux stationes phréatiques. Polskie Archiwum Hydrobiologii,29,425-438.

[87] K. Matsumoto (1976): An introduction to the Japanese groundwater animals with reference to their ecology and hygienic significance. International Journal of Speleology,8,141-155.

[88] M. Mestrov, R. Lattinger-Penko, and V. Tavcar (1976): La dynamique de population de l'Isopode *Proasellus slavus* ssp.n. et les larves de chironomides dans l'hyporheique de la Drave du point de vue de la pollution. International Journal of Speleology ,8,157-166.

[89] J. Notenboom, W. Hendrix, and A.-J. Folkerts (1996): Meiofauna assemblages discharged by springs from a phreatic aquifer system in the Netherlands. Netherlands Journal of Aquatic Ecology,30,1-13.

[90] R.W. Pennak and J.V. Ward (1986): Interstitial faunal communities of the hyporheic and adjacent groundwater biotopes of a Colorado mountain stream. Archiv für Hydrobiologie,3, Suppl.

74,356-396.

[91] D. Ronneberger (1975): Zur Kenntnis der Grundwasserfauna des Saale-Einzuggebietes (Thüringen). Limnologica,9,323-419.

[92] R. Rouch (1980): Les Harpacticides, indicateurs naturels de l'aquifère karstique. Mémoires de la Societé Géologique de France,11,109-116.

[93] A. Seyed-Reihani, J. Gibert, and R. Ginet (1982): Structure et fonctionnement des ecosystemes du Haut-Rhone Francais. XXIII. Ecologie de deux stations interstitielles; influence de la pluviosite sur leur peuplement. Polskie Archiwum Hydrobiologii,29,501-511.

[94] V. Vanek (1982): Fauna of groundwaters of Bohemian karst (Barrandium). Methodology and preliminary results. Polskie Archiwum Hydrobiologii,29,415-424.

[95] J.V. Ward and M.A. Palmer (1994): Distribution patterns of interstitial freshwater meiofauna over a range spatial scales, with emphasis on alluvial aquifer systems. Hydrobiologia,287,147-156.

[96] J.V. Ward, J.A. Stanford, and N.J. Voelz (1994): Spatial distribution patterns of crustacea in the floodplain aquifer of an alluvial river. Hydrobiologia,287,11-17.

[97] R. Wegelin (1966): Beitrag zur Kenntnis der Grundwasserfauna des Saale-Elbe-Einzugsgebietes. Zoologisches Jahrbuch Systematic,93,1-117.

[98] F. Mösslacher (1997): Ein Vorschlag für die zusätzliche Verwendung von Crustaceen zur Qualitätskontrolle von Gundwässern. 32. IAD Tagung, M. Dokulil (ed.), Wien,Band 1,449-452.

[99] T.L. Poulson (1991): Assessing groundwater quality in caves using indices of biological integrity, 495-511. In: U.S. EPA (1991): *Proceedings of the Third Conference on Hydrogeology, Ecology, Monitoring, and Management of Ground Water in Karst Terranes*, Nashville, Tenessee.

[100] T.L. Poulson (1992): Case studies of groundwater biomonitoring on the Mammoth Cave region, 331-340. In: J.A. Stanford and J.J. Simons (1992): *Proceeding of the First Conference on Ground Water Ecology*, American Water Resource Association, Bethesda, MD.

[101] B. Sket (1973): Gegenseitige Beeinflussung der Wasserpolution und des Höhlenmilieus. 5th International Congress on Speleology,253-262.

[102] B. Sket (1993): Cave fauna and speleobiology in Slovenia. Nase jame / Our Caves,35(1),35-45.

[103] D.C. Culver, W.K. Jones, and J.R. Holsinger (1992): Biological and hydrological investigation of the cedars, Lee County, Virginia, and ecologically significant and threatened karst area, 281-290. In: J.A. Stanford and J.J. Simons (1992): *Proceedings of the First International Conference on Ground Water Ecology*, American Water Resource Association, Betheseda, MD.

[104] J.R. Holsinger (1966): A preliminary study on the effects of organic pollution of Banners Corner Cave, Virginia. International Journal of Speleology,2,75-89.

[105] T.M. Iliffe and T.D. Jickells (1984): Organic Pollution of an Inland Marine Cave from Bermuda. Marine Environmental Research,12,173-189.

[106] F. Malard, J.-L. Reygrobellet, J. Gibert, R. Chapuis, C. Drogue, T. Winiarsky, and Y. Bouvet (1994): Sensitivity of underground karst ecosystems to human perturbation - Conceptual and methodological framework applied to the experimental site of Terrieu (Herault - France). Verhandlungen der Internationalen Vereiningung für Limnologie,25,1414-1419.

[107] F. Malard, J. Mathieu, J.-L. Reygrobellet, and M. Lafont (1996): Biomonitoring groundwater contamination: Application to a karst area in Southern France. Aquatic Sciences,58,158-187.

[108] F. Malard, S. Plénet, and J. Gibert (1996): The use of invertebrates in ground water monitoring: a rising research field. Groundwater Monitoring and Remediation,Spring 1996,103-113.

[109] M. Rejic (1973): Biological pollution indicators in underground waters. Bioloski Vestnik Ljubljana,21(1),11-15.

[110] K.S. Simon and A.L. Buikema (1997): Effects of organic pollution on an Appalachian cave: Changes in macroinvertebrate populations and food supplies. The American Midland Naturalist,

138,387-401.

[111] F. Malard, J.-L. Reygrobellet, and T. Winiarsky (1997): Physico-chemical and biological dynamics of a sewage- polluted limestone aquifer. Interntionale Revue der gesamten Hydrobiologie,82,507-523.

[112] V. Kulhavy (1982): Einfluß der Desinfektion auf die chemischen und biologischen Eigenschaften von Brunnenwasser. Polskie Archiwum Hydrobiologii,29,519-526.

[113] M. Boulal, A. Touyer, M. Messouli, M. Yacoubi-Khebiza, and Y. Ait Ichou (1997): Impact de la pollution sur la faune aquatique des puits de la region d'agadir. XIII Int. Symposium of Biospeleology,Abstracts,27.

[114] D.L. Danielopol (1983): Der Einfluss organischer Verschmutzung auf das Grundwasser-Ökosystem der Donau im Raum Wien und Niederösterreich. In: Bunderministerium für Gesundheit und Umweltschutz (1983): *Beiträge Gewässerökologie 5*, Austria, Vienna.

[115] J. Notenboom, R. Serrano, I. Morell, and F. Hernandez (1995): The phreatic aquifer of the "Plana de castellon" (Spain): relationships between animal assemblages and groundwater pollution. Hydrobiologia,297,241-249.

[116] A. Petrova (1982): Pollution organique et autopurification des eaux alluviales et proluviales. Polskie Archiwum Hydrobiologii,29,513-517.

[117] L.W. Sinton (1984): The macroinvertebrates in a sewage-polluted aquifer. Hydrobiologia,119,161-169.

[118] M. Yacoubi-Khebiza, F. de Bovee, N. Coineau, and C. Boutin (1997): Mesoscale Ecology of aquatic interstitial communities in sediments of the Marrakesh High Atlas valleys. XIII Int. Symposium of Biospeleology,Abstracts,70.

[119] J. Marxsen (1988): Investigations into the number of respiring bacteria in groundwater from sandy and gravelly deposits. Microbial Ecology,16,65-72.

[120] C. Zhang, R.M. Lehman, S.M. Pfiffner, S.P. Scarborough, A.V. Palumbo, T.J. Phelps, J.J. Beauchamp, and F.S. Colwell (1997): Spatial and temporal variations of microbial proberties at different scales in shallow subsurface sediments. Applied Biochemistry and Biotechnology,63-65,797-807.

[121] R.W. Harvey and L. George (1987): Growth determination for unattached bacteria in a contaminated aquifer. Applied and Environmental Microbiology,53,2992-2996.

[122] G. Novarino, A. Warren, N.E. Kinner, and R.W. Harvey (1994): Protists from a sewage-contaminated aquifer on Cape Cod, Massachusetts, USA, 143-153. In: J.A. Stanford and H.M. Valett (1994): *Proceedings of the Second International Conference on Ground Water Ecology*, American Water Resource Association, Atlanta.

[123] T. Okubo and K. Matsumoto (1983): Biological clogging of sand and changes of organic constituents during atrtificial recharge. Water Research,17,813-821.

[124] J.L. Sinclair (1990): Eukaryotic microorganisms in subsurface environments, 39-51. In: C.B. Fliermans and Hazen (1990): *Microbiology of deep subsurface*, WSRC, Information Service, South Carolina.

[125] J. Notenboom, A.J. Flokerts, D. De Zwart, and A. Sterkenburg (1997): The shallow groundwater ecosystem below four diary farm on sandy soil in relation to soil use, RIVM Report 67276002, Bilthoven, The Netherlands.

[126] A.Q. Armstrong, R.E. Hodson, H.M. Hwang, and D.L. Lewis (1991): Environmental factors affecting toluene degradation in groundwater at a hazardous waste site. Environmental Toxicology and Chemistry,10,147-158.

[127] E. Arvin, Jensen, Aamand, and Jorgensen (1988): The potential of free living ground water bacteria to degrade aromatic hydrocarbons and heterocyclic compounds. Water Sciences and Technology,20,109-118.

[128] P.M. Bradley, F.H. Chapelle, and J.E. Landmeyer (1994): Microbial transformations of TNT

in contaminated soils and aquifer materials at Weldon Spring, Missouri, 199-207. In: J.A. Stanford and H.M. Valett (1994): *Proceedings of the Second International Conference on Ground Water Ecology*, American Water Resource Association, Atlanta.

[129] W.C. Ghiorse and J.T. Wilson (1988): Microbial ecology of the terrestrial subsurface. Advanced Applied Microbiology,33,107-172.

[130] H.-M. Hwang, J.A. Loya, D.L. Perry, and R. Scholze (1994): Interactions between subsurface microbial assemblages and mixed organic and inorganic contaminant systems. Bulletin Environmental Contamination and Toxicology,53,771-778.

[131] J.L. Sinclair, D.H. Kampbell, M.L. Cook, and J.T. Wilson (1993): Protozoa in subsurface sediments from sites contaminated with aviation gasoline or jet fuel. Applied and Environmental Microbiology,59,467-472.

[132] M. Mestrov and R. Lattinger-Penko (1981): Investigation of the mutual influence between a polltued river and its hyporheic. International Journal of Speleology,11,159-171.

[133] J. Gibert, P. Marmonier, V. Vanek, and S. Plénet (1995): Hydrological exchange and sediment characteristics in a riverbank: relationship between heavy metals and invertebrate community structure. Canadian Journal of Fisheries and Aquatic Sciences,52,2084-2097.

[134] S. Plénet and J. Gibert (1994): Invertebrate community response to physical and chemical factors at the river/aquifer interaction zone I. Upstream from the city of Lyon. Archiv für Hydrobiologie,132,165-189.

[135] S. Plénet, H. Hugueny, and J. Gibert (1996): Invertebrate community responses to physical and chemical factors at the river/aquifer interaction zone II. Downstream from the city of Lyon. Archiv für Hydrobiologie,136,65-88.

[136] C.M. Schmidt, P. Marmonier, S. Plénet, M. Creuzé des Châtelliers, and J. Gibert (1991): Bank filtration and interstitial communities. Example of the Rhone River in a polluted sector (downstream of Lyon, Grand Gravier, France). Hydrogeologie,3,217-223.

[137] D.J. Reasoner (1983): Microbiology of potable water and groundwater. Journal of Water Pollution Control Federation,55,891-895.

[138] M. Rutgers, I. Van't Verlaat, B. Wind, L. Posthuma, and A.M. Breure (1998): Rapid method to assess pollution-induced community tolerance in contaminated soil. Environmental Toxicology and Chemistry,in print.

[139] P. Doelman, E. Jansen, M. Michels, and M. Van Til (1994): Effects of heavy metals in soil on microbial diversity and activity as shown by the sensitivity-resistance index, an ecologically relevant parameter. Biology and Fertility of Soils,17,177-184.

[140] C. Burkhardt, H. Insam, T.C. Hutchinson, and H.H. Reber (1993): Impact of heavy metals on the degradative capabilities of soil bacterial communities. Biology and Fertility of Soils,16,154-156.

[141] R.M. Lehman, F.S. Colwell, and J.L. Garland (1997): Physiological profiling of indigenous aquatic microbial communities to determine toxic effects of metals. Environmental Toxicology and Chemistry,16,2232-2241.

[142] J. Cairns Jr., P.V. McCormick, and B.R. Niederlehner (1992): Estimating ecotoxicological risk and impact using indigenous aquatic microbial communities. Hydrobiologia,237,131-145.

[143] P. van Beelen and P. Doelman (1997): Significance and application of microbial toxicity tests in assessing ecotoxicological risks of contaminants in soil and sediment. Chemosphere,34,455-499.

Environmental Science Forum Vol.96 (1999) pp. 141-160
© 1999 Trans Tech Publications, Switzerland

Protozoa in Polluted Water Biomonitoring

J.R. Pratt and N.J. Bowers

Environmental Sciences and Resources, Portland State University,
P.O. Box 751, Portland, Oregon 97207, USA

Keywords: Protozoa, Ciliates, Amoebae, Saprobian System, Environmental Impacts

Abstract

Protozoa respond rapidly to changing environmental conditions, making them useful in the assessment of ongoing pollution of waters. Since many protozoa are cosmopolitan, understanding their indicator value may have broad applicability to many ecosystems. Three approaches to the use of protozoa in biological monitoring dominate. The first, and oldest, is the use of the indicator species approach based on the tolerance of select species to polluted waters, typically waters polluted by oxygen demanding wastes. The indicator species approach can provide appropriate evidence of traditional, sewage-based water pollution, but indicators of organic pollution may not indicate pollution by toxic materials or physical alterations such as increased temperature. Further, the indicator approach relies on species-level taxonomic precision. Other approaches are less taxonomically precise because they depend on changing patterns of community composition and may not require the same level of taxonomic identification. Such approaches focus on detecting changes in community composition or, more commonly, reductions in the total number of extant taxa as a measure of stress. Similarly, nontaxonomic community measures of biomass or metabolic activity can also be useful in assessing pollution effects. Finally, the use of protozoa as rapid toxicity indicators is also appropriate for evaluating possible water pollution effects. Rapid tests are suitably sensitive and focus on population growth and other responses of particular test species to evaluate pollution effects. Overall, protozoa can quickly provide valuable information on the status of a water body suspected of being polluted. Further research is needed to more fully account for species differences among aquatic habitats and to fully assess tolerance of particular indicators. The time over which changes occur and can be made detectable is of importance. Since protozoa are sensitive to water pollution, they can also be used to estimate recovery from water pollution and other human-influenced changes in streams and rivers.

Environmental indicators have a long, rich, and unfulfilling history. Protozoa and algae were among the first organisms used in efforts to quantify and qualify the effects of environmental change nearly a century ago [25, 26] and even the newest indicators of environmental condition are deeply tied to this history of indicator development and use (see [11]). In some ways, it is not clear whether we lack appropriate indicators or lack faith in older indicators.

The best indicators would be sensitive to local impacts, predictive, and diagnostic without providing false positives or false negatives. Such indicators do not exist, but the enormous number of developed indicators shows that efforts to identify particular environmental changes correlating with human influences on ecosystems have been of recurring interest to environmental biologists, environmental scientists, and policy analysts. Not only do we wish to monitor environmental change, we need to know how policies for managing the human environment work, if at all. To be of value, indicators need to be able to indicate the status of and processes within a water body. Additionally, they should allow comparisons over space and time, they should relate to water uses understood by the public (the so-called "fishable, swimmable, drinkable" test), and they should be suited to data that can actually be obtained with sufficient regularity to make monitoring feasible.

Indicators of the condition of fresh water are based on the close relationship between the chemical composition of waters and the biological communities [11, 20, 32]. Additionally, certain ecosystem changes seem to produce predictable changes in aquatic ecosystems. In this chapter, we examine the use of protozoan indicators in biological monitoring. Protozoa are a diverse and ubiquitous group of single-celled eukaryotes that are an essential component of most aquatic and many terrestrial communities. For example, fresh water protozoan communities are frequently complex and may consist of as many as 1000 species representing several functional groups (e.g., bactivores, producers, algivores, saprobes, omnivores). Their global distribution makes them ideally suited for biomonitoring, since similar communities may be found worldwide. Protozoa may be classified based on their primary mode of movement (e.g., flagellates, ciliates, or amoebae), although this method of classification is artificial due to the tremendous diversity and polyphyletic nature of these groups. The use of protozoa as bioindicators has focused on ciliate taxa, yet all three groups have proven useful as indicators of environmental stress.

There are three basic approaches to the use of protozoa in biological monitoring. First, a great deal of monitoring has focused on identifying the effects of sewage pollution; most effectively evaluated by saprobian methods. Second, effects of sewage and other pollutants such as toxic materials need to be evaluated by community analysis. Third, the special case of toxicity prediction can also be effectively estimated by protozoan indicators. We conclude with some comments about the importance of sampling strategy, taxonomy, and ecological considerations in the use of protozoa in biomonitoring.

Protozoa as pollution indicators

Most protozoan indicators focus on the status of the biological populations or communities in waters receiving a variety of stressors, and the value of using these rapidly responding organisms has been repeatedly stated [3, 6, 15, 47, 48]. Indicators evaluate stressors and their consequences differently depending on the ecosystem context and presumed use: waters in industrial use areas are considered to be a less significant problem than impaired waters in pristine or high-protection areas. Some indicators focus on the assessment of toxicity, while others attempt to link the structure of biological communities to important ecosystem processes.

Indicators based on the presence, absence, or relative abundance of particular species have a long history in applied biology [11]. Regional patterns of water quality influence the species of aquatic life present in a particular ecosystem [24]. Further, aquatic species are influenced by stressors (pollutants) in water and influence these pollutants by absorption, transformation, and degradation. Finally, some types of nuisance organisms (e.g., toxic algae) in polluted waters adversely influence a variety of water uses. In general, conditions which support expected native populations also support other uses of water, while the presence of nuisance organisms (e.g., fecal bacteria, cyanobacteria, aquatic weeds) may prevent other beneficial uses of water. The nature of the aquatic biota is intimately linked to supporting energy sources (autochthonous algal production vs. allochthonous detritus), and the activity of organisms has also been suggested as an important indicator of water quality. Aquatic life indicators have ranged from classic "indicator species" approaches to more complex dominance and diversity indices, community measures of biomass and activity, and specific indicators of the effects of toxic materials.

Protozoa of many kinds are useful indicators. Typically, ciliates were used since they are active under a broad range of conditions and are sufficiently diverse ecologically and systematically to determine environmental patterns. Amoebae with tests and loricae may leave traces (microfossils) of varying age that may be useful in environmental analysis. For example, paleolimnological studies reveal the indicator value of traces (plates, whole tests) of abundant testate amoebae, that can indicate temporally significant changes in peatlands [14]. Examples of the use of this group of species are more common for terrestrial and transition aquatic ecosystems [1].

Saprobian system

Among the oldest, most typical, and important forms of water pollution is the disposal of sewage in surface waters. Sewage and similar oxygen demanding wastes (e.g., food processing wastes, agricultural wastes) alter the supply of nutrients and organic carbon in natural waters, often resulting in oxygen depletion and nuisance growths of bacteria and algae. Protozoa respond rapidly to their bacterial and algal prey [39] which respond to changes in the nutrient content of waters. These responses have been considered predictable, and the changes in the presence and abundance of particular protozoa (and other taxa) have been used as an indicator of pollution. Although attempts have been made to develop lists of indicator taxa for other forms of pollution (toxic pollutants, radionuclides), these have been less successfully and widely applied.

In fact, indices based on saprobity as a measure of oxygen and organic matter content have been most widely applied in Europe [25, 36, 50]. Because of the large amount of taxonomic work and the compartive ease of sampling and handling, the most effective use of Protozoa as indicators has focused on the ciliates [15] with considerably less effective use of amoebae or flagellates. Saprobian indices have limitations, since lists of taxa with particular indicator value are weighted toward indicating moderate pollution: more taxa are classified as indicating beta-mesosaprobic conditions than any other condition, with alpha-mesosaprobic indicators next most common (Fig. 1). Extreme conditions are less well-represented in the lists. That is, fewer taxa are considered to be indicative of highly polluted situations and very few taxa indicate oligosaprobic situations (yet these are the most highly valued waters and are very diverse in the biotas). Further, natural waters vary in oxygen and organic matter content, so the boundary between polluted and non-polluted conditions is relative along a river course. The need for expert taxonomic skills has been demonstrated repeatedly by Foissner and his colleagues, who have recently attempted to provide accessible keys to indicator ciliates [16].

Despite the limitations of indicator lists (and concerns about sampling and taxonomy explained below), protozoan indicators can be quite effective in identifying degraded water conditions and recovery from those conditions. For example, Madoni and Ghetti [30] studied the Torrente Stirone (Northern Italy), a tributary to the Taro River. By sampling ciliates at 14 stations along the stream (and in important tributaries), they showed the adverse biological effects of waste water loaded to the T. Stirone by the Torrente Ghiara (Fig. 2). Their work clearly shows the changing patterns of ciliates of differing saprobian classification [50] in response to waste loads along the stream. Their work also identified

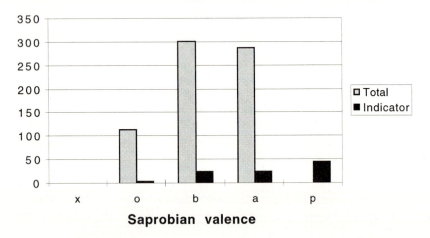

Fig. 1. Distribution of saprobian valences among the 387 indicator species of ciliates (based on [15]). Key: x - xenosaprobity, o - oligosaprobity, b - beta-mesosaprobity, a - alpha-mesosaprobity, p - polysaprobity. Species with mixed classifications appear more than once. Indicators species are those with a valence of 10 in a given category.

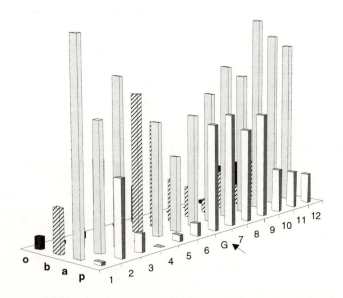

Fig. 2. Analysis of the Torrente Stirone (Italy) using ciliate protozoa. Shown are the distribution of ciliates based on saprobian classification at 12 main stream sampling stations and at a sampling station in the Torrente Ghiara (G), a tributary which carries a significant waste load. The distribution of polysaprobic forms clearly shows the impact of the Torrente Ghiara waste load and downstream recovery. (Redrawn from Madoni and Ghetti [30].)

limitations of such classification systems: not all of the 114 species sampled were classified in the saprobian system, and the saprobic classification of stations based on ciliates differed significantly from saprobity based on the index of Pantle and Buck [36]. In general, indices based on ciliate communities classified stations during different seasons as either beta- (6 of 27 occurrences) or alpha-mesosaprobic (21 of 27 occurrences, a poly saprobic condition was detected only once), while chemical values suggest a distribution of beta-mesosaprobity (4/27), alpha-mesosaprobity (6/27), and polysaprobity (9/27) with mixed classifications accounting for the remainder (8/27).

Such comparisons serve to demonstrate both the usefulness and complexity of applying the saprobian system to analysis of ciliate communities. Rigorous analysis of the environmental requirements and indicator value of ciliate indicators is still needed. Foissner [15] identified a large number of taxa requiring additional analysis. The importance of accurate taxonomy is reflected in the differential distribution of saprobian valences among congeneric taxa (Table 1). Examination of Table 1 reveals many similarities within genera but some obvious differences related to the autecology of the listed species. The distribution of tolerances is variable among species, so accurate taxonomy is essential, and more essential for some genera than others. For example, the listed *Cyclidium* species are distributed across a narrower tolerance range than the other genera shown.

A more extreme example is shown in Table 2, which shows the distribution of tolerances among species of the peritrich ciliate genus *Vorticella*. Peritrich ciliates of the genus *Vorticella* are particularly interesting to examine for pollution tolerance, since they are commonly be found in sewage treatment facilities. Survey data can help show the difficulties in understanding the importance of historical indicator work. For example, Madoni and Ghetti [31] surveyed 84 sewage treatment plants using activated sludge or rotating biological contactors. They found *V. microstoma* (a polysaprobic species, SI=4.0) at only 10% of the activated sludge plants and at 14% of the plants using rotating biological contactors. By contrast, they found *V. convallaria* (an alpha-mesosaprobic species, SI=2.9) at 84% of the activated sludge plants and 57% of the plants using rotating biological contactors.

In studies of predominantly toxic river pollution, Jiang and colleagues [22] estimated "species pollution values (SPV)" for a particular species by averaging a chemical pollution index for any station at which the species occurred. The SPV index has a range (0-5) similar to that of the saprobian index (1-4). Comparison of these indices for difference species shows both concordance and disagreement: *V. microstoma* SI=4.0, SPV=4.1 and *V. picta* SI=1.1, SPV=1.29 but *V. aequilata* SI=4.0, SPV=1.01 and *V. cupifera* SI=2.7, SPV=1.29. These inconsistencies do not invalidate either approach but does show that there is significant divergence in the indicator value of a given species responding to oxygen-demanding wastes and other pollutants. The indicator value of species is best defined in a particular context (i.e., sewage pollution, industrial pollution, etc.).

Community metrics and measures

Saprobian indicators may suffice for monitoring organic pollution (which is very common), but there is not an effective indicator species approach for assessing nutrient pollution in the absence of oxygen-demanding waste, thermal pollution, or toxic pollution. Additionally, for higher organisms, habitat modification may be a significant stress on natural populations. An alternative to the use of indicator species is an assessment based on derived characteristics of the community, which rely on a less thorough knowledge of the autecology of individual species. The definition of a community by ecologists varies from interacting populations in a habitat to definitions based on the species of a certain

Table 1. Distribution of saprobic valences among selected ciliate genera. Data from Foissner [15]. Key: s - consensus classification, x - xenosaprobity, o - oligosaprobity, b - beta-mesosaprobity, a - alpha-mesosaprobity, p - polysaprobity.

Species	s	x	o	b	a	p
Amphileptus						
carchesii	a			1	8	1
claparedii	a			2	8	
meleagris	a				10	
pleurosigma	b-a			5	5	
punctuatus	a			1	9	
rotundus	a			1	8	1
trachellioides	o		7	3		
Genus average		0	1	1.86	6.86	0.29
Cyclidium						
citrullus	a			1	8	1
elongatum	b-a			5	5	
glaucoma	a				9	1
heptatrichium	b			8	2	
oblongum	a-b			4	6	
singulare	a				10	
versatile	a-b		2	3	5	
sp.	a-b			4	6	
Genus average		0	0.25	3.12	6.37	0.25
Litonotus						
anguilla	b-a			5	5	
carinatus	b-a			5	5	
crystallinus	b-a			5	5	
cygnus	b			10		
fasciola	a			1	8	1
fusidens	b-p			3	4	3
hirundo	a			1	8	1
lamella	a			2	8	
proceus	o-b		5	5		
varsavensius	b-a			5	5	
varsavensius						
var. *polysaprobica*	p-I				1	9
sp.	a			1	7	2
Genus average		0	0.42	3.58	4.67	1.33

Table 2. Distribution of saprobic valences among ciliate species of the genus *Vorticella*. Data from Foissner [15]. Also shown are the summary saprobian index (SI) and a species pollution value (SPV, based on the work of Jiang and colleagues [20]. The SI is scaled 0-4 while the SPV is scaled 0-5. Key: s - consensus classification, x - xenosaprobity, o - oligosaprobity, b - beta-mesosaprobity, a - alpha-mesosaprobity, p - polysaprobity.

Species	s	x	o	b	a	p	SI	SPV
Vorticella								
aequillata	p					10	4.0	1.01
alba	a-p				5	5	3.5	
campanula	b	1		6	3		2.2	1.63
campanulata	b-a			5	5		2.5	
citrina	b			8	2		2.2	
communis	b			10			2.0	
convallaria	a			1	9		2.9	2.11
cupifera	b-a			5	3	2	2.7	1.29
elongata	b			10			2.0	
fromenteli	a			2	8		2.8	
hamatella	b-a			4	6		2.6	
hians	p-i					10	4.0	
marginata	b		2	8			1.8	
mayeri	b			10			2.0	
microstoma	p-i					10	4.0	4.10
m. f. elongata	p					10	4.0	
m. f. monilata	p					10	4.0	
m. f. turgescens	p-I					10	4.0	
natans	b		3	7			1.7	
nebulifera	o-b		6	4			1.4	
nutans	b-p			10			2.0	
octava	a-b		1	3	4	2	2.7	1.01
picta	o		9	1			1.1	1.29
p. f. longa	a			2	8		2.8	
similis	o-b		6	4			1.4	2.08
telescopica	a-p				5	5	3.5	
telescopioides	a-p				5	5	3.5	
vernalis	b-p			10			2.0	
vestita	a			2	8		2.8	
sp.	b-p			3	3	4	3.1	
Genus average		0	0.93	3.83	2.47	2.77		
Index range							1.1 to 4.0 5.0	1.01 to 4.10

taxonomic group in a particular habitat (e.g., plankton, periphyton, benthos). This latter definition will be familiar to stream ecologists, limnologists, and pollution biologists. Examples will be found elsewhere in this volume.

Communities are often defined by convenience, familiarity, sampling strategy, and tradition (e.g., periphyton algae, benthic macroinvertebrates) rather than linkages or strong interactions among the species (see also [43]); some of the species may interact strongly while many probably do not. Communities defined at least partly by taxonomic affinity are defined by the investigator and may not form an ecologically meaningful unit. These limitations should be acknowledged openly; they are not necessarily weaknesses of the approach. Communities include large arrays of species, and, typically, with an adequate sampling of the available species, a range of sensitivities will be found. Individual species react to environmental gradients, including those caused by pollution, and there is a great deal of overlap in sensitivities among taxa [38]. Clearly, analyses which examine large portions of many biological communities in an ecosystem will be the most valuable in assessing pollution, but such approaches are excessively costly and possibly unnecessary to detect and monitor environmental effects.

Unlike the indicator species approach, community approaches usually focus on the numbers of extant taxa in a sample and the taxonomic composition of the collected species (see [37]). This approach is largely based on dominance and diversity changes associated with environmental stressors (Fig. 3). The effects of toxic pollutants often eliminate particular sensitive taxa resulting in more simplified communities [35]. Such approaches have been used in the laboratory and the field to predict and evaluate pollutant effects on protozoan communities. Stressors typically reduce species richness and alter the composition of communities, such as the effects of a chlorinated wastewater effluent on protozoa colonizing artificial substrata (Fig. 4). These simple analyses show the effectiveness of using species richness as a surrogate for more complicated measures of diversity. In addition to reductions in species richness, subtle changes in the composition of communities may occur under stress. One approach is to examine the similarity among communities exposed to a stressor [27, 46].

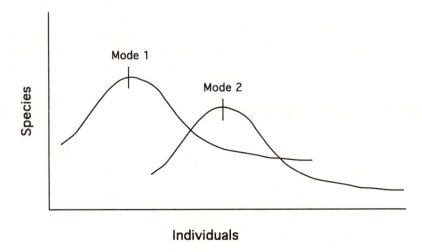

Fig. 3. Hypothetical shift in dominance and diversity of communities under stress (based on [8]).

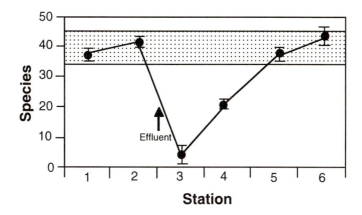

Fig. 4. Effect of wastewater assessed as a change in protozoan species richness upstream (stations 1 and 2) and downstream (stations 3-6) of a chlorinated municipal sewage treatment discharge. The chlorinated effluent entered the stream (Spring Creek, Centre County, Pennsylvania, USA) between stations 2 and 3. Error bars are standard deviations (n=3). The shaded area identifies the 95% confidence interval for the upstream reference stations.

Additionally, the composition of protozoan communities as distinguished by functional groups [41] based on general feeding strategy can reveal differences in the importance of various energy sources (especially algae and bacteria) in the environment. For example, in laboratory studies of copper toxicity (see Table 3) we observed significant changes in the composition of protozoa in microcosms as copper toxicity affected the algal food supply. In this case, protozoan species feeding on bacteria increased in importance as the toxic effects were realized.

The pattern of community composition can also be used to draw inferences about ecosystem forces structuring communities along biological gradients. There are many procedures available for conducting such analyses, although ecologists, applied scientists, and protozoologists need to be careful in using these approaches since sampling may not be sufficient to validate the assumptions of these methods. Recalling that species are "variables," it is necessary to assure that the number of samples taken approaches the number of variables analysed if multivariate procedures are used to deduce pattern in the data set [18].

A good example of the use of pattern analysis (in this case correspondence analysis) is a study by Madoni [29] which shows the link between biological structure and pollution conditions in the streams of the Northern Appenines (Italy). This study identified seven community types plus a group of species without community affiliation. Community types B-D identify high gradient, clean water systems with community types E-H corresponding to lower gradient systems with added pollution. Communities B-D were classified by saprobity as oligo- to beta-mesosaprobic, while communities E-H ranged from beta- and alpha-mesosaprobic to polysaprobic.

Table 3. Effects of copper on protozoan species richness, functional group composition, chlorophyll biomass, nutrient cycling activity and oxygen balance in laboratory microcosms. Microcosms were established with natural periphyton communities and were dosed continuously with copper-amended dechlorinated dilution water. Data show mean measured copper concentrations the mean (SD) microcosm responses (n=3) sampled after 14-d exposure to copper. Units for alkaline phosphatase activity are nmole p-nitrophenol/mg protein/hr.

Treatment	Species richness	Bacterivore species (%)	Chlorophyll (ug/L)	Alkaline Phosphatase Activity	Oxygen (mg/L)
Control	34.0 (1.7)	75 (5.1)	172 (56.5)	13.5 (3.77)	9.87 (0.11)
9.9 ug/L	31.0 (2.6)	79 (3.0)	141 (22.6)	11.1 (1.80)	10.0 (0.10)
19.9	29.0 (5.6)	77 (9.2)	87.7* (16.9)	17.0 (3.46)	9.63* (0.11)
40.0	19.0* (2.6)	84 (15.2)	46.3* (14.6)	36.0 (10.4)	9.53* (0.15)
90.0	10.0* (1.0)	99* (21.7)	22.7* (2.3)	58.7* (12.6)	9.40* (0.20)
205	10.7* (0.6)	99* (24.6)	13.7* (2.3)	130* (44.7)	9.33* (0.06)

Asterisks (*) denote treatments significantly different from control ($p < 0.05$).

Microbial activity

While a great deal of biologically important information can be determined from assessments focusing on the structure of communities, these analyses give only qualitative information on the functioning of a particular system. That is, the rates of nutrient uptake or cycling, respiration, or productivity cannot be estimated by enumerating species or individuals. Methods focused on assessing community activity are also valuable indicators of adverse effects [40]. In general, such approaches, like approaches based on community analysis, are useful when used as comparative measures of adverse effect and/or recovery. Our knowledge rarely permits us to apply an absolute standard for the value of a measure such as community respiration or nutrient cycling rate. Despite the relative nature of the measures,

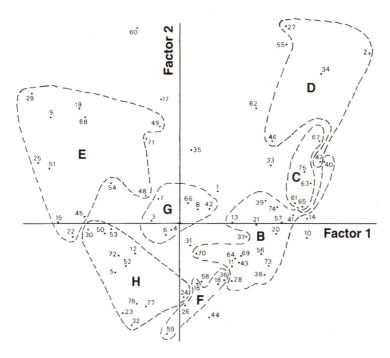

Fig. 5. Community types in Northern Appenine (Italy) streams identified by correspondence analysis. The community types correspond to particular species associations, with communities H and E having some evidence of pollution. Community type F corresponds to lower stream courses with sluggish flow and moderate pollution, and community type G corresponds to common species found at many sites (e.g., *Cinetochilum margaritaceum*). Redrawn from Madoni [29]. Community A consisted of species with a random distribution.

activity estimates can be very valuable in understanding the mechanisms behind adverse effects and the severity of the impacts. Such approaches have been effectively used in the prediction of toxic chemical effects, often in controlled laboratory ecosystems such as mesocosms and microcosms (see [28, 44]).

An example of the use of such measures is given in Table 3 (see above for discussion of functional group data) which presents data from a laboratory microcosm test of the effects of low levels of dissolved copper. Copper is a well-known and common toxicant in industrial wastes and is also used in many parts of the world as an aquatic herbicide. In microcosm tests, copper causes reduction in protozoan species richness, shifts communities toward heterotrophy as photosynthetic forms are eliminated, and compromises the ability of communities to recover essential nutrients as important nutrient cycling organisms are eliminated by direct toxic effects. The increase in the activity of alkaline phosphatase enzyme systems reflects the loss of nutrients from the community and the comparative induction of enzyme systems to recover those nutrients [35, 44].

Toxicity indicators

A final approach in the use of protozoa as bioindicators is the use of protozoa as rapid test organisms for assessing toxic effects. A number of techniques to assess toxicity to protozoa have been developed. They include assessment of lethal toxic effects, assessment of depression of population growth [13, 21, 45], an assessment of overall metabolic activity [33], and assessment of changes in behavior [49], including feeding [23]. Of these, the most effective approaches have been those focused on short-term growth measurements.

Such studies are appropriate for assessing the toxicity of compounds, for assessing wastewaters, for evaluating potential adverse environmental effects, and for evaluating so-called "ambient" or in situ toxicity in water. Increasing interest in rapid biological testing [4] and so-called "battery-of-tests" approaches point directly to the use of protozoa in evaluating toxicity. Protozoa, especially many ciliates, have ideal characteristics for such assessments: (1) they can be cultured easily, often in defined media, (2) they have rapid growth rates, doubling two to three times per day, (3) they can be fed a variety of bacteria as food (or fluorescent beads if feeding is to be assessed), and (4) many produce resting cysts which can be stored, thus eliminating the need for continuous culturing.

Several different examples of model organisms are available. The most commonly tested organisms are *Tetrahymena*, *Colpidium*, and several species of *Colpoda*, although a number of efforts have been addressed at using other species. While the *Colpoda* species used are not truly aquatic (they are soil forms), they can be tested quickly in liquid medium. The most common tests estimate growth rates in controlled experiments and use direct counting (microscopic counts, automated particle counters) to assess growth rate differences between populations exposed to different toxic compounds.

There are several considerations in developing these rapid tests. First, the tested organisms must be obtained from rapidly growing, well-fed cultures and food organisms must be present in the tests. These test organisms must have high growth rates and require appropriate energy sources. Second, the testing conditions must assure the biological availability of the putative toxic compounds. Organic components of growth media often bind both inorganic and organic toxicants and reduce the sensitivity of tests. Most protozoa can be maintained in clean, inorganic media for some period of time if adequate food is provided. For example, we have grown *Colpoda inflata* in Sonneborn's *Paramecium* medium with food bacteria but have subsequently conducted growth tests in a minimal salts medium [17, 45].

As an example, the growth response of ciliates under toxic stress in experiments conducted over a 24-hr period is shown in Fig. 6. In experiments where 100 ciliates/ml were inoculated into test conditions, controls produced 400-800 cells (2-3 doublings) in 24 hr. Further, the monotonic effects of toxic materials on growth demonstrates the sensitivity and appropriateness of protozoa for use in predicting toxicity. It is clear that different strains and species differ in their sensitivity to toxicants. Further, it is feasible to use these kinds of tests in field situations where mixtures of toxic materials may be present. In addition to interspecific differences in sensitivity to toxic compounds, the culture conditions under which testing occurs can have significant effects on the detection of toxicity. As noted above, organic test media can reduce bioavailability and measured toxicity. When we tested ciliates in contrasting media with and without significant amounts of dissolved organic carbon which binds some toxicants, we found significant changes in toxicity (Fig. 7). The presence of nonspecific organic carbon increased the median growth inhibiting concentration from 25 ug/L ($\pm$ 2.5 ug/L) in the minimal salts medium to 75 ug/L ($\pm$ 1 ug/L) in 10% Sonneborn's *Paramecium* medium.

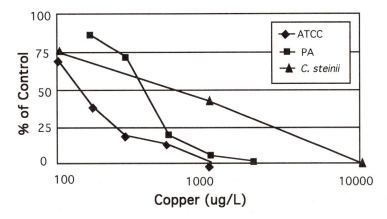

Fig. 6. Effects of copper on different species and strains of *Colpoda inflata* (ATCC and PA) and *C. steinii*. Strains of *C. inflata* are from the American Type Culture Collection (ATCC) and a strain (PA) isolated from garden soil in Boalsburg, Centre County, Pennsylvania, USA. Data for *C. steinii* from [17].

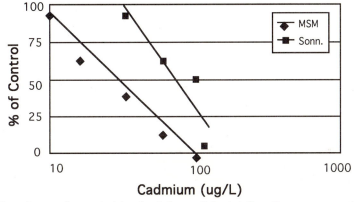

Fig. 7. Effect of test medium on toxicity of cadmium to *Colpoda inflata*. Data are comparative reductions in *C. inflata* growth in the presence of cadmium for tests in minimal salts medium (MSM) and 10% Sonneborn's *Paramecium* medium (approximately 40 mg dissolved organic carbon/L).

One potential drawback with respect to the evaluation of protozoan toxicity tests is the need for microscopic evaluation of test material, either for counting purposes or for the assessment of feeding, morphological changes, or assessment of behaviors. Noevers and Matsos [33] designed a clever macroscopic method for evaluating the effect of heavy metals on ciliates based on bioconvective patterns. Briefly, ciliates exposed to heavy metals tend to concentrate these metals, which in turn alters cellular metabolism. These alterations in metabolic function may be visualized macroscopically by the alteration in the formation of macrocopic, reticulate patterns which are generated when a concentrated culture is allowed to settle. The resulting patterns may be digitized and analyzed for the number of nodal points created, the number of polygons, the area and perimeter of the polygons, and cell-side distribution. In addition, the timing of the creation of these patterns may also be used as a sensitive indicator of toxic stress.

Issues in the use of protozoan indicators

<u>Sampling issues</u>

Efforts to assess environmental conditions such as pollution must pay significant attention to developing a sampling strategy [9, 19]. Of particular importance are the following points.

1. Samples must be taken along the pollution gradient paying special attention to potential sources, including tributaries. Samples should be taken from more than one upstream or reference site where pollution is thought to be minimal. Upstream, reference site variability needs to be characterized.

2. Focus on a particular habitat and expend similar effort in collecting at each site. An alternative to collecting natural substrata is the use of artificial devices such as glass slides, polyurethane foam, fouling plates, and similar devices [7, 48]. Advantages of such devices are knowledge of their history, the sampling of similar (artificial and replicable) habitat at each site, the ability to control placement and the timing of collections. Collection time is often minimal. Disadvantages include the requirement of two trips (placement and retrieval), possible sampler loss or damage, and the selectivity of some devices for particular taxonomic groups [2].

<u>Taxonomic issues</u>

No more eloquent argument for the need for accurate taxonomy can be found than the work of Foissner and colleagues (e.g., [15]). For indicator species approaches, taxonomy must be precise, since similar and related species often show important differences in tolerance. For approaches based on community measures (composition, similarity, functional groups) taxonomy can be less precise (i.e., genus-level), but then all taxa should be identified to the same taxonomic level. An alternative is the use of the so-called "lowest practical taxonomic unit" where taxonomic consistency is the focus (see [8]). Using this approach, differences in resolution may vary among groups based on the available taxonomic references. Here, investigator experience is most important, as this approach is usually used when time does not permit complete analysis of the community.

While conventional taxonomic identification of protozoa is based on observation of both living and stained material, pollution studies do not permit the same level of taxonomic precision that would be expected in systematic or autecological studies. Protozoologists disagree on the use of living material for identification, although several recent references focus on live material [16]. Only a few taxonomists have recognized this problem and developed keys such as the "ciliate atlas" which aids in identifying over 200 indicator species.

Taxonomic identification of protozoa in pollution studies is confounded by several problems. No universal fixative is available to kill and preserve samples. Since the organisms are predominantly naked cells, all fixatives are selective. Some are better than others, but some cells will not survive a particular fixative. Use of multiple fixatives on subsamples might help (but creates more work). Further, dead cells are difficult to separate from algae, fungal hyphae, and debris in all but the cleanest samples (i.e., plankton) and separation is very problematic in samples from periphyton or benthos. Ciliates and testate amoebae are the best known and most effectively analyzed. The zooflagellates are systematically diverse but more difficult to identify, since modern references are insufficient to identify many of the smallest species without recourse to electron microscopy. These procedures are not very useful to ecologists, and little helpful taxonomic work is occurring.

Living material changes rapidly in composition. Within a few hours of sampling, some populations disappear. Given that routine preservation is not possible, living material is more suitable for routine evaluation, but taxonomic precision is not possible using this method. For ongoing pollution assessment, it may be possible to distribute sampling through time to achieve better use of investigator time, but for acute events or varying pollution in small streams, conditions will vary day to day. Sampling in as short a time interval as possible is preferred. This is likely to require the use of living material, coupled with photodocumentation and a healthy set of taxonomic references to complete. In such cases, well-defined sampling protocols, which may involve the use of artificial substrata, are recommended.

Ecological issues

The principal role for protozoa in ecosystems is feeding on bacteria, small algae, some fungi and other animal-like organisms. Whether protozoa for an important link to higher organisms in the food web is uncertain, but the importance of protozoa in recycling nutrients from bacteria and algae is unquestioned. Protozoa form taxonomically large and diverse communities in most surface waters and soils and litters. Therefore, regardless of their known and yet to be discovered ecological importance, these diverse communities contain many different "sensors" of environmental conditions. The structure of protozoan communities can do much to show the nature and extent of environmental influences.

Many protozoa appear to have a cosmopolitan distribution: the same species are found in comparable habitats all over the world. This means that knowledge about many protozoa is universal, unlike our understanding of the autecology of more limited species. Work in our laboratory has shown that the same species from many different areas have similar toxicological profiles. For example, Xu and colleagues [53] found that *Colpoda inflata* sampled from soils on five continents showed very similar responses to standard toxicants. These same strains were also found to be genetically similar based on comparisons of a variable region of the large subunit of the rRNA gene [5]. It is not known how strains of the same species isolated from contaminated soils, which may show more genetic variation, would respond to exposure to toxicants. Protozoa respond quickly to environmental stress and it is possible that a pollution event can come and go without being detected by long-lasting changes in these communities. On the other hand, protozoa are active all year long, so they can be used as environmental monitors when other species are absent, migrating, or otherwise inactive.

Statistical issues

Although space in this chapter is insufficient to amplify the statistical issues in detecting the effects of pollutants on aquatic ecosystems, applied scientists will realize the complexity of decision making in the face of legal challenges to analyses showing adverse effects of waste water, industrial pollution, habitat modification, nonpoint source runoff and other stressors. Approaches to detecting pollution that rely on calculating a derived index (e.g., saprobian index, pollution index, diversity index) need to be examined for the statistical properties of the chosen measure. A valuable analysis of sampling, data recording, and statistical issues in biomonitoring is given by Norris and Georges [34]. Several factors are worth emphasizing: (1) ANOVA is often inappropriate for comparing biotic indices among sites ("treatments") since there may be little evidence that the chosen index is normally distributed and homoscedastic. ANOVA is rarely an appropriate procedure. (2) The multivariate nature of species by samples data sets argues for the use of multivariate statistics or , at least, measures of biological association (similarity, dissimilarity) of sampled communities. Similarity and dissimilarity (including distance measures) are amenable to bootstrapping and Monte Carlo simulation [46] for detecting differences. These measures must be examined carefully since not all of them have desirable statistical properties (see [19, 46, 52]). (3) Several multivariate techniques can reduce the complexity of biomonitoring data sets and allow evaluation of linkages between physicochemical and biological patterns. However, these techniques should be applied judiciously. Diversity indices of many sorts are very problematic since diversity estimates may be from a heterogeneous data set (some species, some genera, some families); these measures have been called "answers for which there are not questions" [19]. Additionally, modern computing techniques make it possible to utilize the more complex, raw data.

Further, when applied scientists use methods that depend on scores or rankings (see [10, 52]), they may lose power to detect some of the changes they think they have observed. That is, the assignment of "scores" derived from real data artificially limits the variability in data by converting real data to a narrower range. For example, a data range for biotic diversity of 0-50 species might be reduced to a set of scores of 0-5 in an artificial index. Further, assigning scores assumes that the scale is arithmetic (the distance from 0-1 should represent the same degree of variation as the difference from 2-3 or 4-5). Such an assumption is rarely validated (see [19]), and scores often seem to be slightly to very qualitative, despite their quantitative appearance.

Summary and research needs

Protozoa are appropriate and useful indicators of environmental pollution. They have been used effectively for nearly a century in understanding the effects of waste water and industrial pollution on streams and rivers and have been effectively used in understanding the mechanisms of adverse effects and in predicting effects of suspected pollutants. Despite this effectiveness, there are a number of needs in applied studies in general and the use of protozoa, in particular, in biomonitoring.

Greater emphasis on taxonomic research and training is essential in most areas of pollution biology if the use of organisms as environmental monitors is to continue. Developing taxonomic lists with organisms incompletely identified is understandable because of the poor state of taxonomy of many groups. However, ciliate taxonomy is reasonably well-developed, especially for more commonly encountered species. Continuing work in ecologically oriented taxonomy - taxonomy that gives cues to ecologists to identify organisms in the living state - is essential. Further, taxonomic work on naked amoebae and flagellates is urgently needed. Suprafamilial taxonomy has shown the polyphyletic nature of the artificial group "protozoa," so it may be useful to restrict future studies to more taxonomically homogenous groups (e.g., ciliates, testate amoebae, euglenid flagellates).

A fuller understanding of the scope and significance of cosmopolitan distribution of many species is needed. Scientists need to know if morphologically (and therefore taxonomically) similar species are ecologically similar in tolerances, behavior, feeding, etc. Clearly, the ability of applied scientists to share information around the world will be enhanced by this knowledge.

Finally, a better quantitative understanding of the ecological importance of protozoa in ecosystems is needed. While we now understand the significance of the microbial loop in the oceans and oligotrophic continental waters, we still do not know if protozoa are critical to ecosystem function. Clearly, the role of protozoa in the soil, sediments, and various waters is important to protozoologists, but is the role important in the ecosystem? The answer from studies of parasitic protozoa in humans is obvious:

malaria, Chagas' disease, Leishmaniasis are all important. The effects of parasites of insects, birds, plants, provide clear evidence of how protozoa affect life on earth. Is the same true of protozoa in the periphyton, the plankton, or in soils and litters?

Protozoa are ubiquitous in ecosystems, diverse, and sensitive to human influences on ecosystem structure and function. They indicate the kind and severity of pollution and can be quickly and effectively used as tools in understanding the effects of human populations on aquatic systems.

Literature cited

[1] V. Balick (1991): The effect of the road traffic pollution on the communities of testate amoebae (Rhizopoda, Testacea) in Warsaw (Poland). Acta Protozoologica, 30, 5-11.

[2] S.S. Bamforth (1982): The variety of artificial substrates used for microfauna, 115-130. In: J. Cairns, Jr. (1982): *Artificial Substrates*, Ann Arbor Science, Ann Arbor, MI.

[3] H. Bick (1957): Beiträge zur ökologie einiger Ciliaten des Saprobiensystems. Vom Wasser, 24, 224-246.

[4] C. Blaise (1991): Microbiotests in aquatic ecotoxicology: characteristics, utility, and prospects. Environmental Toxicology and Water Quality, 6, 145-155.

[5] N. Bowers and J.R. Pratt (1995): Estimation of genetic variation among soil isolates of *Colpoda inflata* (Stokes) (Protozoa: Ciliophora) using the polymerase chain reaction and restriction fragment length polymorphism analysis. Archiv für Protistenkunde, 145, 29-36.

[6] J. Cairns, Jr. (1978): Zooperiphyton (especially Protozoa) as indicators of water quality. Transactions of the American Microscopical Society, 97, 44-49.

[7] J. Cairns, Jr., ed. (1982): *Artificial Substrates*, Ann Arbor Science, Ann Arbor, MI.

[8] J. Cairns, Jr., D.L. Kuhn and J.L. Plafkin (1979): Protozoan colonization of artificial substrates, 34-57. In: R.L. Weitzel (1979): Methods and Measurements of Periphyton Communities: A Review, American Society for Testing and Materials, Philadelphia.

[9] J. Cairns, Jr. and J.R. Pratt (1986): Developing a sampling strategy, 168-186. In: B.G. Isom (1986): *Rationale for Sampling and Interpretation of Ecological Data in the Assessment of Freshwater Ecosystems*, American Society for Testing and Materials, Philadelphia.

[10] J. Cairns, Jr. and J.R. Pratt (1993): A history of biological monitoring using benthic macroinvertebrates, 10-27. In: D.M. Rosenberg and V.H. Resh (1993*): Freshwater Biomonitoring and Benthic Macroinvertebrates*, Chapman & Hall, New York.

[11] J. Cairns, Jr. and J.R. Pratt (1995): The relationship between ecosystem health and the delivery of ecosystem services, 63-76. In: D.J. Rapport and P. Calow (1995): *Evaluating and Monitoring the Health of Large-scale Ecosystems*, Springer-Verlag, Berlin.

[12] R. Costanza, R. d'Arge, R. de Groot, S. Farber, M. Grasso, B. Hannon, K. Limburg, S. Haeem, R.V. O'neill, J. Paruelo, R.G. Raskin, P. Sutton, and M. van den Bett (1997): The value of the world's ecosystem services and natural capitol. Nature, 387, 253-260.

[13] D. Dive, S. Robert, E. Angrand, C. Bel, H. Bonnemain, L. Brun, Y. Demarque, A. LeDu, R. El Bouhouti, M. Fourmaux, L. Guery, O. Hanssens and M. Murat (1989): A bioassay using the measurement of growth inhibition of a ciliate protozoan *Colpidium campylum*. Hydrobiologia, 188/189, 181-188.

[14] M.S.V. Douglas and J.P. Smol (1988): Siliceous protozoan and chrysophycean microfossils from the recent sediments of *Sphagnum* dominated Lake Colden, N.Y., U.S.A. Verh International Verein Limnologie, 23, 855-859.

[15] W. Foissner (1988): Taxonomic and nomenclatural revision of Sládecek's list of ciliates (Protozoa: Ciliophora) as indicators of water quality. Hydrobiologia, 166, 1-164.

[16] W. Foissner, H. Blatterer and F. Kohman (1991-1994): Taxonomische und Ökologische Revision der Ciliaten des Saprobiensystem, Band I-IV, Informationsberichte der Bayerische Landesamt für Wasserwirtschaft, München.

[17] T.A. Forge, M.L. Berrow and J.F. Darbyshire (1995): Protozoan bioassays of soil amended with sewage sludge and heavy metals using the common soil ciliate *Colpoda steinii*. Biology and Fertility of Soils, 16, 282-286.

[18] H.G. Gauch (1970): *Multivariate Analysis in Community Ecology*, Cambridge University Press, London.

[19] R.H. Green (1979): *Sampling Design and Statistical Methods for Environmental Biologists*, John Wiley, New York.

[20] C.W. Hart, Jr. and S.L.H. Fuller, eds. (1974): *Pollution Ecology of Freshwater Invertebrates*,

[21] M.P.M. Jannsen, C. Oosterhoff, G.J.S.M. Heijmans, and H. Van der Voet (1995): Toxicity of metal salts and the populatin growth of the ciliate *Colpoda cucullus*. Bulletin of Environmental Contamination and Toxicology, 54, 597-605.

[22] J. Jiang, F. Weisong, G. Manru, and S. Yunfen (1995): Establishment of Protozoan Pollution Value in Hangjiang River, Institute of Hydrobiology, Chinese Academy of Sciences, Wuhan.

[23] C.M. Juchelka and T. W. Snell (1995): Rapid toxicity assessment using ingestion rate of cladocerans and ciliates. Archives of Environmental Contamination and Toxicology, 28, 508-

[24] J.R. Karr, L.A. Toth and D.R. Dudley (1985): Fish communities of midwestern rivers: a history of degradation. BioScience, 35, 90-95.

[25] R. Kolkwitz and M. Marsson (1908): Ökologie der pflanzlichen Saprobien. Berichte der Deutschen Botanischen Gesellschaft, 26A, 505-519.

[26] R. Kolkwitz and M. Marsson (1909): Ökologie der tierischen Saprobien. Internationale Revue der Gesamten Hydrobiologie und Hydrographie, 2, 126-152.

[27] W.G. Landis, G.B. Matthews, R.A. Matthews, and A. Sergeant (1994): Application of multivariate techniques to endpoint determination, selection and evaluation in ecological risk assessment. Environmental Toxicology and Chemistry, 13, 1917-1927.

[28] W.G. Landis, R.A. Matthews, A.J. Markiewicz, N.J. Shough, and G.B. Matthews (1993): Multivariate analysis of the impact of the turbine fuel Jet-A using a standard aquatic microcosm toxicity test. Environmental Science, 2, 113-130.

[29] P. Madoni (1984): Ecological characterization of different types of watercourses by the multivariate analysis of ciliated protozoa populations. Archiv für Hydrobiologia, 100, 171-188.

[30] P. Madoni and P.F. Ghetti (1981): Ciliated Protozoa and water quality in the Torrente Stirone (Northern Italty). Acta Hydrobiologica, 23, 143-154.

[31] P. Madoni and P.F. Ghetti (1981):The structure of ciliated Protozoa communities in biological sewage-treatment plants. Hydrobiologia, 83, 207-215.

[32] J.L. Metcalfe (1989): Biological water quality assessment of running waters based on macroinvertebrate communities: history and present status in Europe. Environmental Pollution Series A, 60, 101-139.

[33] P.A. Noever and H.E. Matsos (1991): A bioassy for monitoring cadmium based on bioconvective patters. Journal of Environmental Science and Health, 26A, 273-286.

[34] R.H. Norris and A. Georges (1993): Analysis and interpretation of benthic macroinvertebrate surveys, 234-286. In: D.M. Rosenberg and V.H. Resh (1993*): Freshwater Biomonitoring and Benthic Macroinvertebrates*, Chapman & Hall, New York.

[35] E.P. Odum (1985): Trends expected in stressed ecosystems. BioScience, 35, 419-422.

[36] R. Pantle and H. Buck (1955): Die biologische Überwachung der Gewässer und die Darstellung der Ergebnisse. Gas- und Wasserfach, 96, 604-612.

[37] R. Patrick (1949): A proposed biological measure of stream conditions based on a survey of the Conestoga Basin, Lancaster County, Pennsylvania. Proceedings of the Academy of Natural Sciences, Philadelphia, 101, 277-341.

[38] R. Patrick, J. Cairns, Jr. and A. Scheier (1968): The relative sensitivity of diatoms, snails, and fish to twenty common constituents of industrial wastes. Progressive Fish-Culturist, 30, 137-140.

[39] L.E.R. Picken (1937): The structure of some protozoan communities. Journal of Ecology, 25, 324-384.

[40] J.R. Pratt and N.J. Bowers (1992): Variability of community metrics: detecting changes in structure and function. Environmental Toxicology and Chemistry, 11, 451-457.

[41] J.R. Pratt and J. Cairns, Jr. (1985): Functional groups in the Protozoa: roles in differing ecosystems. Journal of Protozoology, 32, 415-423.

[42] J.R. Pratt and J. Cairns, Jr. (1992): Ecological risks associated with the extinction of species, 93-118. In: J.Cairns, Jr. (1992): *Predicting Ecosystem Risk*, Princeton Scientific Publishing, Princeton.

[43] J.R. Pratt and J. Cairns, Jr. (1996): Ecotoxicology and the redundancy problem: understanding efects on community structure and function, 347-370. In: M.C. Newman and C.H. Jagoe (1996): *Ecotoxicology: a Hierarchical Treatment*, Lewis Publishers, Boca Raton.

[44] J.R. Pratt and J.L. Rosenberger (1993): Community change and ecosystem functional complexity: a microcosm study of copper toxicity, 88-102. In: J.W. Gorsuch, F.J. Dwyer, C.G. Ingersoll, and T.W. LaPointe (1993): *Environmental Toxicology and Chemistry*, vol. 2, American Society for Testing and Materials, Philadelphia, PA.

[45] J.R. Pratt, D. Mochan and Z. Xu (1997): Rapid toxicity evaluation using soil ciliates: sensitivity and bioavailability. Bulletin of Environmental Contamination and Toxicology, 58, 387-393.

[46] J.R. Pratt and E.P. Smith (1991): Significance of change in community structure: a new method for testing differences, 91-103. In: *Biological Criteria: Research and Regulation*, EPA-440/5-91-005. Office of Water, Washington DC.

[47] N. Ricci (1991): Protozoa as tools in the pollution assessment. Marine Pollution Bulletin, 22, 265-268.

[48] J.C. Riedle-Lorjé (1981): Investigations on the value of the aufwuchs as an indicator of water quality in the fresh- and brackish-water zones of the Elbe-Estuary with special regard to industrial influences. Archiv für Hydrobiologia, Supplement, 61, 153-226.

[49] R.O. Roberts and S.G. Berk (1990): Development of a protozoan chemoattraction bioassay for evaluating toxicity of aquatic pollutants. Toxicity Assessment, 5, 279-292.

[50] V. Sládecek (1973): System of water quality from the biological point of view. Archiv für Hydrobiologie Ergebnisse der Limnologie, 7, 1-218.

[51] R. Srámek-Husek (1956): Zur biologischen Charakteristik der höheren Saprobitätsstufen. Archiv für Hydrobiologia, 51, 376-390.

[52] H.G. Washington (1984): Diversity, biotic, and similarity indices: a review with special references to aquatic ecosystems. Water Research, 18, 653-694.

[53] Z. Xu, N. J. Bowers, and J.R. Pratt (1997): Variation in morphology, ecology, and toxicological responses of *Colpoda inflata* (Stokes) collected from five biogeographic realms. European Journal of Protistology, 33, 136-144.

Environmental Science Forum Vol.96 (1999) pp. 161-194
© 1999 Trans Tech Publications, Switzerland

Crustaceans as Bioindicators

M. Rinderhagen[1], J. Ritterhoff[1] and G.-P. Zauke[1,2]

[1] Carl von Ossietzky Universität Oldenburg, FB Biologie (ICBM),
Postfach 2503, DE-26111 Oldenburg, Germany

[2] corresponding author: E-Mail: ZAUKE@HRZ1.UNI-OLDENBURG.DE

Keywords: Life History, Reproduction, Amphipods, Isopods, Biotic Indices, Bioaccumulation, Feeding, Growth, Behaviour, Population Studies, Sediment Toxicity, Lake Zooplankton

Abstract

First some special biological features of crustaceans are described, including morphology and anatomy (segmentation, extremities, cuticle), respiration, circulation, excretion and osmotic regulation, reproduction and life history strategies (precopula, breeding, larvae, moulting), distribution and abundance, and modes of life (living in the benthos or plankton). This section is followed by a detailed review of results at different methodological levels, particularly focussing on freshwater amphipods, isopods and Cladocerans (e.g. as components of lake zooplankton). One basis for an ecological assessment is knowledge of ecological preferences and distribution of crustaceans under different environmental conditions. Although information on the abundance of species is essential, the relevance of reproduction and population dynamics is emphasised. Subsequently, the development of biotic indices (including the *Gammarus/Asellus* ratio or the saprobic index involving crustaceans) is discussed. To understand ecological effects of aquatic pollution, information on bioaccumulation of substances is crucial. The applicability of feeding, growth, behaviour (e.g. precopula) and drift as meaningful ecological indicators (or biomarkers) is then discussed at length. Finally, information from field studies on survival and reproduction is presented, regarding aqueous-phase toxicity, sediment toxicity and response and recovery of lake zooplankton communities exposed to acidification and metals. Some details are illustrated in a case study focussing on accumulation strategies, toxicokinetic models and detoxification mechanisms in gammarid amphipods.

Introduction

Crustaceans are frequently used as bioindicators and biomonitors in various aquatic systems. One reason is that they are a very successful group of animals, distributed in a number of different habitats including marine, terrestrial and freshwater environments. They are thus interesting candidates for comparative investigations. Some of the special features of crustaceans, particularly of reproduction strategies, may be highly important for the interpretation of data from bioindicator studies using these organisms, and for the development of ecotoxicological endpoints. The focus of this presentation is on the utilisation of crustaceans, mainly freshwater species, as bioindicators or biomonitors. The two terms are not always clearly distinguished. We use the term (i) *bioindicator* to define a collective of organisms from the field (in a statistical sense) which give information about the environmental state, with effect variables being their mere presence or absence, their life history status or population dynamics (e.g. regarding age structure, abundance, genetic structure or condition index [1] and the term (ii) *biomonitor* to define an organism which can be used to establish geographical and temporal variations in the bioavailability of contaminants by measuring the accumulated concentrations of chemicals in the whole body or in specific tissues [2]. Thus, results of standard toxicity tests are not within the scope of this section. Some recent developments of such tests are critically reviewed in [3]and [4]. In marine systems, crustaceans are not frequently used as bioindicators as yet. Here the emphasis is on other groups, including bivalves, polychaetes and echinoderms, mainly because much of the effort is spent on the benthos. We start with a brief description of special biological features of crustaceans which might be relevant to the topic, followed by a review of results obtained at different methodological levels. Some important issues are illustrated in more detail by describing a case study. It is not intended and would be almost impossible to present a complete review of results published in this field. This is the goal of more specialised review papers, which are published regularly [5-12]. Furthermore, the selection of organisms is restricted in accordance with the main goals mentioned above. While the Cladocera, for example, play an important role in standard toxicity tests, the literature on bioindicators and biomonitors in freshwaters is clearly dominated by amphipods and isopods.

Special biological features of crustaceans

Both the number of different morphological, physiological and behavioural adaptations and the immense variation in life history patterns and reproductive traits have facilitated the evolutionary success of crustaceans [13]. Only a brief outline of some special biological features and adaptations of crustaceans relevant to the topic is given in this section (for more detailed information see e.g. [14-19]). Some of these special features of crustaceans, particularly of reproduction strategies, may be highly important for the interpretation of data from bioindicator studies using these organisms.

Morphology and anatomy

Basically, crustaceans are arthropods that are equipped with mandibles, and have two pairs of antennae and respire by means of gills. The body of crustaceans is segmented. It consists of the head (cephalon), various numbers of segments (thorax and abdomen) and the end of the body (telson), but the physique of crustacean species may vary widely [18-20]. There are long-limbed animals with numerous segments and species with few segments. The number of segments can vary even within a subclass. In some species the body is shortened or elongated. Thorax and abdomen (pleon) cannot always be distinguished in recent crustaceans. Often, some of the thoracic segments are fused with the cephalon to form a cephalothorax. The remaining thoracic segments are then called the peraeon [19]. Thus, crustaceans consist of either a cephalon, thorax and a pleon or a cephalothorax, a peraeon and a pleon. Those segments bearing legs that are not smaller or different in some other respect from those of the more anterior segments are regarded as belonging to the abdomen.

Many crustacean species have developed a dorsal shield (carapace). This is a duplication of the integument, originating from the far end of the cephalon. It may cover only certain segments of the body and can even be fused to these segments. However, it may also cover the body laterally or even surround the body, to greater or lesser extent, like a shell. Other crustacean species are completely surrounded by a shell [18].

The typical limbs of crustaceans are biramous [18, 20]. They consist of a broad, flattened basal stem (protopodite) and two attached branches, the exopodite (the outer one) and the endopodite (the inner one). There exist many variations on this basic structure. In many Malacostraca (isopods, amphipods and most decapods) the exopodite is completely reduced, producing a so-called walking leg (stenopodium). Another limb variant is the leaf-like phyllopodium. Normally, the extremities are named after either the section of the body to which they are attached or after their function (thoracopods, pereiopods, pleopods only in Malacostraca). Extremities which are attached to the cephalon are involved in ingestion of food (maxillipeds). A fusion of pleon and telson segments can be found in some groups of crustaceans. In these cases, the extremities are often modified and then called uropods. All these morphological structures are highly relevant in the identification of crustacean species, which is a necessary precondition for using these animals as bioindicators.

Considerable amounts of calcium minerals are deposited in the chitin-protein cuticle of numerous crustaceans, and make the shell hard and durable. The calcium occurs as calcite, vaterite or hydroxyapatite [18, 21]. Calcium is mainly taken up from the environment (e.g. from water or food). It can also be stored in the haemolymph, in the hepatopancreas and in some crustacean groups in special organs. In these cases, the stored calcium often is excreted (e.g. in *Carcinus* to 90%). There exists a strong relationship between the metabolism of calcium and that of other essential and non-essential metals.

Respiration

The respiration occurs mostly via gills. These are often modified appendages of the thoracopods (epipodites), having a thin cuticle and large surface area and promoting thus a high rate of gas exchange. Often the interior surface of the carapace is also involved in respiration. The extremities, mainly the thoracopods, move so as to send a current of oxygen-rich water past the gills. In some little crustaceans no special respiratory organs have been developed. Here the respiration occurs through the whole body surface (cutaneous respiration [18, 19, 22]). The respiration mechanisms make crustaceans susceptible to water-borne chemicals.

Circulation

The circulatory system of the crustaceans is open, and the development of the arterial blood vessel system varies greatly among the different groups of crustaceans [18, 19, 22]. Generally, a long heart lies dorsal to the gut within a pericardial sinus. Because of the open circulatory system, a mixture of blood and coelomic fluid, which is called haemolymph, flows around the organs. Oxygen carriers in crustaceans are either the ferruginous (iron-containing) haemoglobin or the cupriferous (copper-containing) haemocyanin. Special adaptations have been evolved to provide crustaceans with the essential element copper; among these molecules, the metallothioneins serve a dual purpose, being also involved in the detoxification of potentially toxic metals like cadmium [23-26].

Excretion and osmotic regulation

Crustaceans have relatively slight tolerance to changes in salinity, with the exception of species which live in estuaries or tidal zones. Generally, crustaceans are stenohaline. The physiological mechanisms of ionic and osmotic regulation keep water and salt balance nearly constant [18, 19]. The essential adaptations of ionic and osmotic regulations are depending on the particular habitat. Marine crustaceans are isotonic to sea water; thus osmotic regulation is no problem for these species. However, since the ionic composition of the haemolymph and the cells is not identical with that of sea water, the ionic balance needs to be regulated. On the other hand, the ionic concentration in the haemolymph of crustaceans living in brackish water or fresh water is higher than in the surrounding medium. These animals must therefore regulate both the ionic concentration and the osmotic pressure. Losses of salts have to be compensated or, at least, reduced to a minimum. Too high inflows of water into the body must be prevented and a surplus of water eliminated from the body. Active processes are involved in these regulatory mechanisms, so that osmotic regulation in limnetic crustaceans requires some energy investment. Furthermore, all the processes are closely related to the metabolism of essential and non-essential trace elements.

Different organs and mechanisms are involved in ionic and osmotic regulation. The primary organs for excretion and osmotic regulation in crustaceans are the antennal glands and maxillary glands. These are nephridia, opening at the basal segment of the second antenna and at the base of the second maxilla. They always consists of three sections: an end sac, regarded as a vestige of the coelom (sacculus), a channel (nephridial channel) of various lengths in the different groups of crustaceans, and a short distal terminal duct (exit duct), which opens into the nephropore. The primary urine is formed in the end sac by means of membrane filtration. A reabsorption of ions occurs in the distal sections. Normally, the antennal gland functions in the early larval stages; later it regresses and the maxillary gland is formed, which is responsible for the excretion in adult animals.

Other organs are involved in excretion, including the body surface, the gut, the epithelia of the hepatopancreas and, above all, the gills. The gills are important exchange organs and are able to take up ions from the water. Therefore, often a special epithelium with osmotic regulatory activity exists in the gills [18]. In limnetic crustaceans, the body surface is less permeable to water and ions than in marine species. A more detailed description of osmotic regulation in crustaceans is given by Pequeux [27].

Reproduction, development and life history strategies

Most crustaceans are dioecious. However, within the different groups of crustaceans both hermaphroditism and parthenogenesis (e.g. in most Cladocera) can be found [13, 18, 19]. Cirripedia are originally dioecious, but most of the species are now hermaphrodites. Sex is primarily genetically determined, although environmental factors often have considerable influence on it. For example, in Malacostraca the gonads are hermaphroditic during early development. Later in the juvenile stage, genetic factors determine whether an androgen gland will be induced, producing males, or whether it will atrophy, resulting in females. Effective environmental factors include temperature, day length and salinity. For example, in *Gammarus duebeni* (Amphipoda) the ratio of sexes is determined by day length. There also exist cytoplasmic sex factors which are regarded as single-celled parasites. In some amphipods and isopods, they can suppress the androgen glands, so that all animals become females, which in turn produce only female young, despite reproduction with both sexes [18]. Such effects may be misunderstood, for example, as being caused by environmental pollution.

Copulation occurs in almost all crustaceans. In a number of cases, a precopula precedes the copulation. Special organs responsible for the care of the brood are formed during the maturity-moult of the females, and the eggs become prepared for oviposition. The male clings to the female with particular, sometimes specially formed extremities. Very rarely, the eggs are released into the water or attached to solid materials. Instead, brood care predominates. The eggs are carried on the body until the larvae hatch. A brooding chamber may be formed by a section of the carapace cavity or a specially developed epipodite of the thoracopods. Some species produce a large number of small eggs which hatch to planktotrophic larvae; they rely on phytoplankton or bacteria for nourishment and remain in the pelagic environment for several weeks or months. Those species with direct development produce only a few large eggs which hatch to lecithotrophic larvae; they usually do not feed on planktonic food, depending instead on yolk in the egg for their growth and development [13]. The features discussed here may be relevant in studies using crustaceans as bioindicators. The disruption of precopula pairs is frequently used as a toxicological endpoint. The development of many crustaceans, including amphipods, isopods and decapods, offers the opportunity to use reproductive traits as indicators of environmental quality or integrity.

According to Sastry [13] it is difficult to classify the life history traits of crustaceans as being clearly r- or K-selected. Rather, particular components of life history and reproduction strategies are more representative of r- or K-characteristics over an r-K-continuum. There seems, however, to be a greater tendency toward K- than towards r-selected strategists. In the aquatic environment, only crustaceans living in the lower mesopelagic, bathypelagic and deep-sea benthic environments and some freshwater decapods show characteristics that clearly correlate with those of K-selected strategies.

The development appears to take two basic forms, being either direct or indirect [13]. Between these two major types, there are many gradations. During indirect development, the larva has to undergo several moults after hatching from the egg, to reach its final form (metamorphosis). The characteristic larva of crustaceans is the nauplius. However, there exist numerous different types of larvae (for a detailed description see [18-20]). In many cases, the larvae hatch at a more advanced stage of development (e.g. metanauplius, zoëa larva, megalopa larva). In direct development, the juveniles have the same number of segments and are of the same constitutional types as the adults. Their development consists of growth and maturation, without metamorphosis. Especially planktonic larvae may be subject to environmental constraints; for example, those of barnacles and decapods are influenced by increased UV-B radiation [28, 29].

The growth of all crustaceans occurs in stages, each combined with a moult, at least until sexual maturity is reached [18, 19]. Moulting is controlled by hormones. The moult is induced by the synthesis of the moulting hormone α-ecdysone. The synthesis of α-ecdysone is prevented by an antagonistic hormone. The time between sequential moults becomes longer with increasing age. The beginning of a moult is influenced by temperature and amount of food available.

The life span of crustaceans varies widely. According to Sastry [13] even individuals of different generations within a population may differ in life span. For example, an overwintering generation may live some months longer than the spring generation. Generally, smaller crustaceans produce several generations per year, while larger species may live quite a long time (e.g. *Astacus astacus* for 20 years [18]). The life span of organisms is relevant, for example, in estimating exposures to environmental stressors under field conditions.

Distribution and abundance

Crustaceans are a very successful group with an immense variety of ecological types living in a great number of different habitats. Many crustaceans live in marine environments, others in various freshwater systems, including lakes and rivers, but also in the ground water or in cave waters. In both the marine and freshwater environment, crustaceans may live in the benthos as well as in the pelagic zone. All habitats are occupied, from tropical to arctic waters, from the deep sea to the marine littoral and from lowland freshwater systems to the mountains [13, 30]. Crustaceans even live in small pools of water in the phytotelmata of bromeliads. The extreme diversity of adaptations in morphology, physiology, way of life and reproduction to the different conditions enable crustaceans to establish themselves in almost any body of water (for details of adaptations see for example ref. [14, 30] and literature cited therein). Because of special adaptations, numerous species can also be found in terrestrial habitats. In particular, these include the terrestrial isopods (woodlice; Oniscidea). Furthermore, some crustacean species are specialised to live as parasites. Thus, for any question of interest it will be possible to find some crustacean species sufficiently relevant to be used for the ecological comparison of a wide range of habitats.

Modes of life

The locomotion of crustaceans greatly depends on the habitat in which they are living [17, 19]. Benthic crustaceans either walk or combine walking and swimming. Different limbs are used for the various types of locomotion (e.g. thoracopods for walking, pleopods for swimming). Other crustacean species are permanently swimming, above all the smaller planktonic crustaceans (Cladocera, Copepoda), but also some bigger species such as prawn-like or shrimp-like decapods and some amphipods. Changes in locomotor behaviour are also good indicators of environmental stress as, for example, outlined below for amphipods. Other species exhibit a completely sessile mode of life (e.g. marine Cirripedia), which would be a great advantage in biomonitoring studies, although these species are seldom used [2, 31].

The variety of adaptations of crustaceans to different habitats is also reflected in a variety of feeding mechanisms [17-19]. Some crustaceans are filter feeders (e.g. Anacostraca, Phyllopoda, many Copepoda and most Cirripedia). Other groups seize their food directly. They may behave as selective or non-selective deposit feeders (grazer), scavengers or predators. There also exist some sucking parasites. However, there is almost no strict division between either feeding mechanisms or kind of food. Many crustaceans employ an alternate or secondary feeding mechanism and many of them ingest a variety of foods. In marine environments, many crustaceans in the lower mesopelagic zone are omnivores with a tendency to carnivory, while in bathypelagic zones most crustaceans are carnivorous. Only a few crustaceans are specialised feeders (for more details see ref. [32] and literature cited therein).

Review of results obtained at different methodological levels

In this section we shall give an overview of results of investigations (since about 1990) at different methodological levels that have used crustaceans as bioindicators or biomonitors, referring to the most important biological features discussed in the previous section. This review will focus mainly on freshwater organisms, particularly on amphipods, isopods and Cladocerans, which are often used in freshwater studies as bioindicators and biomonitors [33]. In contrast, crustaceans have seldom been used for that purpose in marine studies, which will thus be only briefly addressed. Some arguments are provided by Schulz [34] for the coastal zone of the Baltic Sea. Bioindicators should be benthic species, distributed over the whole region of interest, and should live in the littoral zone to enable easy sampling. Further, they should be of ecological importance. Thus, he recommends the barnacle *Balanus improvisus* as the only crustacean to be used as a bioindicator for the Baltic Sea. Accordingly, mainly benthic littoral bivalves or polychaetes are used as biomonitors and, to lesser extent, as bioindicators in marine systems [35-39]. Regarding pelagic marine systems, crustaceans are also used as biomonitors [40-43], but not yet as bioindicators.

Ecological preferences and distribution

The micro-distribution of various gammarid species depends on particular ecological preferences, such as habitat structure (including water flow, substrate and food) as well as differences in physiological tolerance to oxygen content, temperature and water quality. The presence or absence of predators also plays an important role. Dahl and Greenberg [44] showed that *Gammarus pulex*, for example, is able to evaluate differences in habitat quality and to respond to it. When artificial streams were modified by adding a coarse substrate, for example, the abundance of amphipods at these sites was increased.

Crane [45] investigated the population characteristics of *Gammarus pulex* in five English streams as sampled during 1992. The results showed a great variability of *G. pulex* under different lotic conditions. Significant differences between populations of different streams involve population density, standing crop biomass, individual size and sex ratio. It was suggested that some of these differences were caused by pollutants or by the physical structure of the stream bed.

Invasion of neozoic species may be caused by different tolerance to environmental pollution [46]. Toxic effects can result in the elimination of local taxa, and neozoic species which are probably more resistant to pollution may fill the empty ecological niches. Streit and Kuhn [47] suggest that this may be a reason for the invasion of *Gammarus tigrinus* in the River Rhine after elimination of *Gammarus fossarum* (see also section describing toxic effects).

Meijering [48] studied the occurrence of *Gammarus fossarum*, *Gammarus pulex* and *Gammarus roeseli* in Hessian running waters (Germany) from 1968 to 1988. After considering the ecological preferences of the particular species with respect to temperature, water flow rate, food and geological conditions, the lack of oxygen mainly caused by organic pollution and low pH were recognised as the most important factors which determine the distribution of these *Gammarus* species. For example, *G. pulex* is more robust against pollution, low oxygen and low pH. Thus, sometimes it may be the dominant or the only *Gammarus* species in an area affected by these environmental constraints. For the isopod *Asellus aquaticus* it was reported that it not only responds directly to pollution by developing an increased tolerance [49]; it also exhibits an indirect adaptation of life history strategies, as demonstrated by less investment in reproduction, producing fewer, but larger offspring at the polluted site of a stream. Laboratory studies indicated that these phenomena probably have a genetic basis.

Laboratory experiments by Pilgrim and Burt [50] proved that the North American amphipod *Hyalella azteca* is acid-sensitive like the *Gammarus* species mentioned above, being also influenced by water temperature. At lower temperatures, the amphipods are able to tolerate an acid stress for a longer period of time than at higher temperatures. These results suggest that tolerance to acidification probably varies seasonally and that the summer months are the most sensitive ones. Most of the natural habitats of *H. azteca* (which prefers ponds to lakes) show seasonal variations of water temperature. Thus, when using *H. azteca* in laboratory bioassays, it is more realistic to use a variety of temperatures simulating the range in the field and acclimatise *H. azteca* to them correspondingly, before pH adjustments are made. The temperature-dependent sensitivity to acidification has then to be taken into account for the prediction of ecological effects.

Reproductive aspects such as egg survival and brood development (each time-related to optimal temperatures), reproductive potential and fecundity of populations of *Gammarus fossarum* and *Gammarus roeseli* in Austria were intensively investigated [51, 52]. No significant intra-specific differences were found between populations of either species, but there were differences between the two species, which appear to have different temperature requirements. The optimum temperature for *G. fossarum* is a little bit lower than that for *G. roeseli*; moreover, *G. fossarum* has a wider optimal temperature range for egg survival and reproductive success (50% survival or more, with 3.6-19.2 °C for egg survival and 4.5-19.1 °C for reproductive success) compared to *G. roeseli* (egg survival: 11.9-15.1 °C; reproductive success: 7.0-21.0 °C). These attributes may well influence the distribution patterns of both species. The reproductive period of *G. fossarum* extends from December to September, and that of *G. roeseli* from March to September. Conditions will be optimal for *G. fossarum* in streams that stay cool in summer, whereas for *G. roeseli* optimal conditions will occur in summer-warm streams, from which *G. fossarum* is ordinarily absent. Only *G. fossarum* is able also to breed in winter-cold streams, with development times of more than 3 months (at <10 °C) and thus long generation times.

A specialised investigation of the presence and habitat preferences of the amphipod *Gammarus minus* is provided by Glazier et al. [53]. The presence or absence of this amphipod was evaluated in several cold springs in central Pennsylvania, showing differences in pH and ionic content. The amphipods were missing in springs with pH<6.0 and conductivity <25 $\mu S \cdot cm^{-1}$, and there was a gradual linear decrease in population density with decreasing alkalinity and water hardness (Ca^{2+} and Mg^{2+}). Brood size and brood mass were independent of spring pH and ionic content, but correlated with maternal body size. Furthermore, the percentage of females breeding was independent of pH and ionic content in the various springs. Most of the observed differences between the populations in the different springs could not be clearly explained. It was suggested that the population density is possibly more a function of food availability than of direct physiological effects on the amphipods of alkalinity and hardness.

As outlined by Meurs and Zauke [54], it is not sufficient to examine the abundance of organisms in a given ecological system to obtain a complete picture of potential ecological processes. To achieve this goal, detailed investigations into the population dynamics are required, as demonstrated for the amphipod *Gammarus zaddachi* in the Weser estuary and a major freshwater tributary (River Hunte). At freshwater sites in the River Hunte, the length distribution of gammarids shows distinct seasonal fluctuations, with small juveniles appearing in late spring (April-May) and larger adults appearing in winter (December-February). Most interestingly, although precopula pairs were frequently observed in winter, no mature females carrying juveniles in the brood pouch occurred at all, and the abundance of the population at these sites approached zero between February and March. The authors inferred that the gammarid populations at these freshwater sites were not maintained by autochtonous reproduction, but by invasion of juveniles from other parts of the river system. *G. zaddachi* is known to reproduce in brackish waters and this was also observed at a site in the Weser estuary. Females carrying juveniles in the marsupium were abundant in December and in March-April. An analysis of water movements during ebb and flood revealed that it might well be possible for juveniles (released in the brackish water region of the estuary) to move upwards and reach the freshwater sites of the River Hunte as predicted by the corresponding seasonal analysis. The authors stress the necessity of such detailed investigations in complex river systems before the ecological situation can be fully assessed — for example, in the context of environmental impact assessments.

Gammarus/Asellus ratio and other biotic indices

The occurrence of the amphipod *Gammarus pulex* and/or the isopod *Asellus aquaticus* has been shown to be a useful tool in the assessment of water quality. *G. pulex* and *A. aquaticus* are widely distributed in Europe. Although they have similar trophic niches (with respect to life history patterns and utilisation of organic matter as food), they generally occur in different sections of rivers. The ecological niche differentiation of *G. pulex* and *A. aquaticus* was explored by Graca et al. [55, 56], who analysed their microdistribution, life history patterns and food preferences. *A. aquaticus* mainly occurs in the lower stretches of water courses [57, 58], while *G. pulex* commonly occurs in the upper stretches. Because of its higher tolerance to pollution and hypoxic conditions, *A. aquaticus* dominates in habitats under anthropogenic pressure [59], whereas *Gammarus* species normally inhabit well oxygenated, unpolluted stretches. However, there are also coexisting populations of both species. For the coexisting populations observed, Graca et al. [55] found no inter-specific differences in the microdistribution, but intra-specific differences in the distribution of juveniles and large adults. The first predominantly inhabited algae/bryophyte mats, while the latter preferred coarse-grained substrates. Data on population dynamics of the two species gave no clear results concerning whether or not competition is a important factor in determining their distribution. For example, Hargeby [60] did not find any indication supporting the hypothesis that the disappearance of *G. pulex* during acidification of streams is caused by competition with *A. aquaticus*. Moreover, coexisting populations did not appear to have other food preferences than allopatric populations. Both species feed on fungal-conditioned leaf material and preferred the same fungal species. If there is a choice, however, *A. aquaticus* preferred mycelia, while *G. pulex* preferred fungal-conditioned leaf material [59, 61]. In contrast to *A. aquaticus*, *G. pulex* is able to tolerate deterioration in food quality, compensating the reduced energy intake by reduction of the energy expenditure. A high overlap in trophic niches, indicating that competition for food plays an important role in the distribution patterns of *G. pulex* and *A. aquaticus*, was reported by Graca et al. [55]. It was suggested that other biotic and/or abiotic factors than competition must be responsible for the spatial separation of these two species.

Dissolved oxygen contents and ammonia concentrations are further important abiotic factors in determining the distribution of amphipods and isopods (as well as of other freshwater invertebrates) in the aquatic environment. *A. aquaticus*, for example, is five times more resistant to hypoxia and two times more resistant to un-ionisised ammonia than *G. pulex* [62]. Juveniles are less susceptible than adults to both substances, as indicated by examination of respiratory surfaces, ventilation rates and haemolymph. From the results and literature data it was suggested that the different susceptibility of the two species to low-oxygen conditions may be caused by interspecific differences in ventilation rates and blood characteristics (such as haemocyanin concentration and its oxygen affinity). The reasons for the differential inter-specific tolerance to ammonia are less clear.

The different tolerances of the two species to pollution and low-oxygen conditions are routinely used as a simple tool for the assessment of water quality [63]. In this method, the abundance of *G. pulex* in relation to the abundance of *A. aquaticus* is used to generate a *Gammarus/Asellus* ratio. It was shown for various rivers that changes in the distribution of both species were consistent with changes in water quality. Highly significant negative correlations between the *Gammarus/Asellus* ratio and BOD, concentrations of ammonia-nitrogen, nitrite-nitrogen and phosphate-phosphorous were reported. For some pollutants significant correlations between the abundance of *Asellus* and the concentration of water-borne chemicals could be established. However, the converse did not hold. For example, increasing chloride concentrations are positively correlated with the abundance of *A. aquaticus*, but the number of *G. pulex* does not increase with decreasing concentrations of chlorides [63].

A disadvantage of this methods method lies in the fact that the *Gammarus/Asellus* ratio is, to some extent, influenced by the sampling techniques. When Standard Aufwuchs Units and airbrick substrate colonisation samplers were used, the data were in good agreement with results obtained with the aid of Surber samplers (for details of the particular samplers see ref. [64] and literature cited therein). Conversely, basket samplers produced biased results, because they favour colonisation by *Gammarus*.

While the *Gammarus/Asellus ratio* is a rather specialised biological indicator, some crustaceans are also included in different general biotic indices which cover a broader range of taxa. In Germany and some other countries of Central Europe, the 'saprobic system' is adopted by law in official water quality assessment programs [33]. The saprobic index (S) can be regarded as the weighted mean of the abundance (A_i) of the i-th species found in a river system or stretch. The weights are given by the saprobic value (s_i) of the i-th species (characterising the water quality for which they are regarded as an indicator; range 1-4, with 1=clean water and 4=polluted water) and the indicative weight (G_i) (characterising the tolerance of a species against different levels of water quality, ranging 1-16; with 1=poor indicator and 16=good indicator). The saprobic index is defined as [33, 65]:

$$S = \frac{\sum_{i=1}^{n} s_i * A_i * G_i}{\sum_{i=1}^{n} A_i * G_i}$$

Among the crustaceans, 7 species are listed in the German saprobic system, viz. *Asellus aquaticus* (L.) with s=2.7 and G=4; *Atyaephyra desmaresti* (Millet) with s=1.9 and G=8; *Gammarus fossarum* (Koch) with s=1.6 and G=8; *Gammarus pulex* (L.) with s=2.1 and G=4; *Gammarus roeseli* Gervais with s=2.0 and G=8; *Gammarus tigrinus* Sexton with s=2.4 and G=4; and *Proasellus coxalis* Dollfus with s=2.8 and G=4. The enormous importance of amphipods and isopods in the assessment of the quality of freshwaters is highlighted by this list.

Although the saprobic system is adopted by law in official water quality assessment programs, it may be criticised from an ecological point of view [66-68]. Some important issues involve the problem of calibration and validation of the saprobic values and indicative weights, the fact that this measure reflects almost exclusively degradable organic pollution, and the unsolved problem of the comparability of different lotic systems. Braukmann [69], for example, proposed a regional differentiation of this concept, taking into account an intrinsic saprobic index of corresponding rivers. He suggested that the saprobic index of alpine rivers be reduced by 0.7; of montane rivers by 1.0 and of lowland rivers by 1.7 to make the results comparable. Wuhrmann [67] advocated comparing biotic communities from a polluted stretch of a river only to those in similar micro-habitats at uncontaminated sites upstream of the source of pollution. Such a concept may be greatly facilitated by utilising artificial substrates (e.g. brick material) for a standardised sampling of macroinvertebrates [70]. A broader perspective for indicator development involving the zonation, the river continuum, the stream hydraulics, the resource spiraling, the flood pulse, the riverine productivity and the catchment concept was recently reviewed [71].

Regarding contamination of groundwaters, crustaceans could be also used as bioindicators as shown by their mere presents or absence. Malard et al. [72] investigated groundwater contamination in a karst area in southern France. A sewage-polluted site was dominated by oligochaetes, exhibited low stygobite richness and its groundwater fauna consisted mainly of stygoxene taxa. On the other hand unpolluted sites showed fauna assemblages dominated by crustaceans, with a high number of stygobite species.

In addition to organic pollution (e.g. domestic sewage effluents), crustaceans have been used as bioindicators of other environmental variables such as anthropogenic salinity or naturally occurring physicochemical properties. In general, the abundance and diversity of benthic microcrustaceans could be used as a bioindicator for the ionic content, physical variables and habitat availability in freshwaters. This has been shown by Rundle and Ramsay [73] for streams in two physiographically contrasting regions of Britain. Only 33% of the forty-three copepod and Cladoceran species were found both in lowland south England and upland Wales. The authors stressed that it is important to understand species ecology and biogeography before assessing pollution impacts on stream communities.

Utilisation of ostracods as potential bioindicators was investigated by Milhau et al. [74] in the Slack River basin, northern France. The abundance of ostracods, represented by seventeen species, was related to different water-quality levels and physicochemical variables. In the Aral Sea, ostracods were used by Boomer et al. [75] as indicators of the anthropgenic salinity increase of the past three decades. Of the eleven species of ostracods living in the Aral Sea in 1960 only one is still present. Furthermore, the modern ostracod distribution in lakes of the north-central USA shows how ostracod abundance is influenced by the concentrations of major ions like calcium, sulphate and bicarbonate [76]. Again different ostracod species were associated with different water qualities: for instance, *Limnocythere sappaensis* and *Heterocypris glaucus* were found in bicarbonate-enriched sulphate-dominated waters, whereas *Limnocythere staplini* was abundant in bicarbonate-depleted sulphate-dominated waters. *Candona ohioensis* and *Limnocythere itasca* were found in freshwaters, and *Candona rawsoni* in both bicarbonate-enriched and bicarbonate-depleted sulphate-dominated waters. Smith recommended ostracods as an aid to identify changes in both ionic composition (solutes) and ionic concentrations. Llano [77] investigated the distribution of marine benthic ostracods and its modification by upwelling phenomena in marine epibathyal and continental shelf areas. He concluded that ostracods can be used as indicators in hydrologic studies and geological surveys. Ostracods can also be used in palaeoecological studies. Griffiths et al. [78] stated that *Psychrodromus olivaceus* may serve as an indicator of cool, solute-rich waters and coarse substrata in modern and subfossil faunas.

Bioaccumulation in experiments and field studies

Crustaceans, particularly amphipods and isopods, are often used in bioaccumulation experiments and in field studies, as the large number of papers on this topic attest. Within the scope of this chapter, only a few important aspects will be mentioned. Some details are illustrated in the case study (see below).

Results for the amphipod *Gammarus fasciatus* are described by Amyot et al. [79]. The authors inferred that *G. fasciatus* is a suitable biomonitor for Cd, Ni and Pb, but not for Cu and Zn, which seem to be regulated. Significant variations in concentrations of Ni, Cd and Pb in field samples were reported for different seasons. It should be mentioned, however, that metals were only measured two times a year, so that the data base might be too small to infer 'seasonal' variations. No significant differences were found for Mn, Cu, Zn and Cd. It was hypothesised that these variations could be related to moulting or food contamination.

In another study, Amyot et al. [80] used *G. fasciatus* to investigate the distribution of some metals in various organs of these animals. This distribution was strongly dependent on the metal under consideration. For Cd and Cu, highest levels were found in the hepatopancreas, while for Cr, Fe, Mn and Ni concentrations were highest in the gut (one order of magnitude higher than in the other organs). Zn showed rather invariable concentrations throughout the body. An important fraction of Pb was associated with the exoskeleton. Depuration did not alter the mean body concentration of Cd, Cu and Zn, whereas Pb significantly decreased. Furthermore, the gut content had a significant effect on concentrations of Cr, Fe, Mn, Ni and Pb. Defecation (which was called depuration in ref. [80]) caused an important decrease in whole-body metal concentrations, pointing to the fact that the gut content should not be included in biomonitoring studies. Thus, depuration should be employed in any such investigations.

Regarding the river Rhône, Plenet [81] used the amphipods *Gammarus fossarum* (detrivorous animals, living on the bottom of rivers; epigeans) as well as *Niphargus* cf. *rhenorhodanensis* and *Niphargopsis casparyi* (omnivorous animals, living in the interstitial of the sediment; hypogeans) as biomonitors of zinc and copper pollution. He found differences in the ability to accumulate Zn and Cu between the epigean and hypogean amphipods, with the epigean *G. fossarum* showing lower body concentrations of Zn or Cu compared to *N.* cf. *rhenorhodanensis* or *N. casparyi*. A positive correlation between Zn and Cu body burdens and metal concentrations in sediments was noted for *G. fossarum*. For the hypogean amphipods such relationships were not found. Apparently, *N.* cf. *rhenorhodanensis* could regulate its body concentration of Zn and Cu up to a certain threshold level. Because of this, *G. fossarum* was suggested to be a useful biomonitor of metal pollution, in contrast to both hypogean amphipods.

Bioaccumulation of As, Cd, Cu, Pb and Zn was investigated in two groups of the amphipod *Hyalella azteca* [82]. One control group was exposed for 28 d under laboratory conditions to sediments collected from a depositional area of a river. Another group of organisms was collected from riffles adjacent to the depositional area and their metal accumulation under field conditions was measured. Although the patterns of accumulated metals were similar in the two groups, metal concentrations were 50-75% lower in organisms from the control group, exposed in laboratory experiments. It was concluded that long-term monitoring of contaminated rivers should not only rely on short-term sediment bioassays, but should also include the evaluation of long-term bioaccumulation in the field.

Bioaccumulation by the non-deposit-feeding amphipod *Rheopoxynius abronius* of polycyclic aromatic hydrocarbons (PAHs) from contaminated sediments was examined by Meador et al. [83]. Sediments were collected at the Hudson-Raritan estuary and the organisms were exposed to them for 10 d. Analyses of the tissue concentrations of 24 PAHs and comparison of the data for *R. abronius* with that for the non-selective, deposit-feeding polychaete *Armandia brevis* did not yield significant differences for the low molecular weight PAHs (LPAHs). In contrast, bioaccumulation of high molecular weight PAHs (HPAHs) differed markedly between the two species. After analysing correlations between concentrations in tissues and environmental substrates as well as partition coefficients, the authors concluded that the bioavailability of HPAHs was significantly lower for *R. abronius,* probably due to their partitioning to dissolved organic carbon, while it was still high for the deposit-feeding polychaete.

The preceding examples show that bioaccumulation is a very complex process, influenced by several environmental factors. For example, the uptake route plays an important role, and it may be that no simple correlations can be established between the concentrations of pollutants in the substrate and in the organism, or that more than one uptake route is involved.

Effects on feeding activity and growth

Feeding activity and growth are two sensitive sub-lethal indicators at the level of the individual organism which can themselves affect higher levels of biological organisation (the population, community or ecosystem).

Taylor et al. [84] developed a non-destructive experimental procedure for the measurement of feeding in the amphipod *Gammarus pulex*. It allows feeding activity to be monitored under experimental conditions involving a time-response analysis. The amphipods are placed into cylindrical pots (basal area 18 cm^{-2}) together with 10 shell-less eggs of *Artemia salina* as a standard food source for adult individuals. The number of eggs eaten in each pot was recorded frequently and the time at which 50% of the eggs have been consumed was defined as median feeding time (FT$_{50}$). The authors found that the feeding activity decreased under copper exposure, with a positive linear correlation with the metal concentration and exposure time. For example, the FT$_{50}$ increased from 1.6 h in the control to 11.5 h following a 3-h exposure to 101 µg Cu l^{-1}. A further development of this approach revealed that the feeding activity of juvenile *G. pulex* was a sensitive response criterion for use in assessing the sub-lethal toxicity of copper, lindane, and 3,4-dichloroaniline (3,4-DCA) [85]. Reductions in gammarid feeding activity were detected following 96 hours exposure at 12.1 µg Cu l^{-1} or 8.4 µg lindane l^{-1} and 240 hours exposure at 918 µg l^{-1} 3,4-DCA. However, a significant increase was observed in the feeding rate of gammarids that had been exposed for 240 h at 0.09 µg lindane l^{-1} in comparison with control values. The increase in feeding rate may be interpreted as a possible stimulatory effect associated with the toxicant action of lindane. The feeding bioassay was also used as a tool in an investigation of species interactions in toxicant systems. The feeding responses of *G. pulex*, which had been maintained in the presence of *Asellus aquaticus* (as interacting pairs) and exposed to a range of concentrations of lindane or 3,4-DCA, illustrate the complex nature of test systems that integrate the stresses of toxicant and competition.

In a series of field studies, it was found that the feeding rate of *Gammarus pulex* is a sensitive indicator of water quality [86, 87]. In these studies, individuals of different populations of *G. pulex* were deployed in cages at different stream sites above and below metalliferous discharges. By comparing feeding rates with the measured bioaccumulation, Maltby and Crane [87] found that the two were negatively correlated for, and manganese as well as iron was taken up by the animals. However, this effect could not be explained by adsorption of these substances alone. Laboratory experiments were carried out to validate the results from the field study. It was shown that exposure to iron concentrations similar to those measured in the field experiment did indeed reduce the feeding rate of *G. pulex*, while manganese had no additional effect. Differences in sensitivity between populations of different sites were reported and the authors concluded that animals originating from a polluted site might have developed a greater metal tolerance. Thus, results obtained for different *Gammarus* collectives must be regarded with some caution. Moreover, Crane and Maltby [86] found differences between the feeding rates measured in two different laboratories. These differences did not seem to be systematic, but the reasons remained unclear, pointing to the necessity of appropriate intercalibration procedures. It should be noted, however, that a reduction of feeding rate under metal exposure is not restricted to crustaceans, as shown, for example, for Fe and larvae of the ephemerid *Leptophlebia marginata* [88].

Moore and Farrar [89] examined the relationship between food rations and growth rates of *Hyalella azteca*. Growth rates decreased significantly with reduced food rations, whereas survival was not influenced after 10 days. However, reproduction was also reduced with low food rations. Effects like those reported in this study might be confused with toxic effects in corresponding experiments and should be carefully considered.

Scope for Growth

Scope for growth is a measure of the energy budget, viz. the difference between energy absorbed (from food intake) and energy metabolised. It gives an indication of the metabolic condition of an organism. Scope for growth is frequently used as an indicator of pollution stress in marine systems (e.g. regarding mussels), but rarely in freshwater systems. If implemented properly, it provides a highly relevant ecological indicator [90]. Maltby et al. [91], for example, investigated the sensitivity of scope for growth in the benthic freshwater crustacean *Gammarus pulex*. The organisms were exposed to zinc, 3,4-dichloraniline, oxygen and ammonia under laboratory conditions. It was shown that scope for growth provides a sensitive assay, with a response to the different stressors. Scope for growth was reduced by any of the four substances applied. A field deployment of a scope for growth assay using *G. pulex* is described by Maltby et al. [92]. The animals were caged up- and downstream of a pollution source. Compared to the upstream reference site, scope for growth was reduced at all sites downstream of the pollution source.

To evaluate the sensitivity of the '*Gammarus* scope for growth assay', Maltby and Naylor [93] used breeding females of *G. pulex* and compared the effect of zinc on scope for growth with that on reproduction. Size and the number of offspring of the current and a subsequent brood were measured over a 2-month period as reproductive parameters, to evaluate the effects of past and present pollution. Comparison of the results with that from the former study [91] revealed that breeding females are more sensitive to stress than males. At a concentration of 0.3 mg $Zn \cdot l^{-1}$, scope for growth was significantly reduced as a result of decreased absorbed energy. The size of offspring in the subsequent brood was likewise decreased. This decrease was observed over the entire juvenile size range and was independent of the number of offspring in the brood, but the number of broods aborted increased as a result of both present and past zinc exposure. These effects result in a longer developmental time and a reduced fecundity, two variables which are important for the survival or continued existence of the population. However, the authors emphasised that in ecologically meaningful ecotoxicological tests only those species should be used which are widely distributed and which play an important role in the functioning of ecosystems [93]. Additionally, they pointed out that tests based on physiological, cellular or subcellular responses are useful only if the observed responses can be related to effects at higher levels of biological organisation (whole organism, population, community or ecosystem).

Effects on moulting and regeneration

Another kind of sub-lethal response to contamination, viz. the effect of pollutants on moulting and regeneration in crustaceans, is described for several species [94]. Moulting and regeneration are controlled by the neuroendocrine system, on which the toxicants act. Heavy metals commonly retard the regeneration of limbs and affect the moult cycle by causing delay in ecdysis. Retardation of regeneration and delayed ecdysis are also caused by the organometal compound tributyltin (TBT), which is additionally related to anatomical abnormalities, for example in the major chela of male crabs *Uca pugilator*. Aromatic hydrocarbons and dioxins decrease the growth increment per moult. In contrast, DDT can accelerate regeneration and moulting.

Behavioural changes

Behavioural changes have been shown to be a very sensitive, non-destructive and rapid sub-lethal response to a wide range of pollutants. They basically represent an integrated response of an individual organism, although some behavioural changes such as precopula disruption (see above) may lead to changes at the community or population level.

Poulton and Pascoe [95] described a behavioural bioassay to evaluate environmental stress using the freshwater amphipod *Gammarus pulex*. Here the disruption of precopulatory pairing is taken as a signal of the presence of pollutants or parasites. The induced separation time, viz. the time from introduction of a pair to the test solution to the release of the female by the male, was measured using cadmium as pollutant. The mean induced separation time decreased with increasing cadmium concentrations, ranging from 8.3 $\mu g \cdot l^{-1}$ to 46.7 $\mu g \cdot l^{-1}$. Mean separation times ranged from 944 s in the control to 471 s at the highest cadmium concentration. The bioassay has also been successfully used in the field. Disruption of precopula has also been described as a result of exposure to various concentrations of γ-hexachlorocyclohexane (lindane) [96]. An exposure to 0.5; 1.0 and 2.0 $mg \cdot l^{-1}$ for 20 min or 1.0 and 2.0 $mg \cdot l^{-1}$ for 10 min or 2.0 $mg \cdot l^{-1}$ for 2 min resulted in separation of the pairs. Within 4 to 72 h after exposure, recovery of precopula occurred. Malbouisson et al. [97] further investigated the use of feeding rate and re-pairing of precopula to assess lindane toxicity. Feeding rate was significantly reduced either by long exposure to low concentrations of lindane (48 h at 5.0 μg l^{-1}) or by short exposure to high concentrations (2-20 min at 2.0 $mg \cdot l^{-1}$ or 20 min at 1.0 $mg \cdot l^{-1}$). Re-pairing of precopula was only affected by combining higher concentrations and longer exposures.

Precopula disruption of *G. pulex* as a sub-lethal bioassay was evaluated by Pascoe et al. [98] in the laboratory and in the field. In the laboratory test, the precopula pairs were exposed to various concentrations of 3,4-dichoroaniline, atrazine, copper and lindane. Median induced separation times were determined and both no-observed-effect concentrations (NOEC) and lowest-observed-effect concentrations (LOEC) were derived from these data and compared with LC_{50} values. Separation time was clearly reduced with increasing toxicant concentrations. In the field, precopula pairs of *G. pulex* were kept in a river for 24 h at 9 sites at various distances below discharges of treated sewage, farm runoff, mint effluent and domestic waste. Precopula disruption occurred more quickly for animals exposed to polluted waters than for those kept at a reference site above the particular discharge. For example, induced mean separation time was 630 s at the site of treated sewage influx, in comparison to 992 s at the unpolluted site. Disruption of precopula was also observed by McCahon et al. [99] as an effect of both increased ammonia and reduced dissolved oxygen concentrations, designed to simulate farm waste pollution. From these results, it can be suggested that precopula disruption is a very sensitive and rapid method of detecting environmental stress for a wide range of pollutants, both in the laboratory and in the field.

Borlakoglu and Kickuth [100] evaluated some behavioural changes as an effect of exposure to a chlorophenolic compound (3,5-dichloro-4-hydroxyphenyl-α,ß-dichloropropionic acid). Escape reaction, swimming behaviour, copulation behaviour and activity were measured. At levels around 5% of the LC_{50} value, behavioural changes were observed, demonstrating the enormous sensitivity of these methods employing *Gammarus* species.

A novel technical approach for the measurement of behavioural changes was developed by Gerhardt et al. [101] and subsequently optimised and used to monitor responses of *Gammarus pulex* to metal pollution [102-104]. This method is based on an impedance conversion technique, which is capable of recording different kinds of behaviour simultaneously. Two electrodes, conducting an alternating current of 0.3 mA, are placed on the side walls of the test chamber. The electrical field between them is modified by movements of the test organisms, and the resulting alteration in the current is measured by another pair of electrodes, recorded on-line and processed by a computer system. *G. pulex* was exposed to various acutely toxic concentrations of Pb (0.01, 0.05, 0.1 and 0.5 mg $Pb \cdot l^{-1}$) and sub-lethal concentrations of Cu (0.05 mg $Cu \cdot l^{-1}$). Both initial early stress responses during short term exposure (1 h) and toxic effects of the metals caused by chronic exposure (Pb: 7 d, Cu: 35 d) could be detected by the system. Mortality due to Pb-exposure was 30%, whereas exposure to Cu did not result in any mortality over 35 days. *G. pulex* showed increased ventilation and decreased locomotion as initial stress reactions within 30 min (Cu) or 1 h (Pb) after metal exposure. After several days of exposure to sub-lethal concentrations, increased ventilation and decreased locomotion were observed. It was concluded that *G. pulex* is a sensitive bioindicator of metal stress and that behavioural changes of ventilation and locomotion are useful indicators of acute and chronic toxicity, especially for the detection of early responses.

Another behaviour in *Gammarus* species, cannibalism, was studied by Dick [105], but has not yet been related to pollution. It should be stressed, however, that cannibalism can bias toxicity tests to a great extent. Thus, test vessels must be supplied with appropriate gauze to prevent cannibalism [106, 107].

Drift

Drift is a special active behavioural response of lotic animals to environmental stress. Liess [108] used *Gammarus pulex* as an indicator of pesticide stress related to surface water runoff in small rural rivers. An overview of these concepts is provided by Liess and Schulz [109]. *G. pulex* showed an increased drifting behaviour as a result of pesticide pollution (e.g. due to lindane, permethrine or fenvalerate), following enhanced farmland runoff. The animals give very sensitive responses to low amounts of fenvalerate (100% drift at concentrations of fenvalerate two orders of magnitude below the LC_{50} value of 18 $\mu g \cdot l^{-1}$). Furthermore, *G. pulex* is capable of recolonising the site within some days or weeks. Other species (e.g. the trichopterean larvae *Limnephilus lunatus*) did not show any drift behaviour and, in staying at contaminated sites, may become less abundant due to toxicants. The authors developed a tentative indicator system for inputs from surface runoff. On the other hand, the possibility that *G. pulex* will adapt to pesticide stress has to be taken into account. An investigation of the reproductive behaviour of two populations showed that the population living in a habitat exposed more to pesticides may become adapted to it, as indicated by a higher precopula rate.

Matthiessen et al. [110] studied *G. pulex* that had been caged for an in situ bioassay in a headwater stream draining treated farmland and thus could not avoid pollution by drifting. They found that the animals stopped feeding and died after heavy rainfall. The farmland was treated with 3 kg carbofuran ha^{-1} as broadcast granules, and concentrations up to 26 $\mu g \cdot l^{-1}$ were measured in the headwater stream after a heavy rainfall. These concentrations exceeded the *G. pulex* 24 hr LC_{50} of 21 $\mu g \cdot l^{-1}$. It was further reported that a concentration as low as 4 $\mu g \cdot l^{-1}$ of this substance could reduce the feeding rate.

Taylor et al. [111] used the migration of *G. pulex* from the sediment into the water column as a behavioural measure of the presence of copper in non-acidified and acidified sediments. The drift response was affected by both types of sediment, to a degree related to the copper concentration. Acidification increased the behavioural response to copper.

Effects on survival, reproduction / mixed effects

Many tests using organisms focus on acute toxicity and provide LC_{50} values. Various sub-lethal effects are often evaluated together with effects on survival. In this section research on tests that produce several responses is reviewed. Amphipods, isopods and Cladocerans are the crustaceans mainly used in toxicity tests (bioassays). *Gammarus pulex* is one of the amphipods most frequently used in Europe, particularly in aqueous-phase toxicity tests, whereas *Hyalella azteca* is the most frequently used amphipod in the USA, particularly for sediment toxicity tests. Both the aqueous-phase and the sediment can be an important source of contaminants for organisms in natural systems. For both media various methodological approaches are used in bioassays: static or flow-through conditions may be employed for the aqueous-phase test, and the test organisms may be exposed to whole sediment, interstitial water, eluates or extracts. Ref. [112] and [113] provide an overview of test procedures with benthic metazoans for the assessment of sediment toxicity, including freshwater and marine crustaceans.

Aqueous-phase toxicity

McCahon et al. [99] simulated the effects of farm waste effluent (increased ammonia, reduced dissolved oxygen concentrations) in two second-order streams. The feeding rate of *Gammarus pulex* was significantly reduced by exposure to ammonia (max. $4.7 \, \text{mg·l}^{-1}$) but recovered post-exposure. A reduced feeding rate and an increased mortality were also induced by parasitism with the acanthocephalan *Pomphorhynchus laevis*. In addition, disruption of precopula was observed as an effect of both increased ammonia and hypoxia (see also behavioural changes). It was concluded that farm waste pollution incidents can be realistically simulated in the field and that it is important to use sub-lethal response criteria such as feeding rate in the case of short-term peaks of pollution. Despite the low acute mortality observed under exposure, McCahon et al. [99] suggested that the disruptions of feeding and reproductive behaviour were so significant as to produce effects even at the ecosystem level.

Toxicity of ammonia was also studied in the amphipod *Hyalella azteca* by Borgmann [106] under laboratory conditions. Young (0-1 week old) and adult (4-5 weeks old) *H. azteca* were exposed to various concentrations of ammonia for up to ten weeks. Mortality was similar for juveniles and adults and growth was not reduced at concentrations below that causing mortality, i.e. 1 mM total ammonia. However, reproduction was reduced at concentrations as low as 0.32 mM. Additional effects of pH on ammonia toxicity were also evaluated.

Statistical methods may help to detect correlations between cause and response in large sets of environmental data. Mulliss et al. [114, 115] exposed *Gammarus pulex* (Amphipoda) and *Asellus aquaticus* (Isopoda) from non-polluted sites to urban aquatic discharges for 36 d and measured the mortality and the bioaccumulation of heavy metals of both species. A number of different chemical and hydrological characteristics were also measured. The variables influencing mortality were assessed by multivariate statistical methods. Some variables influenced the mortality rate of both species (viz. BOD_5, aqueous copper concentration, flow rate and suspended solids). Additionally, the mortality of *G. pulex* was altered by ammonia, total aqueous lead concentration and dissolved concentrations of copper and zinc, whereas the mortality of *A. aquaticus* was affected by some other variables, including the body concentrations of zinc, lead, cadmium and copper and the dissolved aqueous concentration of cadmium. As a result some predictive equations were presented. Maltby [49] showed that *A. aquaticus* not only responds directly to pollution, by developing an increased tolerance, but also modifies its life history strategy in an indirect way by investing less in reproduction, producing fewer, but larger offspring at the polluted site of a stream. Results from de Nicola Giudici et al. [116] did not suggest a significant reduction of recruitment after exposure to copper (see below). Laboratory studies indicate that these differences may have a genetic basis.

Differences in tolerance to environmental pollution may result in the invasion of neozoic species, which fill empty ecological niches after local taxa have been eliminated by toxic effects. Streit and Kuhn [47] showed that elimination of *Gammarus pulex* as well as *Gammarus fossarum* in the river Rhine and the immigration of *Gammarus tigrinus* could be explained by the greater tolerance of the latter to organophosphorous insecticides. Acetylcholinesterase inhibition was used as a parameter of sub-lethal toxicity in that study. Crane et al. [117] also used acetylcholinesterase activity in *G. pulex* in an in-situ experiment to determine the effects of the organophosphorous insecticide Malathion 60 on aquatic invertebrate communities. After application of Malathion 60, acetylcholinesterase activity was significantly decreased, while mortality was higher only at one experimental station, and there were no clear effects on feeding rates at any station. Thus, it was concluded that the use of

Malathion 60 on watercress beds presents little risk to *G. pulex* if the recommended dose is used and if the effluent is diluted, for example by passing through a settling pool. However, discharging the effluent directly into the aquatic environment may cause sub-lethal or possibly lethal effects.

Crane [118] found differences in tolerance to zinc between four field-caught populations of *Gammarus pulex*. Mortality was lower in one population during the first 2 days of exposure, whereas after 6 days of exposure mortality as well as sub-lethal effects such as feeding rate were similar for all four populations. An LC_{50} for zinc after 6 days is given as 0.75 mg·l^{-1}. The author concluded that only short-term processes could be responsible for the variation in tolerance to zinc. Thus when organisms collected from the field are used in toxicity tests, longer-term tests would be more appropriate. During short-term tests it may be that some populations of a given species may be more able to resist or tolerate toxic concentrations than others. This idea is supported by results of Gerhardt et al. [104], showing that *G. pulex* from an unpolluted brook was more sensitive to Cu exposure (≤ 50 µg·l^{-1}) than a local population from the Cu-contaminated brook.

Maund et al. [119] investigated the effects of copper on population structure and recruitment of *Gammarus pulex*. Exposure to copper reduced the population density, recruitment and growth. LOEC for recruitment of *G. pulex* was 14.6 µg·l^{-1}. It was suggested that the sub-lethal effects of copper at this concentration are caused by reduction of the amount of metabolic energy available for reproduction (e.g. by reducing feeding rate as described above for several substances and organisms, or by disruption of various physiological functions). Effects of chronic exposure to copper observed in *G. pulex* were compared to results from ref. [116] for *Asellus aquaticus*. Some effects were similar, such as the decrease of growth and survival, whereas the reduction of recruitment was not observed in *A. aquaticus* (in contrast to results from Maltby [49] discussed above, which indicate that pollution leads to differences in reproduction strategies as an indirect adaptation). Maund et al. [119] discussed whether or not these differences may be due to differences in life history strategies of the two species investigated. The influence of life history of the brackish-water isopod *Idothea balthica* on effects of acute and long-term exposure to Cd and Cu was reported in ref. [120, 121].

Toxicity of 3,4-dichloroaniline (DCA), atrazine, copper and lindane to the juvenile stage of *Gammarus pulex* (2nd or 3rd moult) was evaluated by Taylor et al. [122] and compared with that obtained for the 2nd larval instar of the insect *Chironomus riparius*. For *G. pulex*, lindane was the most toxic substance, followed by copper, atrazine and DCA, following exposure for 240 h (corresponding LC_{50} values in the same order: 0.07; 0.33; 4.4 and 5.0 mg·l^{-1}). Collyard et al. [123] showed that there were no significant age-specific differences in the sensitivity of *Hyalella azteca* to contaminants with various toxic modes. LC_{50} values were determined by exposure of *H. azteca* (with ages ranging from <1 to 26 d) to the organophosphate pesticide dazinon, a mixture of alkylphenol ethoxylates, copper, cadmium and zinc for 96 h.

The sensitivity of *H. azteca* (together with two other non-crustacean benthic macroinvertebrates) to five metals (Cu, Pb, Zn, Ni and Cd) and five pesticides (chlorpyrifos, dieldrin, p,p'-DDD, p,p'-DDE and p,p'-DDT) was evaluated by Phipps et al. [124]. Exposure occurred in a water-only, flow-through test system using natural lake water under laboratory conditions. *H. azteca* was the organism most sensitive to metals (10 d LC_{50}; Cu: 31 µg·l^{-1}, Zn: 73 µg·l^{-1}, Ni: 780 µg·l^{-1}; 8 d LC_{50}; Cd: 2.8 µg·l^{-1}, Pb: 16 µg·l^{-1}), but was less sensitive to pesticides (10 d LC_{50}: chlorpyrifos: 0.086 µg·l^{-1}, dieldrin: 7.6 µg·l^{-1}, p,p'-DDD: 0.19 µg·l^{-1}, p,p'-DDE: 1.66 µg·l^{-1}, p,p'-DDT: 0.07 µg·l^{-1}). It was emphasised that testing of multiple species in sediment assessments is very important and that sensitivity is not only dependent on test species but also on the test regime. Static conditions rather than flow-through tests or the use of nominal concentrations rather than measured concentrations may indicate effective concentrations higher than the actual, unknown effective concentrations.

McCahon et al. [125] studied the lethal and sub-lethal effects of acid, aluminium and lime on *Gam-*

marus pulex. By introducing measured doses of sulphuric acid, aluminium sulphate and lime into a soft-water stream, they produced zones polluted with aluminium at low pH, some of which were limed and others not. Concentrations of aluminium were lower in the limed zone, probably owing to adsorption to the suspended lime particles. Mortality was increased in the polluted zones, but the addition of lime reduced mortality. Median lethal times were 89.4 h in the unlimed zone and 169.4 h in the limed zone. Individuals parasitised with the acanthocephalan *Pomphorhynchus laevis* showed a significantly higher mortality in both zones (median lethal times 63.6 h and 100.8 h). More than 90% of precopula pairs separated in the polluted zones during an exposure period of 3 h, but they began to re-establish precopula after transfer into the unpolluted control zone.

A bioassay for determining the effects of toxicants on the growth of a sensitive life stage of *Gammarus pulex* (juveniles, body length 3-4 mm, wet weight ca. 2 mg) was described by Blockwell et al. [126]. Lindane (γ-hexachlorocyclohexane) was used as toxicant at mean concentrations of 0.19, 0.88, 2.67 and 6.11 $\mu g \cdot l^{-1}$ over 14 days under laboratory conditions. Effects on growth were detected by weighting the animals. Mortality was less than 10% at the lower concentrations and 30% at 6.11 $\mu g \cdot l^{-1}$. Growth of juvenile *G. pulex* was reduced during the exposure period. The no-observed-effect concentration (NOEC) is given as 2.67 $\mu g \cdot l^{-1}$, the lowest-effect concentration as 6.11 $\mu g \cdot l^{-1}$.

Sediment toxicity

Toxicity of aqueous-phase and sediment-bound copper to the crustaceans *Ceriodaphnia dubia* (Cladocera), *Daphnia magna* (Cladocera) and *Hyalella azteca* (Amphipoda) as indicated by survival and reproduction was investigated by Suedel et al. [127]. The organisms were exposed in static systems for different periods up to 14 d. Sensitivity decreased in the order *C. dubia* > *D. magna* > *H. azteca*. Mortality and growth effects increased with increasing duration of the tests in the case of *H. azteca*. When copper-contaminated sediments were used in the experiments, effects on mortality, growth and reproduction were correlated with the concentrations of copper in the overlying water rather than with the concentration of copper in bulk sediment or pore water. The effects of copper-contaminated sediments on the same organisms were also studied by Kubitz et al. [128]. In addition, they compared the sensitivity of the *H. azteca* toxicity test with a fluorescent *D. magna* test, based on in-vivo inhibition of enzymatic processes as a rapid screening tool. The results were similar to those of Suedel et al. [127]. *D. magna* and *C. dubia* were equally sensitive to copper-contaminated sediment (LC$_{50}$ 22 and 24 mg$\cdot l^{-1}$), whereas *H. azteca* was less sensitive (LC$_{50}$ 43 mg$\cdot l^{-1}$). It was shown that amphipod growth was a more sensitive bioassay endpoint than lethality. Growth of *H. azteca*, *Daphnia magna* fluorescence, mortality of *C. dubia* and of *D. magna* were equally sensitive upon exposure to copper-contaminated sediments. However, results of experiments using spiked sediments did not necessarily predict toxicity in field-collected sediments containing the same causative agent, in this case copper.

Toxicity of sediments to *D. magna* was also studied by Bridges et al. [129]. The animals were exposed 21 d to harbour sediment eluates. Endpoints of the bioassay were survival, age at first reproduction, number of broods produced and the total number of young produced per adult. Survival and reproduction data from each sediment were parameterised with a stochastic matrix model to integrate all test end-points. Diminished survival in turn had negative effects on reproduction. Population growth was mainly influenced by the first age of reproduction, followed by fecundity and survival. There were also some indications that toxicity evaluations using Cladocerans may be confused by the presence of suspended sediments in eluates.

Mortality and sub-lethal effects of copper in *Gammarus pulex* was also investigated by Taylor et al. [111]. The individuals were exposed to non-acidified and acidified copper-treated sediments. Mortality and a behavioural responses (drift into the water column, see above under behavioural changes) were measured. Mortality was low under natural conditions (untreated sediments). Addition of copper (initial aqueous concentrations of 136-942 $\mu g \cdot l^{-1}$) did not cause a significant increase of mortality, while acidification increased mortality up to 90%. Kubitz et al. [130] designed a screening and a confirmatory bioassay using *H. azteca* to assess the toxicity of river sediments. Survival and growth were tested in the first step of the bioassay. In the second step only growth was investigated. Endpoint of this bioassay was 25% growth inhibition. Because of its size-specific fecundity, growth is a good surrogate of reproductive fecundity in *H. azteca*.

Response and recovery of lake zooplankton communities to acidification and metals

Many studies have been performed on effects of acidification and metal emissions on phyto- and zooplankton communities and subsequent recoveries in manipulated (viz. lime treated) lakes — for example, in the region of Sudbury, Ontario, Canada which was heavily contaminated by copper and nickel smelters (see ref. [131, 132] and the literature cited therein). In these studies crustacean zooplankton species were used as bioindicators at different levels of consideration. In contrast to the concept of biotic indices (especially regarding the saprobic system, see above), the assessment of the results of this suite of experiments requires comparison to either a pre-contamination status (for which information is seldom available) or to temporal and spatial reference lakes in the same region.

One approach involved investigations on the population level with the Cladoceran *Daphnia galeata mendotae*, taking into account paleolimnological surveys of ephippial densities in lake sediments, actual standing stocks of *D. g. mendotae*, genetic analyses (including Hardy-Weinberg deviations or reduced clonal diversity) and laboratory bioassays quantifying survival and brood production upon exposures to actual and historical water-born Cu and Ni levels [132]. The authors conclude that *D. g. mendotae* has recovered after manipulation of the experimental lakes and that decreasing metal concentrations most probably account for that recovery. The second approach involved univariate metrics for the crustacean zooplankton (abundance, species richness, diversity and evenness) and multivariate metrics based on correspondence analysis [131]. The multivariate metrics incorporating the abundances of taxa were the best overall performers. In the first three axes 50% of the variance in the zooplankton data was summarised. Axis I separated zooplankton communities of small, nutrient-rich and dystrophic lakes, axis II communities of deep and nonacidic lakes. The strongest correlate of axis III scores was lake elevation.

Koivisto et al. [133] addressed the question whether or not cadmium pollution changes trophic interactions in experimental foodwebs composed of phytoplankton, small-bodied zooplankton (*Chydorus spharicus*, *Cyclops* sp. and rotifers), *Daphnia magna* and *Notonecta* sp. as zooplanktivorous predator. Mazumder [134] suggested the absence or near absence of large *Daphnia* and planktivorous fish as an indicator of the functional dominance of odd or even links in pelagic lake ecosystems. Thus the study of Koivisto et al. [133] links ecotoxicology to the theoretical concept of the trophic cascade (see further literature cited therein). Their results indicated that Cd (ranging from 10-20 $\mu g \cdot l^{-1}$) negatively affected phytoplankton and zooplankton abundance, whereas *Notonecta* reduced the *Daphnia* population and hence indirectly increased phytoplankton biomass in the dominant odd-linked food web. Cadmium and *Notonecta* predation had an interaction effect on phytoplankton but not on zooplankton.

Influence of experimental conditions / validation

Experimental conditions can be very important in toxicity tests using crustaceans as bioindicators. Nipper and Roper [135] showed for *Chaetocorophium* cf. *lucasi*, an amphipod living in estuarine muddy sediments, that survival was largely dependent on the type of sand used in the trials. This indicates that some amphipods may be very sensitive to sediment features, a finding which can lead to incorrect interpretations of toxicity tests. Similar results were found by Chappie and Burton [136]. Survival of *H. azteca* during in situ tests was reduced to less than 50% by limiting the period of acclimation to cold water (e.g. 2-3 h in field experiments instead of 3-5 h in laboratory experiments for acclimation to <15 °C). An interlaboratory comparison among 10 laboratories, examining the precision of sediment toxicity tests using *H. azteca* as test organism, showed that the test methods had relatively low variance and mainly acceptable levels of precision [137]. The necessity for validation procedures of experimental data is stressed by Chapman [138-140].

Case study

The suitability of gammarid amphipods as biomonitors was investigated at the University of Oldenburg (ICBM) in a suite of field studies (measuring experiments in a statistical sense) and manipulative experiments under laboratory and semi-field conditions. The main goal was to obtain statistical information about the temporal and spatial heterogeneity of metal levels in gammarids, to learn more about accumulation strategies and potential detoxification mechanisms and to develop toxicokinetic compartment models as a tool for calibration of biomonitors.

Although many pure and applied ecological studies deal with time, the problem of temporal autocorrelations is rarely addressed. In many cases, classical regression analysis is applied, completely ignoring the fact of temporal autocorrelation. Adequate procedures would involve, for example, transfer-function noise models [141-144]. Using such an approach it was possible, for example, to relate monthly Cd concentrations in the amphipod *Gammarus tigrinus* from the River Weser to the (relative) monthly water temperature of the previous month [145]. Most interestingly, this delay parameter in the transfer-function noise model was in good agreement with timescales derived from toxicokinetic experiments (summarised in ref. [146]).

In assessing the quality status of ecological systems it is necessary to compare results of different biomonitoring studies. One important problem arising in that attempt is that many biomonitoring studies operate on different spatial and/or temporal scales. Proper definitions of a hierarchy of units of investigation and the corresponding independent replicates in the sense of Hurlbert [147] are important issues for the development and optimisation of experimental designs. An example of this topic is outlined by Zauke et al. [148] for cadmium in gammarid amphipods. The smallest spatial scale (and grain) taken into account was at the level of single localities within an estuary, requiring independent subsamples from each locality as replicates; on the largest scale were European coastal waters.

Taking life history traits of gammarid collectives into account, it was possible to detect linear effects of fecundity and growth status on Cd body burdens of gammarids within the Weser estuary [149]. Such studies may lead to normalisation procedures to reduce the variability of metal concentrations in organisms on the level of single localities, but at the cost of much experimental effort and the great biological expertise necessary to determine the life history traits of the gammarid collectives. This will not always be justified. The alternative is to analyse gammarid collectives without paying attention to measures of normalisation. Under these cicumstances it was, for example, not possible to detect differences of Cd in gammarids within the Weser estuary due to an enlarged variability of Cd in these samples. Because of the trade-off between grain and effort it is the experimenters' responsibility to select an appropriate experimental design.

Assuming homogeneity of Cd in gammarids within an estuary (see above) and similar accumulation strategies (see below), the temporal variability within the Weser estuary and the spatial variability between the estuaries of the Elbe, Weser and Ems were investigated in more detail [148]. On basis of the 95% confidence intervals, no substantial variability of Cd in gammarids from the Weser estuary was obvious within the time interval 1984 - 1991, despite a considerable heterogeneity between estuaries. On the scale of European coastal waters preliminary results did not indicate substantial differences between the Weser estuary and the Island of Helgoland, but showed increased Cd concentrations in gammarids of Arctic waters (Barents Sea, unpublished data).

Accumulation strategies vary between pure regulation and net accumulation (either unlimited or showing depuration), depending on the biological species and the chemical element considered [150-153]. Only a few general conclusions are possible, for example, that barnacles are net accumulators for Zn or that many decapod crustaceans are regulators for Cu and Zn. On the other hand, there may be pronounced interspecific variability even among related species, calling for detailed evaluation at the species level. Only after appropriate validation and calibration procedures have successfully been performed is it acceptable to use collectives of different species composition in biomonitoring studies [146].

Accumulation strategies of organisms may be evaluated by exposure to various external metal concentrations, either under controlled laboratory or under semi-controlled field conditions. In toxicokinetic studies, the time course of uptake and clearance (depuration) is followed and data evaluation involves, for example, compartment modelling, which provides rate constants and bioconcentration factors (BCFs) for the theoretical equilibrium [146, 154-158]. A finding that metal concentrations do not readily reach a plateau phase is indicative of a net accumulation strategy. Though such experiments are necessary for a calibration, they are rather time consuming and expensive.

Considering amphipod crustaceans, for example, net accumulation strategies have been reported for Cd and Cu in *Echinogammarus pirloti* [152], for Cd in *G. pulex* [159, 160], for Pb in *G. pseudolimnaeus* [161] and for Cd, Pb and Hg in the North American species *Hyalella azteca*, where Cu is apparently regulated and for Zn an intermediate strategy is most likely [157, 162, 163]. Application of compartment modelling has confirmed net accumulation of Zn in *G. pulex* [155] and Cd in mixtures of *G. salinus* and *G. zaddachi* [146, 164], not only showing uptake but also elimination of these elements.

Some examples of this toxicokinetic approach are compiled for the element cadmium in Table 1 and Fig. 1. The approximate coefficients of determination suggest that the models derived explain a great proportion of the measured variance and thus give a fair description of the data. Nevertheless, more attempts at model validation are required before the bioavailability of metals for these organisms can be fully assessed. Validation is a process carried out by comparison of model predictions with independent field experiments and experimental measurements. An example of such a validation is given in Fig. 2. The solid lines represent model predictions based on the parameters summarised in Table 1, while the symbols represent independent measurements from uptake studies. Results compiled in Fig. 2 indicate an increased variability; nevertheless, model predictions and measurements are in fairly good agreement, at least within the first days of the experiments. The models are calculated until day 40 to illustrate the predictive behaviour when approaching theoretical equilibrium. We do not infer, however, that a unique set of k_1 and k_2 values exists that might be suitable as a prognostic tool, but rather that there is a whole family of k_1 and k_2 combinations as given by the different models in Fig. 2. Clearly, the differences become more pronounced when approaching the theoretical equilibrium, but even then the variability seems to be acceptable in terms of what we expect from 'ecological models'.

To summarise the results of these case studies, the gammarids investigated appear to be suitable as biomonitors at least for the element cadmium, and the data provide a good basis for the calibration of these biomonitors as well as for normalisation procedures to be employed in routine programmes.

Table 1. Toxicokinetics of Cd in gammarid amphipods from the Rivers Weser, Hunte and Ems. Summary of statistical information to assess the goodness-of-fit of the two-compartment models applied*.

Col	exp	BCF	k_1	SE_{k1}	k_2	SE_{k2}	R^2
J_1	17	425[a]	25.9	2.4	0.061	0.013	0.878
Q_1	17	377[a]	24.9	1.7	0.066	0.011	0.940
P_1	17	542[a]	65.8	4.2	0.121	0.011	0.959
J_2	23	947[a]	52.8	2.6	0.056	0.007	0.968
Q_2	23	1180[a]	70.6	5.2	0.060	0.010	0.950
P_2	23	1190[a]	98.4	6.9	0.083	0.011	0.945
Field	0.5	680[b]					
K	51.5	409[c]	103	26.2	0.253	0.069	0.785
Q	51.5	416[c]	87.7	22.5	0.211	0.061	0.796
Hu	51.5	576[c]	73.2	7.0	0.127	0.013	0.874
Ga	51.5	465[c]	61.4	7.9	0.132	0.020	0.792

Notes

* Statistical computations by SYSTAT for Windows, Version 5, NONLIN subroutine and piecewise regression option (ref. [165], p. 443 ff.).

a ref. [146], flow-through tests under semi-field conditions;

b Weser estuary (summer 1985);

c ref. [164], static tests under laboratory conditions;

Col: gammarid collectives, River Weser: J (Nordenham, 100% *Gammarus zaddachi*), K (Einswarden, 100% *G. zaddachi*), Q (Langlütjen-1, 100% *G. salinus*), P (Bremerhaven, 45% *G. zaddachi* and 55% *G. salinus*); River Hunte: Hu (Huntebrück, 100% *G. zaddachi*); River Ems: Ga (Gandersum, 100% *G. zaddachi*);

exp: exposures [$\mu g l^{-1}$];
BCF: Bioconcentration factors = k_1 / k_2; [d.w.]; w.w./d.w.-ratio = 4;
k_1: rate constant for uptake phases [d^{-1}];
k_2: rate constant for clearance phases [d^{-1}];
SE_{k1} and SE_{k2}: approximate standard errors;
R^2: approximate coefficient of determination;

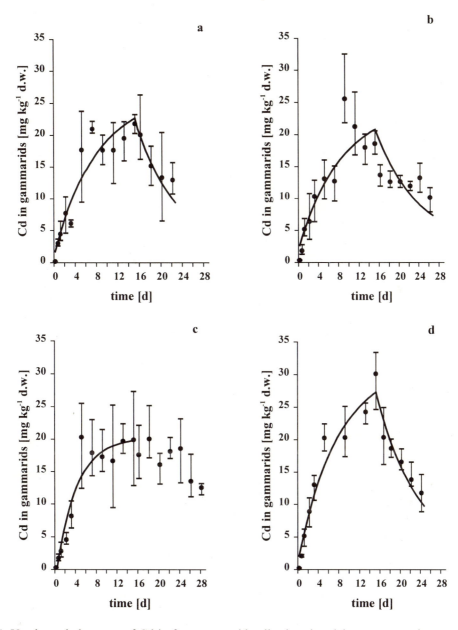

Fig. 1. Uptake and clearance of Cd in four gammarid collectives in a laboratory experiment under constant Cd exposure (51.5 ± 1.3 µg l^{-1}): (a) *G. zaddachi*, locality Gandersum, Ems, (b) *G. salinus*, locality Langlütjen-1, Weser, (c) *G. zaddachi*, locality Einswarden, Weser and (d) *G. zaddachi*, locality Huntebrück, River Hunte. Comparison of observed (●) mean values (with vertical bars = range of triplicate subsamples) and predicted results (———) from two-compartment models (kinetic data in Table 1); day 15 = end of uptake phases.

From Ritterhoff et al. [164]; © Elsevier Science B.V.

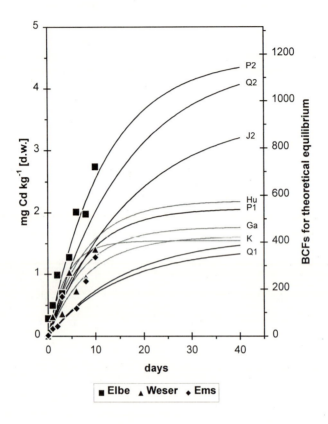

Fig. 2. Validation of compartment models (————) with experimental results on bioaccumulation of Cd in gammarid collectives (*G. zaddachi*, *G. salinus*). Models: Q_1, J_1, P_1, Q_2, J_2, P_2 from Zauke et al. [146]; Hu, Ga, K, Q from Ritterhoff et al. [164] (kinetic data in Table 1). Experimental data (Elbe, Weser, Ems) from Zauke et al. [107]; exposure: 3.8 µg Cd l^{-1}.

Conclusions

As has been shown in numerous investigations on different methodological levels, crustaceans offer excellent opportunities to derive sensitive and ecologically relevant indicators of environmental stress. This is mainly due to their specific biological attributes, including their life history strategies. Because the females often carry eggs or juveniles in specialised structures on the body (e.g. in the marsupium), the reproductive behaviour, for example, can be assayed in laboratory and semi-field experiments as well as in biological field surveys. If it can be proved in such field surveys that the population can reproduce under the given environmental constraints, this finding can be regarded as a true ecological indicator. On the other hand, if life history traits like fecundity deteriorate in laboratory or semi-field experiments, this finding provides a meaningful endpoint which allows some prediction of the potential future development of the populations under study. Not only reproduction, but also other behavioural responses of crustaceans, such as changes in feeding, drifting, locomotion or precopula behaviour have proved to provide sensitive endpoints with respect to bioindication. In all these aspects, crustaceans have great advantages compared to other groups, for example, bivalves or gastropods. It is striking, however, that almost no explicit indicator system has been developed for the marine environment as yet. One reason might be that strong pollution gradients on a local or even regional scale have been documented mainly for freshwater systems, e.g. in relation to degradable domestic waste waters (leading to different biotic indices) or to acidification and metal emissions in lakes. Another reason might be that limnology is the leading discipline in the development of theoretical concepts, e.g. the trophic cascade.

In the last few years we have seen increasing numbers of papers dealing with exposures to mixtures of contaminants or to contaminated sediments collected in the field. Such approaches are particularly relevant to this chapter, more so than the application of classical standard toxicological protocols. Nevertheless it is surprising to note that most studies deal either with ecotoxicological effects or with bioaccumulation, but not with both aspects in combination. A combined approach would be particularly appropriate when animals are exposed to mixtures of contaminants or under field conditions, because we can expect that especially those substances that are taken up can produce adverse effects. However, in this context it would also be necessary to evaluate potential detoxification mechanisms — for example, the binding of metal ions to metallothioneins or granules and the metabolism of organic xenobiotics, probably resulting in less toxic products. Thus, in the future more integrated studies are required, taking into account all these different aspects. Furthermore, we need the development of predictive models, relating bioaccumulation (e.g. on basis of compartment models) to effects, as well as subsequent verification and field validation of these models. To assess the significance of bioindicators, stringent methods for a calibration are required, to test whether a finding from a contaminated area is statistically significantly different from a finding derived for a reference site.

Routine monitoring activities in running waters to date mainly rely on some chemicals in the water phase, oxygen budget and some sort of biotic indices, depending on the country involved. In Germany, for example, the saprobic system is still in use, although it has been frequently criticised. The advantage of these methods is that results can be condensed to water quality maps, from which information is readily accessible even for non-scientist. Thus, more emphasis must be placed on linking integrated approaches of bioindication and biomonitoring with approaches commonly employed in landscape ecology, such as those using geostatistics or geographical information systems. To achieve this goal, we need closer co-operation among different scientific groups, now working in isolation in their particular classical fields, and a funding policy which favours innovative integrated approaches instead of stabilising the classical concepts.

References

[1]   C.A.M. Gestel, van and T.C. Brummelen, van (1996): Incorporation of the biomarker concept in ecotoxicology calls for a redefinition of terms. Ecotoxicology, 5, 217-225.

[2]   P.S. Rainbow (1995): Biomonitoring of heavy metal availability in the marine environment. Marine Pollution Bulletin, 31, 183-192.

[3]   G. Persoone (1998): Development and validation of toxkit microbiotests with invertebrates, in particular crustaceans, 437-449. In: P. G. Wells, K. Lee and C. Blaise (1998): *Microscale Testing in Aquatic Toxicology*, CRC Press, Boca Raton, FL, USA.

[4]   P.M. Chapman (1998): Death by mud: Amphipod sediment toxicity tests, 451-463. In: P. G. Wells, K. Lee and C. Blaise (1998): *Microscale Testing in Aquatic Toxicology*, CRC Press, Boca Raton, FL, USA.

[5]   R. Castro, A. Bosquez, M.K. Ali and A.N. Ernest (1997): Aquatic sediments. Water Envi-

ronment Research, 69, 749-777.

[6]   J.F. Davis and T.W. Kratzer (1997): Fate of environmental pollutants. Water Environment Research, 69, 861-869.

[7]   C.R. Lange and K.E. Lambert (1995): Biomonitoring. Water Environment Research, 67, 738-749.

[8]   C.R. Lange, S.R. Scott and M. Tanner (1996): Biomonitoring. Water Environment Research, 68, 801-818.

[9]   C.R. Lange and S.R. Lange (1997): Biomonitoring. Water Environment Research, 69, 900-915.

[10]   D.J. Reish, P.S. Oshida, A.J. Mearns, T.C. Ginn, E.M. Godwinsaad and M. Buchman (1997): Effects of pollution on saltwater organisms. Water Environment Research, 69, 877-892.

[11]   D.J. Reish, P.S. Oshida, A.J. Mearns and T.C. Ginn (1996): Effect of pollution on marine organisms. Water Environment Research, 68, 784-796.

[12]   J.F. Davis and T.W. Kratzer (1996): Fate of environmental pollutants. Water Environment Research, 68, 737-755.

[13]   A.N. Sastry (1983): Ecological aspects of reproduction, 179-270. In: F. J. Vernberg and W. B. Vernberg (1983): *The biology of crustacea, Vol. 8: Environmental adaptations*, Academic Press, New York.

[14]   F.J. Vernberg and W.B. Vernberg (1983): *The biology of crustacea, Vol.8: Environmental adaptations,* Academic Press, New York .

[15]   F.J. Vernberg and W.B. Vernberg (1983): *The biology of crustacea, Vol. 7: Behavior and ecology,* Academic Press, New York .

[16]   L.H. Mantel (1983): *The biology of crustacea: Internal anatomy and physiological regulation,* Academic Press, New York .

[17]   R.L. Dorit, J.W.F. Walker and R.D. Barnes (1991): *Zoology.* Saunders College Publishing, Philadelphia, PA, USA.

[18]   H.-E. Gruner (1993): 1. Klasse Crustacea, 448-1030. In: H.-E. Gruner, M. Moritz and W. Dunger (1993): *Lehrbuch der speziellen Zoologie (begr. von A. Kaestner), Wirbellose Tiere, 4. Teil: Arthropoda (ohne Insecta)*, Gustav Fischer, Jena.

[19]   E.E. Ruppert and R.D. Barnes (1994): *Invertebrate Zoology.* Saunders College Publishing, Philadelphia, PA, USA.

[20]   M. Stachowitsch (1992): *The Invertebrates: An Illustrated Glossary.* Wiley-Liss, New York.

[21]   R. Stevenson (1983): Dynamics of the integument, 2-42. In: F. J. Vernberg and W. B. Vernberg (1983): *The biology of crustacea, Vol. 9: Integument, pigments and hormonal processes*, Academic Press, New York.

[22]   P. McLaughlin (1983): Internal anatomy, 1-52. In: F. J. Vernberg and W. B. Vernberg (1983): *The biology of crustacea, Vol. 5: Internal anatomy and physiological regulation,* Academic Press, New York.

[23]   J. Ritterhoff and G.P. Zauke (1998): Potential role of metal-binding proteins in cadmium detoxification in *Themisto libellula* (Mandt) and *Themisto abyssorum* Boeck from the Greenland sea. Marine Environmental Research, 45, 179-191.

[24]   M.G. Cherian and H.M. Chan (1993): Biological functions of metallothionein - a review, 87-109. In: K. T. Suzuki, N. Imura and M. Kimura (1993): *Metallothioneins III: Biological roles and medical implications*, Birkhäuser, Basel.

[25]   R. Dallinger (1995): Metabolism and toxicity of metals: metallothioneins and metal elimination, 171-190. In: M. P. Cajaraville (1995): *Cell biology in environmental toxicology*, Servicio Editorial de la Universidad del Pais Vasco, Bilbao.

[26]   G. Roesijadi (1994): Metallothionein induction as a measure of response to metal exposure in aquatic animals. Environmental Health Perspectives, 102, 91-95.

[27]   A. Pequeux (1995): Osmotic Regulation in Crustaceans. Journal of Crustacean Biology, 15, 1-60.

[28]   D.M. Damkaer and D.B. Dey (1983): UV-damage and photoreactivation potentials of larval shrimp, *Pandalus platyceros* and adult euphausiids, *Thysanoessa raschii*. Oecologia, 60

[29]   D. Wübben and E. Vareschi (1998): *UV-B sensitivity of freshwater and marine zooplankton in a sunshine simulator.* Water of Life, XXVII Congress SIL, Dublin.

[30]   H.-E. Gruner, M. Moritz and W. Dunger (1993): *Lehrbuch der speziellen Zoologie (begr. von A. Kaestner), Wirbellose Tiere, 4. Teil: Arthropoda (ohne Insecta).* Gustav Fischer, Stuttgart.

[31]   G.-P. Zauke, J. Harms and B.A. Foster (1992): Cadmium, lead, copper and zinc in *Elminius modestus* Darwin (Crustacea, Cirripedia) from Waitemata and Manukau harbours, Auckland, New Zealand. New Zealand Journal of Marine and Freshwater Research, 26, 405-415.

[32]   J. Graham (1983): Adaptive aspects of feeding mechanisms, 65-107. In: F. J. Vernberg and W. B. Vernberg (1983): *The biology of crustacea, Vol. 8: Environmental adaptations*, Academic Press, New York.

[33]   DIN (1991): Biologisch-ökologische Gewässeruntersuchung: Bestimmung des Saprobienindex (M2), 1-18. In: Normenausschuß Wasserwesen (NAW) im DIN Deutsches Institut für Normung e.V. (1991): *Deutsche Einheitsverfahren zur Wasser-, Abwasser- und Schlammuntersuchung*, Beuth Verlag, Berlin.

[34]   S. Schulz (1995): *Proposal for some potential indicator species of the Baltic Sea littoral community.* 32.-33. Sitting of the International Working Group on the Project 'Species and Its Productivity in the Distribution Area' for the UNESCO Programme 'Man and the Biosphere', Vilnius (Lithuania), Lithuanian Academy of Sciences.

[35]   K. Naes, E. Oug and J. Knutzen (1998): Source and species-dependent accumulation of polycyclic aromatic hydrocarbons (PAHs) in littoral indicator organisms from Norwegian smelter-affected marine waters. Marine Environmental Research, 45, 193-207.

[36]   A. Jernelov (1996): The international mussel watch: A global assessment of environmental levels of chemical contaminants. Science of the Total Environment, 188, S37-S44.

[37]   B. Beliaeff, T.P. Oconnor, D.K. Daskalakis and P.J. Smith (1997): US Mussel Watch data from 1986 to 1994: Temporal trend detection at large spatial scales. Environmental Science & Technology, 31, 1411-1415.

[38]   D.J. Reish and T.V. Gerlinger (1997): A review of the toxicological studies with polychaetous annelids. Bulletin of Marine Science, 60, 584-607.

[39]   D. Bernds, D. Wübben and G.-P. Zauke (in press): Bioaccumulation of trace metals in polychaetes from the German wadden sea: Evaluation and verification of toxicokinetic models. Chemosphere.

[40]   N.A. Scarlato, J.E. Marcovecchio and A.E. Pucci (1997): Heavy metal distribution in zooplankton from Buenos Aires coastal waters (Argentina). Chemical Speciation and Bioavailability, 9, 21-26.

[41]   S.W. Fowler (1990): Critical review of selected heavy metal and chlorinated hydrocarbon concentrations in the marine environment. Marine Environmental Research, 29, 1-64.

[42]   J. Ritterhoff and G.P. Zauke (1997): Bioaccumulation of trace metals in Greenland Sea copepod and amphipod collectives on board ship: verification of toxicokinetic model parameters. Aquatic Toxicology, 40, 63-78.

[43]   J. Ritterhoff and G.P. Zauke (1997): Trace metals in field samples of zooplankton from the Fram Strait and the Greenland Sea. Science of the Total Environment, 199, 255-270.

[44]   J. Dahl and L. Greenberg (1996): Effects of habitat structure on habitat use by *Gammarus pulex* in artificial streams. Freshwater Biology, 36, 487-495.

[45]   M. Crane (1994): Population characteristics of *Gammarus pulex* (L.) from five English streams. Hydrobiologia, 281, 91-100.

[46]   H.-G. Meurs and G.-P. Zauke (1996): Neozoen und andere Makrozoobenthos-Verän-derungen, 208-213. In: J. L. Lozan and H. Kausch (1996): *Warnsignale aus Flüssen und Ästuaren*, Parey Buchverlag, Berlin.

[47]   B. Streit and K. Kuhn (1994): Effects of organophosphorous insecticides on authochthonous introduced Gammarus species. Water Science and Technology, 29, 233-240.

[48]   M.P.D. Meijering (1991): Lack of oxygen and low pH as limiting factors for *Gammarus* in Hessian brooks and rivers. Hydrobiologia, 223, 159-169.

[49]   L. Maltby (1991): Pollution as a probe of life history adaptation in *Asellus aquaticus* (Iso-poda). Oikos, 61, 11-18.

[50]   W. Pilgrim and M.D.B. Burt (1993): Effect of acute pH depression on the survival of the freshwater amphipod *Hyalella azteca* at variable temperatures: field and laboratory studies. Hydrobiologia, 254, 91-98.

[51]   M. Pöckl and U.H. Humpesch (1990): Intra- and inter-specific variations in egg survival and brood development time for the Austrian populations of *Gammarus fossarum* and *G. roeseli* (Crustacea: Amphipoda). Freshwater Biology, 23, 441-455.

[52]   M. Pöckl (1993): Reproductive potential and lifetime potential fecundity of the freshwater amphipods *Gammarus fossarum* and *G. roeseli* in Austrian streams and rivers. Freshwater Biology, 30, 73-91.

[53]   D.S. Glazier, M.T. Horne and M.E. Lehman (1992): Abundance, body composition and re-productive output of *Gammarus minus* (Crustacea: Amphipoda) in ten cold springs differing in pH and ionic content. Freshwater Biology, 28, 149-163.

[54]   H.-G. Meurs and G.-P. Zauke (1996): Populationsbiologische Untersuchungen an Gammari-den als Grundlage zur Abschätzung von Eingriffen (UVS) in Flußmündungen. Zeitschrift für Ökologie und Naturschutz, 5, 107-113.

[55]   M.A.S. Graca, L. Maltby and P. Calow (1994): Comparative ecology of *Gammarus pulex* (L.) and *Asellus aquaticus* (L.) I: population dynamics and microdistribution. Hydrobiolo-gia, 281, 155-162.

[56]   M.A.S. Graca, L. Maltby and P. Calow (1994): Comparative ecology of *Gammarus pulex* (L.) and *Asellus aquaticus* (L.) II: fungal preferences. Hydrobiologia, 281, 163-170.

[57]   W.D. Williams (1962): The geographic distribution of the isopod *Asellus aquaticus* (L.) and Asellus meridianus Rac. Proceedings of the zoological Society of London, 39, 75-96.

[58]   P. Maitland (1966): Notes of the biology of *Gammarus pulex* in the River Endrick. Hydro-biologia, 28, 141-151.

[59]   M.A.S. Graca, L. Maltby and P. Calow (1993): Importance of fungi in the diet of *Gammarus pulex* and *Asellus aquaticus* I: feeding strategies. Oecologia, 93, 139-144.

[60]   A. Hargeby (1990): Effects of pH, humic substances and animal interactions on survival and physiological status of *Asellus aquaticus* (L.) and *Gammarus pulex* (L.): A field experiment. Oecologia, 82, 348-354.

[61]   M.A.S. Graca, L. Maltby and P. Calow (1993): Importance of fungi in the diet of *Gammarus pulex* and *Asellus aquaticus* II: effects on growth, reproduction and physiology. Oecologia, 96, 304-309.

[62]   L. Maltby (1995): Sensitivity of the crustaceans *Gammarus pulex* (L) and *Asellus aquaticus* (L) to short-term exposure to hypoxia and unionized ammonia: Observations and possible mechanisms. Water Research, 29, 781-787.

[63]   I.T. Whitehurst (1991): The *Gammarus:Asellus* ratio as an index of organic pollution. Water Research, 25, 333-339.

[64]   I.T. Whitehurst (1991): The effects of sampling techniques on the *Gammarus:Asellus* ratio. Water Research, 25, 745-748.

[65]   M. Zelinka and P. Marvan (1961): Zur Präzisierung der biologischen Klassifikation der Reinheit fließender Gewässer. Archiv für Hydrobiologie, 57, 389-407.

[66]   K. Wuhrmann (1969): *Selbstreinigung in Fließgewässern.* Europäisches Abwassersymposium München 1969, München.

[67]   K. Wuhrmann (1974): Some problems and perspectives in applied limnology. Mitteilungen der Internationalen Vereinigung für Limnologie, 20, 324-402.

[68]   G.-P. Zauke and H.-G. Meurs (1996): Kritische Anmerkungen zum Einsatz des Saprobiensystems bei der Gewässerüberwachung, 329-331. In: J. L. Lozan and H. Kausch (1996): *Warnsignale aus Flüssen und Ästuaren*, Parey Buchverlag, Berlin.

[69]   U. Braukmann (1987): Zoozönologische und saprobiologische Beiträge zu einer allgemeinen regionalen Bachtypologie. Archiv für Hydrobiologie: Beihefte Ergebnisse zur Limnologie, 26, 1-355.

[70]   N. De Pauw, D. Roels and A.P. Fontoura (1986): Use of artificial substrates for standardized sampling of macroinvertebrates in the assessment of water quality by the Belgian Biotic Index. Hydrobiologia, 133, 237-258.

[71]   C.M. Lorenz, G.M. Vandijk, A.G.M. Vanhattum and W.P. Cofino (1997): Concepts in river ecology: Implications for indicator development. Regulated Rivers - Research & Management, 13, 501-516.

[72]   F. Malard, J. Mathieu, J.L. Reygrobellet and M. Lafont (1996): Biomonitoring groundwater contamination: Application to a karst area in Southern France. Aquatic Sciences, 58, 158-184.

[73]   S.D. Rundle and P.M. Ramsay (1997): Microcrustacean communities in streams from two physiographically contrasting regions of Britain. Journal of Biogeography, 24, 101-111.

[74]   B. Milhau, N. Dekens and K. Wouters (1997): Assessing the use of ostracods as potential bioindicators of pollution. Application on the Slack River (Boulonnais, France). Écologie: revue trimestrielle / Société Française d'Écologie, 28, 3-12.

[75]   I. Boomer, R. Whatley and N.V. Aladin (1996): Aral Sea Ostracoda as environmental indicators. Lethaia: an international journal of palaeontology and stratigraphy; official journal of the International Palaeontological Association, 29, 77-85.

[76]   A.J. Smith (1993): Lacustrine ostracodes as hydrochemical indicators in lakes of the North-central United States. Journal of Paleolimnology, 8, 121-134.

[77]   M. Llano (1987): Utilization de los estracodos bentonicos marinos como herramientas para el conocimiento hidrologico de las plataformas continentales y su aplicacion en la prospeccion geologica. Memorias de la Sociedad de Ciencias Naturales La Salle, 47, 105-124.

[78]   H.I. Griffiths, K.E. Pillidge, C.J. Hill, J.G. Evans and M.A. Learner (1996): Ostracod gradients in a calcareous stream: Implications for the palaeoecological interpretation of tufas and travertines. Limnologica, 26, 49-61.

[79]   M. Amyot, B. Pinelalloul and P.G.C. Campbell (1994): Abiotic and seasonal factors influencing trace metal levels (Cd, Cu, Ni, Pb, and Zn) in the freshwater amphipod *Gammarus fasciatus* in two fluvial lakes of the St. Lawrence River. Canadian Journal of Fisheries and Aquatic Science, 51, 2003-2016.

[80]   M. Amyot, B. Pinelalloul, P.G.C. Campbell and J.C. Desy (1996): Total metal burdens in the freshwater amphipod *Gammarus fasciatus*: Contribution of various body parts and influence of gut contents. Freshwater Biology, 35, 363-373.

[81]   S. Plenet (1995): Freshwater amphipods as biomonitors of metal pollution in surface and interstitial aquatic systems. Freshwater Biology, 33, 127-137.

[82]   C.G. Ingersoll, W.G. Brumbaugh, F.J. Dwyer and N.E. Kemble (1994): Bioaccumulation of metals by *Hyalella azteca* exposed to contaminated sediments from the upper Clark Fork River, Montana. Environmental Toxicology and Chemistry, 13, 2013-2020.

[83]   J.P. Meador, E. Casillas, C.A. Sloan and U. Varanasi (1995): Comparative bioaccumulation of polycyclic aromatic hydrocarbons from sediment by two infaunal invertebrates. Marine Ecology - Progress Series, 123, 107-124.

[84]   E.J. Taylor, D.P.W. Jones, S.J. Maund and D. Pascoe (1993): A new method for measuring the feeding activity of *Gammarus pulex* (L.). Chemosphere, 26, 1375-1381.

[85]   S.J. Blockwell, E.J. Taylor, I. Jones and D. Pascoe (1998): The influence of fresh water pollutants and interaction with *Asellus aquaticus* (L.) on the feeding activity of *Gammarus pulex* (L.). Archives of Environmental Contamination and Toxicology, 34, 41-47.

[86]   M. Crane and L. Maltby (1991): The lethal and sub-lethal response of *Gammarus pulex* to stress: sensitivity and variation in an in situ bioassay. Environmental Toxicology and Chemistry, 10, 1331-1339.

[87]   L. Maltby and M. Crane (1994): Responses of *Gammarus pulex* (Amphipoda, Crustacea) to metalliferous effluents: Identification of toxic components and the importance of interpopulation variation. Environmental Pollution, 84, 45-52.

[88]   A. Gerhardt (1992): Effects of subacute doses of iron (Fe) on *Leptophlebia marginata* (Insects: Ephemeroptera). Freshwater Biology, 27, 79-84.

[89]   D.W. Moore and J.D. Farrar (1996): Effect of growth on reproduction in the freshwater amphipod, *Hyalella azteca* (Saussure). Hydrobiologia, 328, 127-134.

[90]   C. Naylor, L. Maltby and P. Calow (1989): Scope for growth in *Gammarus pulex*, a freshwater benthic detritivore. Hydrobiologia, 188/189, 517-523.

[91]   L. Maltby, C. Naylor and P. Calow (1990): Effect of stress on a freshwater benthic detritivore: Scope for growth in *Gammarus pulex*. Ecotoxicology and Environmental Safety, 19, 285-291.

[92]   L. Maltby, C. Naylor and P. Calow (1990): Field deployment of a scope for growth assay involving *Gammarus pulex*, a freshwater benthic invertebrate. Ecotoxicology and Environmental Safety, 19, 292-300.

[93]   L. Maltby and C. Naylor (1990): Preliminary observations on the ecological relevance of the *Gammarus* 'scope for growth' assay: Effect of zinc on reproduction. Functional Ecology, 4, 393-397.

[94]   J.S. Weis, A. Cristini and K. Ranga-Rao (1992): Effects of pollutants on molting and regeneration in crustacea. American Zoologist, 32, 495-500.

[95]   M. Poulton and D. Pascoe (1990): Disruption of precopula in *Gammarus pulex* (L.) - Development of a behavioural bioassay for evaluating pollutant and parasite induced stress. Chemosphere, 20, 403-415.

[96]   J.F.C. Malbouisson, T.W.K. Young and A.W. Bark (1994): Disruption of precopula in *Gammarus pulex* as a result of brief exposure to gamma-hexachlorocyclohexane (lindane). Chemosphere, 28, 2011-2020.

[97]   J.F.C. Malbouisson, T.W.K. Young and A.W. Bark (1995): Use of feeding rate and repairing of precopulatory *Gammarus pulex* to assess toxicity of gamma-hexachlorocyclohexane (lindane). Chemosphere, 30, 1573-1583.

[98]   D. Pascoe, T.J. Kedwards, S.J. Maund, E. Muthi and E.J. Taylor (1994): Laboratory and field evaluation of a behavioural bioassay - The *Gammarus pulex* (L.) precopula separation (GaPPS) test. Water Research, 28, 369-372.

[99]   C.P. McCahon, M.J. Poulton, P.C. Thomas, Q. Xu, D. Pascoe and C. Turner (1991): Lethal and sub-lethal toxicity of field simulated farm waste episodes to several freshwater invertebrate species. Water Research, 25, 661-671.

[100]   J.-T. Borlakoglu and R. Kickuth (1990): Behavioral changes in *Gammarus pulex* and its significance in the toxicity assessment of very low levels of environmental pollutants. Bulletin of Environmental Contamination and Toxicology, 45, 258-265.

[101]   A. Gerhardt, E. Svensson, M. Clostermann and B. Fridlund (1994): Monitoring of behavioral patterns of aquatic organisms with an impedance conversion technique. Environment International, 20, 209-219.

[102]   A. Gerhardt (1995): Monitoring behavioural responses to metals in *Gammarus pulex* (L)

(Crustacea) with impedance conversion. Environmental Science and Pollution Research, 2, 15-23.

[103] A. Gerhardt (1996): Behavioural early warning responses to polluted water - Performance of *Gammarus pulex* L (Crustacea) and *Hydropsyche angustipennis* (Curtis) (Insecta) to a complex industrial effluent. Environmental Science and Pollution Research, 3, 63-70.

[104] A. Gerhardt, A. Carlsson, C. Ressemann and K.P. Stich (1998): New online biomonitoring system for *Gammarus pulex* (L.) (Crustacea): In situ test below a copper effluent in south Sweden. Environmental Science & Technology, 32, 150-156.

[105] J.T.A. Dick (1995): The cannibalistic behaviour of two *Gammarus* species (Crustacea: Amphipoda). Journal of Zoology, 236, 697-706.

[106] U. Borgmann (1994): Chronic toxicity of ammonia to the amphipod *Hyalella azteca*; importance of ammonium ion and water hardness. Environmental Pollution, 86, 329-335.

[107] G.-P. Zauke, R. von Lemm, H.-G. Meurs, D. Todeskino, H.-P. Bäumer and W. Butte (1986): *Bioakkumulation and Toxizität von Cadmium bei Gammariden im statischen Labortest und dynamischen Test unter in-situ Bedingungen.* Umweltbundesamt Berlin, Forschungsbericht Wasser 102 05 209, Teil 2, 147 pp., Oldenburg, Germany.

[108] M. Liess (1993): *Zur Ökotoxikologie der Einträge von landwirtschaftlich genutzten Flächen in Fließgewässer.* Cuvillier, Göttingen.

[109] M. Liess and R. Schulz (1995): Ökotoxikologische Bewertung von Pflanzenschutzmittel-Einträgen aus landwirtschaftlich genutzten Flächen in Fließgewässer, V-3.3.5 / 1-44. In: C. Steinberg, H. Bernhardt and H. Klapper (1995): *Handbuch angewandte Limnologie*, ecomed Verlagsgesellschaft, Landsberg am Lech, Germany.

[110] P. Matthiessen, D. Sheahan, R. Harrison, M. Kirby, R. Rycroft, A. Turnbull, C. Volkner and R. Williams (1995): Use of a Gammarus pulex bioassay to measure the effects of transient carbofuran runoff from farmland. Ecotoxicology and Environmental Safety, 30, 111-119.

[111] E.J. Taylor, E.M. Rees and D. Pascoe (1994): Mortality and drift-related response of the freshwater amphipod *Gammarus pulex* (L.) exposed to natural sediments, acidification and copper. Aquatic Toxicology, 29, 83-101.

[112] W. Traunspurger and C. Drews (1996): Toxicity analysis of freshwater and marine sediments with meio- and macrobenthic organisms: A review. Hydrobiologia, 328, 215-261.

[113] W. Ahlf (1995): Biotests an Sedimenten, V-3.6.1 / 1-43. In: C. Steinberg, H. Bernhardt and H. Klapper (1995): *Handbuch angewandte Limnologie*, ecomed Verlagsgesellschaft, Landsberg am Lech, Germany.

[114] R.M. Mulliss, D.M. Revitt and R.B.E. Shutes (1996): A statistical approach for the assessment of the toxic influences on *Gammarus pulex* (Amphipoda) and *Asellus aquaticus* (Isopoda) exposed to urban aquatic discharges. Water Research, 30, 1237-1243.

[115] R.M. Mulliss, D.M. Revitt and R.B.E. Shutes (1996): The determination of the toxic influences to *Gammarus pulex* (Amphipoda) caged in urban receiving waters. Ecotoxicology, 5, 209-215.

[116] M. de Nicola Giudici, L. Migliore, C. Gambardella and A. Marotta (1988): Effect of chronic exposure to cadmium and copper on *Asellus aquaticus* (L.) (Crustacea, Isopoda). Hydrobiologia, 157, 265-269.

[117] M. Crane, P. Delaney, S. Watson, P. Parker and C. Walker (1995): The effect of malathion 60 on *Gammarus pulex* (L) below watercress beds. Environmental Toxicology and Chemistry, 14, 1181-1188.

[118] M. Crane (1995): Effect of zinc on four populations and two generations of *Gammarus pulex* (L). Freshwater Biology, 33, 119-126.

[119] S.J. Maund, E.J. Taylor and D. Pascoe (1992): Population responses of the freshwater amphipod crustacean *Gammarus pulex* (L.) to copper. Freshwater Biology, 28, 29-36.

[120] M. de Nicola Giudici, L. Migliore, S.M. Guarino and C. Gambardella (1987): Acute and

long-term toxicity of cadmium to *Idothea balthica* (Crustacea, Isopoda). Marine Pollution Bulletin, 18, 454-458.

[121] M. de Nicola Giudici and S.M. Guarino (1989): Effects of chronic exposure to cadmium or copper on *Idothea balthica* (Crustacea, Isopoda). Marine Pollution Bulletin, 20, 69-73.

[122] E.J. Taylor, S.J. Maund and D. Pascoe (1991): Toxicity of four common pollutants to the freshwater macroinvertebrates *Chironomus riparius* Meigen (Insecta: Diptera) and *Gammarus pulex* (L.) (Crustacea: Amphipoda). Archives of Environmental Contamination and Toxicology, 21, 371-376.

[123] S.A. Collyard, G.T. Ankley, R.A. Hoke and T. Goldenstein (1994): Influence of age on the relative sensitivity of *Hyalella azteca* to diazinon, alkylphenol ethoxylates, copper, cadmium, and zinc. Archives of Environmental Contamination and Toxicology, 26, 110-113.

[124] G.L. Phipps, V.R. Mattson and G.T. Ankley (1995): Relative sensitivity of three freshwater benthic macroinvertebrates to ten contaminants. Archives of Environmental Contamination and Toxicology, 28, 281-286.

[125] C.P. McCahon and M.J. Poulton (1991): Lethal and sub-lethal effects of acid, aluminium and lime on *Gammarus pulex* during repeated simulated episodes in a Welsh stream. Freshwater Biology, 25, 169-178.

[126] S.J. Blockwell, D. Pascoe and E.J. Taylor (1996): Effects of lindane on the growth of the freshwater amphipod *Gammarus pulex* (L). Chemosphere, 32, 1795-1803.

[127] B.C. Suedel, E. Deaver and J.H. Rodgers (1996): Experimental factors that may affect toxicity of aqueous and sediment-bound copper to freshwater organisms. Archives of Environmental Contamination and Toxicology, 30, 40-46.

[128] J.A. Kubitz, E.C. Lewek, J.M. Besser, J.B. Drake and J.P. Giesy (1995): Effects of copper-contaminated sediments on *Hyalella azteca*, *Daphnia magna*, and *Ceriodaphnia dubia*: Survival, growth, and enzyme inhibition. Archives of Environmental Contamination and Toxicology, 29, 97-103.

[129] T.S. Bridges, R.B. Wright, B.R. Gray, A.B. Gibson and T.M. Dillon (1996): Chronic toxicity of Great Lakes sediments to *Daphnia magna*: Elutriate effects on survival, reproduction and population growth. Ecotoxicology, 5, 83-102.

[130] J.A. Kubitz, J.M. Besser and J.P. Giesy (1996): A two-step experimental design for a sediment bioassay using growth of the amphipod *Hyalella azteca* for the test end point. Environmental Toxicology and Chemistry, 15, 1783-1792.

[131] N.D. Yan, W. Keller, K.M. Somers, T.W. Pawson and R.E. Girard (1996): Recovery of crustacean zooplankton communities from acid and metal contamination: Comparing manipulated and reference lakes. Canadian Journal of Fisheries and Aquatic Sciences, 53, 1301-1327.

[132] N.D. Yan, P.G. Welsh, H. Lin, D.J. Taylor and J.M. Filion (1996): Demographic and genetic evidence of the long-term recovery of *Daphnia galeata mendotae* (Crustacea: Daphniidae) in Sudbury lakes following additions of base: The role of metal toxicity. Canadian Journal of Fisheries and Aquatic Sciences, 53, 1328-1344.

[133] S. Koivisto, M. Arner and N. Kautsky (1997): Does cadmium pollution change trophic interactions in rockpool food webs? Environmental Toxicology and Chemistry, 16, 1330-1336.

[134] A. Mazumder (1994): Patterns in algal biomass in dominant odd- vs. even-linked lake ecosystems. Ecology, 75, 1141-1149.

[135] M.G. Nipper and D.S. Roper (1995): Growth of an amphipod and a bivalve in uncontaminated sediments: Implications for chronic toxicity assessments. Marine Pollution Bulletin, 31, 424-430.

[136] D.J. Chappie and G.A. Burton (1997): Optimization of in situ bioassays with *Hyalella azteca* and *Chironomus tentans*. Environmental Toxicology and Chemistry, 16, 559-564.

[137] G.A. Burton, T. Norbergking, C.G. Ingersoll, D.A. Benoit, G.T. Ankley, P.V. Winger, J. Kubitz, J.M. Lazorchak, M.E. Smith, E. Greer, F.J. Dwyer, D.J. Call, K.E. Day, P. Kennedy and M. Stinson (1996): Interlaboratory study of precision: *Hyalella azteca* and *Chironomus tentans* freshwater sediment toxicity assays. Environmental Toxicology and Chemistry, 15, 1335-1343.

[138] P.M. Chapman (1995): Extrapolating laboratory toxicity results to the field. Environmental Toxicology and Chemistry, 14, 927-930.

[139] P.M. Chapman (1995): Do sediment toxicity tests require field validation? Environmental Toxicology and Chemistry, 14, 1451-1453.

[140] P.M. Chapman, A. Fairbrother and D. Brown (1998): A critical evaluation of safety (uncertainty) factors for ecological risk assessment. Environmental Toxicology and Chemistry, 17, 99-108.

[141] G.E.P. Box, G.N. Jenkins and G.C. Reinsel (1994): *Time series analysis: Forcasting and control*. Englewood Cliffs, Prentice Hall.

[142] J. Malgras and D. Debouzie (1997): Can ARMA models be used reliably in ecology? Acta Oecologica - International Journal of Ecology, 18, 427-447.

[143] S. Ohnishi, Y. Matsumiya, M. Ishiguro and K. Sakuramoto (1995): Construction of time series analysis model effective for forecast of fishing and oceanographic conditions. Fisheries Science, 61, 550-554.

[144] K.I. Stergiou, E.D. Christou and G. Petrakis (1997): Modelling and forecasting monthly fisheries catches: Comparison of regression, univariate and multivariate time series methods. Fisheries Research, 29, 55-95.

[145] H.-P. Bäumer and G.-P. Zauke (1987): Modelling the dynamic relationship between Cd-concentration in *Gammarus tigrinus* and water temperature. Environmental Technology Letters, 8, 529-544.

[146] G.-P. Zauke, R. von Lemm, H.-G. Meurs and W. Butte (1995): Validation of estuarine gammarid collectives (Amphipoda: Crustacea) as biomonitors for cadmium in semi-controlled toxicokinetic flow-through experiments. Environmental Pollution, 90, 209-219.

[147] S.H. Hurlbert (1984): Pseudoreplication and the design of ecological field experiments. Ecological Monographs, 54, 187-211.

[148] G.-P. Zauke, G. Petri, J. Ritterhoff and H.-G. Meurs (1996): Theoretical background for the assessment of the quality status of ecosystems: lessons from studies of heavy metals in aquatic invertebrates. Senckenbergiana maritima, 27, 207-214.

[149] H.-P. Bäumer, A. van der Linde and G.-P. Zauke (1991): Structural equation models, applications in biological monitoring. Biometrie und Informatik in Medizin und Biologie, 22, 156-178.

[150] D.J.H. Phillips and P.S. Rainbow (1989): Strategies of trace metal sequestration in aquatic organisms. Marine Environmental Research, 28, 207-210.

[151] P.S. Rainbow (1993): The significance of trace metal concentration in marine invertebrates, 4-23. In: R. Dallinger and P. S. Rainbow (1993): *Ecotoxicology of metals in invertebrates.*, Lewis Publishers, Boca Raton, USA.

[152] P.S. Rainbow and S.L. White (1989): Comparative strategies of heavy metal accumulation by crustaceans: Zinc, copper and cadmium in a decapod, an amphipod and a barnacle. Hydrobiologia, 174, 245-262.

[153] P.S. Rainbow, D.J.H. Phillips and M.H. Depledge (1990): The significance of trace metal concentrations in marine invertebrates, a need for laboratory investigation of accumulation strategies. Marine Pollution Bulletin, 21, 321-324.

[154] M.P.M. Janssen, A. Bruins, T.H. De Vries and N.M. Van Straalen (1991): Comparison of cadmium kinetics in four soil arthropod species. Archives of Environmental Contamination and Toxicology, 20, 305-312.

[155]   Q. Xu and D. Pascoe (1993): The bioconcentration of zinc by *Gammarus pulex* (L.) and the application of a kinetic model to determine bioconcentration factors. Water Research, 27, 1683-1688.

[156]   B.v. Hattum, P. de Voogt, L. van den Bosch, N.M. Van Straalen and E.N.G. Joosse (1989): Bioaccumulation of cadmium by the freshwater isopod *Asellus aquaticus* (L.) from aqueous and dietary sources. Environmental Pollution, 62, 129-151.

[157]   M. Stephenson and M.A. Turner (1993): A field study of cadmium dynamics in periphyton and in *Hyalella azteca* (Crustacea: Amphipoda). Water Air and Soil Pollution, 68, 341-361.

[158]   K.R. Timmermans, E. Spijkerman, M. Tonkes and H. Grovers (1992): Cadmium and Zinc uptake by two species of aquatic invertebrate predators from dietary and aquaeous sources. Canadian Journal of Fisheries and Aquatic Sciences, 49, 655-662.

[159]   D.A. Wright (1980): Cadmium and calcium interactions in the freshwater amphipod *Gammarus pulex*. Freshwater Biology, 10, 123-133.

[160]   A. Stuhlbacher and L. Maltby (1992): Cadmium resistence in *Gammarus pulex* (L.). Archives of Environmental Contamination and Toxicology, 22, 319-324.

[161]   R.L. Spehar, R.L. Anderson and J.T. Fiandt (1978): Toxicity and bioaccumulation of cadmium and lead in aquatic invertebrates. Environmental Pollution, 15, 195-208.

[162]   B.W. Kilgour (1991): Cadmium uptake from cadmium-spiked sediments by four freshwater invertebrates. Bulletin of Environmental Contamination and Toxicology, 47, 70-75.

[163]   U. Borgmann, W.P. Norwood and C. Clarke (1993): Accumulation, regulation and toxicity of copper, zinc, lead and mercury in *Hyalella azteca*. Hydrobiologia, 259, 79-89.

[164]   J. Ritterhoff, G.-P. Zauke and R. Dallinger (1996): Calibration of the estuarine amphipods, *Gammarus zaddachi* Sexton (1912), as biomonitors: toxicokinetics of cadmium and possible role of inducible metal binding proteins in Cd detoxification. Aquatic Toxicology, 34, 351-369.

[165]   L. Wilkinson, M.A. Hill, J.P. Welna and G.K. Birkenbeuel (1992): *Systat for Windows, Version 5*. Systat Inc., Evanston, IL, USA.

Environmental Science Forum Vol.96 (1999) pp. 195-232
© 1999 Trans Tech Publications, Switzerland

Biomonitoring and Ecotoxicology: Fish as Indicators of Pollution-Induced Stress in Aquatic Systems

L. Cleveland, J.F. Fairchild and E.E. Little

Columbia Environmental Research Center, Biological Resources Division, U.S. Geological Survey, 4200 New Haven Road, Columbia, MO 65201, USA

Keywords: Ecotoxicology, Fish, Pollution-Induced Stress, Aquatic Systems

INTRODUCTION

During the past three decades environmental quality trends have been linked closely to global growth and development. Since 1970 global population has increased by approximately two billion people and commercial consumption of oil increased by three billion tons annually.[1] The increase in global population and energy consumption has been accompanied by increased travel and population mobility, vehicles usage, agricultural activities, and overall improvements in global social and economic status.[1] These human advances have been accompanied by several well-documented international, regional, and local environmental quality issues and concerns. These environmental issues and concerns include global climate change and global warming[1, 2,3]; ultraviolet radiation and stratospheric ozone depletion.[4, 5, 6,7,8]; accelerated loss of biological diversity[9] across all hierarchal levels;[3, 10,11] acidic deposition;[1,2,12,13] and pollution from agricultural and industrial activities.[14,15,16,17,18,19,20,21,22,23]

Even with improved environmental quality over the past three decades, environmental burdens continue to increase primarily because of rising consumption and population growth.[1] Environmental degradation is occurring continually as a result of millions of metric tons of pollutants being released annually from point and non-point sources.[24] Each contaminant that enters the environment has the potential to elicit adverse ecological effects, either singly or in concert with co-contaminants. Also, many new chemicals are developed annually for a wide variety of agricultural, industrial, and other applications. The environmental safety of these chemicals are determined through the use of single chemical and single species tests with a limited number of species[25], which cannot account for adverse ecological impacts that can result from the interaction of multiple species and multiple chemicals in natural systems. Continued and expanded monitoring of environmental quality is necessary to address current environmental concerns and issues and to maintain the currently improved status of the environment compared to earlier decades. Also, environmental monitoring is needed to insure that ecosystem services that benefit natural systems and human society, e.g., maintenance of atmospheric gas balance, flood control, carbon storage, and water quality maintenance, to mention a few, continue indefinitely.[26]

The protection of aquatic systems from pollution-induced stress requires a thorough integration of biomonitoring and ecotoxicology research.[27,28,29] Ecotoxicology is applied to many different species across all environmental compartments; however, this chapter will focus only on the aquatic environmental compartment and species of fish. Also, the biosurvey and chemical and physical monitoring elements of biomonitoring is beyond the scope of this chapter, which will focus entirely on toxicological testing and biomonitoring with fish. Fish are excellent tools for ecotoxicological monitoring for the following reasons:[30,31]

(1) the life-cycle of many species are well-documented,

(2) early life stages are highly sensitive to contaminants,

(3) fish remain in intimate contact with their habitat and avoidance of exposure is difficult,

(4) numerous toxicity test methods for fish have been standardized,

(5) fish provide a natural integrating function for multiple aquatic stressors,

(6) fish can serve as a direct measure of the bioavailability of contaminants and may bioconcentrate contaminants up to 106 times above ambient levels, and

(7) fish can serve as early warning systems for detecting exposure

REGULATORY REQUIREMENTS AND APPLICATIONS OF BIOMONITORING

United States

Environmental regulation of water quality in the United States can be categorized into three areas: 1) chemical registration, 2) regulatory compliance, and 3) Natural Resource Damage Assessments. Human, mammalian, avian, and aquatic tests are required in each of these regulatory areas. Aquatic data is heavily based on the results of toxicity and bioaccumulation studies with fish.

Chemical Registration and Compliance Monitoring. Chemical registration primarily occurs under three Federal statutes: The Toxic Substances Control Act (TSCA; 1976); the Federal Insecticide, Fungicide, and Rodenticide Act (the FIFRA; passed in 1947 and subsequently amended in 1959, 1961, 1972, 1978, and 1988); and the Food and Drug Administration (FDA).

The TSCA is used to regulate the production, transportation, use, and disposal of chemicals of commerce.[32,33] Although not directly applied to the environment these chemicals may enter aquatic systems indirectly during registered uses. Under the TSCA chemical manufacturers are required to submit data under a pre-manufacturing notice. This data concerns anticipated levels of uses; predicted chemical fate; and toxicity. Those chemicals that are predicted to enter the environment in significant amounts are subsequently tested under more rigorous conditions. There are 31 sections under the TSCA; Sections 2 and 4 refer primarily to the conduct of toxicity and bioaccumulation tests with fish. Primary testing guidelines for fish include the conduct of acute toxicity tests, bioconcentration tests, and early lifestage tests. Rainbow trout, bluegill, fathead minnow, sheepshead minnow, and silversides are the primary species used in testing under the TSCA.

A different approach is taken under the FIFRA.[34] The FIFRA regulates chemicals that are manufactured specifically for toxicity and that are intended to be directly applied to the environment (e.g. insecticides, miticides, herbicides, fungicides, etc.). The FIFRA has numerous subdivisions related to human worker exposure, food tolerances, environmental fate, etc. The portion specifically related to the protection of non-target aquatic organisms including fish are provided under 40 CFR Section 158.145 (Subdivision E). Regulatory testing under the FIFRA is conducted under a 4-tiered testing system: tier 1, consisting of limited fate and acute toxicity studies; tier 2, consisting of multiple acute toxicity studies of both freshwater and marine species; tier 3, consisting of early lifestage and chronic testing; and tier 4, consisting of controlled field (microcosm or mesocosm) or field monitoring studies.[35,36] Primary, toxicity tests with fish are emphasized and frequently based on the same standardized tests with species such as rainbow trout, fathead minnow, bluegill, sheepshead, and silversides. However, the FIFRA, unlike the TSCA primarily utilizes a population approach to chemical regulation in which the sensitivity of limited number of species and life-stages are tested in relation to the expected environmental contaminant concentrations. Safety factors are used within each test tier to determine if risk is inferred, at which point additional testing at higher tiers are required to continue the registration process. Recently, strong efforts have been made within the regulatory and research communities to harmonize the approaches for both TSCA and FIFRA in order to develop a probablistic, community-based approach to risk assessment.[37,38]

The FDA, under the requirements of the Federal Food, Drug, and Cosmetic Act of 1938, is responsible for insuring that food, biological products, medicines, and cosmetics are safe and effective.[39] Testing must be conducted to insure that exposure and adverse effects do not occur from manufacturing or hospital, farm, and household use of these chemicals. This includes the registration of fishery chemicals used in the husbandry and veterinary care of fishes. Fish tests similar to those used within the FIFRA are required by the FDA within its jurisdiction.

The Clean Water Act (originally the Federal Water Pollution Control Act 1972; amended in 1977 and 1987) is the primary regulatory authority used in the United States for compliance monitoring.[40] The U.S. EPA requires effluent testing of point source discharges within the National Pollution Discharge and Elimination System (NPDES) process. Both chemical and biological testing are required. Protective concentrations, known as water quality criteria, are developed based on a multi-species dataset containing at least 8 species distributed among representative fish, inverterbrate, and aquatic plant species. Safe concentrations are estimated based on a cumulative response distribution that is used to interpolate a concentration that is protective of 95% of the aquatic species (i.e. a community-based approach to regulation). States are allowed to adopt these standards, or in some cases, to adopt standards that are more stringent than Federal criteria. In addition to these chemical-based criteria, many NPDES permits require the regular conduct of short-term toxicity tests with either freshwater (e.g. fathead minnow) or saltwater (e.g. silversides or sheepshead minnow) fish, and representative invertebrates to determine the toxicity of effluents and receiving systems.[41] Such effluent tests are used to determine the presence of non-measured chemical constituents and to account for potential additivity of chemicals that can be difficult to interpret from chemical-based criteria alone. The CWA also mandates the development of community-based biocriteria using fish and invertebrates to monitor stream quality. The majority of states rely on invertebrate-based biomonitoring. However, fish community criteria are used in many states and are under development in many others. Compliance monitoring also occurs under the FIFRA in the form of tier 4 testing of fish populations in experimental or natural field studies, in addition to pesticide incident reporting that occurs under normal post-registration pesticide use.

Natural Resource Damage Assessments. The Comprehensive Environmental Response, Compensation, and Liability Act of 1980 (CERCLA) provided for the recovery of damages for injury to natural resources, including fish, as a result of the release of hazardous substances to the environment that are covered under either the CERCLA or Sections 311(4) and Sections 311(5) of the Clean Water Act. "Damage" refers to the monetary value of lost resources whereas "injury" refers to a measurable adverse change, loss, or destruction of natural resources. The actual measure of injury to aquatic resources is covered under The Type B Technical Information Document[42] within Section 301 of the CERCLA. Fish assessments, in addition to other forms of assessment, may be used to measure injury in the NRDA process but must meet four criteria: 1) the biological response measured must be frequently observed as the result of exposure to contaminants, 2) exposure to contaminants is known to cause the biological response in free-ranging organisms, 3) exposure to contaminants is known to cause the biological response in controlled experiments, and 4) the biological response must be practical to perform and must produce scientifically valid results. Numerous biological responses in fish are accepted as meeting these criteria and include 1) mortality in natural environments, in-situ cages, and laboratory toxicity tests; 2) reproductive effects, 3) skin and liver neoplasia, 4) brain cholinesterase inhibition, 5) external fin erosion, and 6) behavioral avoidance. Other endpoints, including other biochemical and immunological indicators, were not included in the original set of acceptable endpoints. However, the Type B Technical Document is currently being reviewed for possible revision and/or inclusion of other endpoints for Natural Resource Damage Assessments.

Europe

Environmental regulation in Europe is coordinated by the Organization for Economic Cooperation and Development (OECD), which is an intergovernmental organization composed of 25 democratic European countries.[43] The OECD was formed in 1960 with the primary goal of coordinating, stimulating, and harmonizing the economic development of European countries. Although the OECD is not a governing body it does formulate decisions which are legally binding among member countries; OECD recommendations, although not legally binding, carry considerable political and moral obligation.[43] The principle environmental governing committee of the OECD is the Environmental Policy Committee (EPOC) which is responsible for pollution control and prevention; natural resource management; and environmental health, safety, and assessment.[43] The overall goals of the OECD in regulation of industrial chemicals, pesticides, food additive, and pharmaceuticals is similar to that in the United States and is based on the assessment of hazard, exposure, risk assessment, and risk management. Again, ecotoxological assessments using fish rely on acute toxicity tests, chronic (14-d tests), early lifestage tests, and bioaccumulation studies. Many of the OECD Guidelines for Testing of Chemicals are based on internationally agreed upon methods developed in 1981 following the "Decision Concerning Mutual Acceptance of Data in the Assessment of Chemicals" (e.g. the MAD decision of 1981) which is legally binding among OECD member countries. In this same spirit there is now an intercontinental effort to harmonize test guidelines and allow the acceptance of fish toxicity tests for all uses within the OECD and the U.S. regulatory entities (i.e. TSCA, FIFRA, and FDA).[43]

LABORATORY APPROACHES

Laboratory toxicity tests have been used extensively in ecotoxicology and biomonitoring. These tests involve the exposure of various species in the laboratory to single chemicals, chemical mixtures, water quality variables (e.g., pH, O_2), environmental conditions (e.g. UV radiation) and environmental samples (e.g., effluents, sediments) under static, static-renewal, or flow-through conditions.[30,44,45] These tests are used to evaluate the concentrations of chemicals and the duration of exposure required to produce an effect. All toxicity testing is based on the direct relationship that exists between exposure concentration and the observed effects produced in organisms. Simply stated, as the dose or concentration of a toxicant increases the observed effects in the exposed organism increase. Some of the advantages associated with the use of laboratory toxicity tests in ecotoxicological biomonitoring include: (1) many tests have been standardized, allowing toxicity results to be easily compared across testing facilities, (2) laboratory tests provide a cost effective means of rapidly screening large numbers of chemicals for toxicity, (3) the test variables, e.g., water quality, temperature, photoperiod, etc., can be controlled within finite limits, (4) laboratory tests allow measurement of the direct effects of toxicants on species or life stages within species, and (5) laboratory tests allow rapid determination of the relative sensitivities to toxicants of species and life stages within species. The advantages of laboratory toxicity testing are accompanied by several formidable disadvantages including, (1) laboratory tests ignore many interactions occurring at the community and ecosystem level of biological organization that ultimately control the fate and effects of contaminants in aquatic systems;[45] (2) laboratory single species and single chemical tests are low in environmental realism and predicted effects thresholds may be artefacts of the test procedure and may not exist in natural systems;[46] and (3) laboratory tests often do not incorporate chemical, physical, and biotic factors such as water quality, adsorption onto particulates and sediments, photolysis, hydrolysis, fluctuating exposure regimes, and the nature of the aquatic system that may modify the responses of organisms to toxicants in field situations.[47] Whereas the limitations associated with laboratory tests are critical with respect to their predictive value as ecotoxicological tools, laboratory tests still constitute the basic foundation of ecotoxicological research.

Laboratory Exposure Systems. Laboratory toxicity testing utilize static, static-renewal, and flow-through exposure systems which can be used to conduct acute, intermittent exposure, or chronic toxicity tests. Static systems are less desirable compared to static renewal or flow-through systems because of problems associated with maintaining acceptable exposure conditions, e.g., the concentrations of the test substance may decline as the exposure progresses due to processes such as adsorption, decomposition, volatilization, and bioconcentration. Thus, static systems are used primarily to conduct short-term exposures of 24 to 96 hours.[30] These laboratory exposure systems accommodate a broad range of experimental designs and can be used to generate numerous types of toxicological data on fish.

Acute toxicity tests procedures, uses, and limitations have been thoroughly discussed and reviewed.[30,48, 49,50, 51,52,53,54,55] Acute toxicity tests provide a rapid means of screening large numbers of single compounds and complex mixtures for toxicity, and of determining the relative sensitivity of different fish species. The objective of an acute toxicity test is to determine the concentration of a test material (e.g., a chemical or effluent) or the level of an agent (e.g., temperature or pH) that produces a deleterious effect on a group of test organisms during a short-term exposure under controlled conditions. Groups of fish usually are exposed to several different concentrations of the material mixed in water, although exposure can be accomplished by injection of the material into the organism or by incorporating the material into feed.[30] The 96-hour median lethal concentration (96-h LC50), or the concentration of a substance in which 50% of the test organisms are killed in a 96-hour period, is the most frequently estimated acute toxicity value. The LC50 is used most often because it is more reproducible than other quantile values; however, the LC10, LC95, or other lethal concentration values may be of more interest in particular instances. Various computer programs are available that will estimate LC50 values and utilize the method that fits the response data best. Acute toxicity tests have limited ecological utility because they do not provide information on long-term effects on survival, growth, and reproduction and acutely toxic concentrations of contaminants occur infrequently in natural systems.

In intermittent exposure toxicity tests organisms are exposed repeatedly to contaminant concentrations for various periods of time. Each contaminant exposure period is followed a by period of exposure to conditions similar to the control treatment during which the organisms are allowed to recover. The duration of the exposure period and recovery period is dependent on the study objectives and the field scenario that one wishes to simulate.

In chronic toxicity tests organisms are exposed to toxicant concentrations or a stimulus that is lingering or continues for an extended period of time. Chronic exposures may last weeks to years, depending on the test species and life stage. Other terms used to describe chronic toxicity tests are partial chronic, rapid chronic, subchronic, or early life stage. Various types of proportional diluters (e.g., Mount and Brungs[56]) are used to conduct chronic toxicity tests. Factors that modify toxicity and should be controlled during chronic tests include (1) species and strain of test organism, (2) the life stage and size of test organisms, (3) nutrition, health and parasitism, (4) photoperiod, and (5) water quality (e.g., temperature, dissolved oxygen, pH, dissolved salts, suspended solids, and dissolved organic carbon).[30] The objective of a chronic test is to determine the concentration of a test material (e.g., chemical or effluent) that produces deleterious effects on a group of test organisms during a long-term exposure under controlled conditions. Estimates of toxicity are described in terms of (1) no-observable-effect concentrations (NOEC), (2) lowest-observable-effect concentrations (LOEC), (3) maximum acceptable toxicant concentrations (MATC), or effective concentrations (e.g., EC50s or EC20s, etc.). Results of chronic toxicity tests are most often used to predict effects of toxicants on fish populations under field conditions; however, because chronic tests do not incorporate many field variables that modify toxicity, Type I and Type II errors[46, 57] associated with predictions continue to limit their ecological applications.

Biological Endpoints

The definitive purpose of conducting laboratory toxicity tests with fish is to amass data and information on real or potential effects that natural fish populations may incur when exposed under similar conditions. The information can then be used by resource managers for regulatory purposes, to support mitigative action, post-mitigation monitoring, or to optimize overall management strategies for aquatic resources. Although laboratory procedures available to measure chemically-induced stress in fish are well-developed, widely used, and are thoroughly validated [30, 58,59,60,61], laboratory toxicity tests are low in environmental realism and do not account for many of the biotic and abiotic factors that modify responses to chemical stressors in natural systems. Despite these limitations laboratory tests still remain the backbone of contaminant hazard assessments. Biological endpoints or indicators of chemically-induced stress monitored during laboratory toxicity tests depend on the ecotoxicological objectives of the tests. Measurement of three general types of biological indicators have been proposed [62] : compliance indicators, diagnostic indicators, and early warning indicators. Generally, compliance indicators, e.g., fish population attributes, are measured at the population, community, or ecosystem levels of biological organization and are focused on issues such as the sustainability of target populations or communities as a whole.[62] Measurements such as enzyme levels or activity and chemically-induced changes in biochemical parameters performed on individual organisms are used as diagnostic and early warning indicators.[62] Used in concert, compliance, diagnostic, and early warning indicators provide information that can be used to judge whether a system has been restored, whether a targeted environmental quality is being maintained, to identify probable causes of declines in the quantity or quality of natural systems, and to support intervention and the implementation of actions before population or community level responses are realized or aquatic systems become severely impaired.[62] During laboratory tests with fish, mortality, growth, reproduction, and behavior are measured more often compared to more subtle responses such as biochemical and physiological changes, histological changes, and genetic damage. All of these responses in fish integrate the independent and interactive effects of many chemical stressors and are better indicators of ecological status compared to the concentrations and loadings of individual chemicals.

  <u>Mortality, Growth, and Reproduction</u>. The responses most often measured during exposure of fish to contaminants in the laboratory are effects on mortality, growth, and reproduction. Measurement of these responses requires less time and is less expensive compared to the measurement of other endpoints. Laboratory tests to determine the effects of water pollutants on fish were performed as early as the late 1800's and early 1900's in Europe[63] and by 1937 in the United States.[64] The focus of these early studies were primarily to determine the effects on fish and other aquatic biota of several water pollution problems, including toxic materials in industrial wastes, heavy metals in mine drainage, organic pollution from sewer releases and effluents, and individual compounds. More recently, an extensive data-base on the acute toxicity of 410 chemicals to aquatic organisms, including fish, was developed by Mayer and Ellersieck.[65] A comparison of fish survival to other endpoints, including growth, reproduction, histopathology, and biochemical parameters, among seven species of fish exposed to 28 chemicals in chronic toxicity tests showed that survival was equal to or more sensitive than all the other endpoints 56 to 69% of the time.[66] Individual endpoints measured for fishes were more sensitive than survival 19 to 61% of the time, but

reproduction was always more sensitive than survival.[66] Results of 173 tests in which fish were exposed in full life cycle, partial chronic or early life stage test to metals, pesticides, unclassified organics, inorganic compounds, detergent chemicals, and complex effluents showed that survival was significantly reduced in 57% of the tests at the lowest effect concentration, growth was reduced in 36% and egg hatching in 19% of the tests.[67] Mortality and growth continue to be the primary endpoints measured on fish in the laboratory in support of natural resource damage assessment and contaminant hazard assessment activities. [42,68,69,70,71]

Fish Behavior. Behavioral functions of fish can be deleteriously affected by sublethal concentrations of pollutants, and alterations in these behavioral functions can influence survival of natural populations through impaired performance of critical life functions.[72] Behavioral responses have been shown to be highly sensitive indicators of contaminant-induced stress of fish and the responses can be evaluated during laboratory testing.[73]   Qualitative evaluations of behavioral responses of fish are accomplished through direct observations, whereas quantitative evaluations are done by use of video recordings.[74] The methods used to assess contaminant-induced changes in fish behavior can be incorporated into standard test protocols, they are readily applicable to different fish species,   and they provide rapid, sensitive, and ecologically relevant assessments of sublethal exposure.[74]

The ecological relevance of fish behavioral responses and the ease at which the measurement of the responses can be incorporated into standardized laboratory tests have resulted in their utilization to characterize sublethal exposure. The feeding responses of fish including reduced feeding,[70,75,76,77,78] prey capture and strike frequency,[79] and efficiency of feeding[80] have been evaluated in laboratory tests and shown to be consistent, highly sensitive indicators of sublethal exposure to a range of environmental contaminants. The locomotory responses of fish have been evaluated as indices of toxicosis.[72,81] Measures of swimming activity are commonly used to assess contaminant-related changes in fish locomotion[82] and includes the evaluation of such variables as frequency and duration of movements,[70] speed and distance traveled during movement,[82] frequency and angle of turns[83], position in the water column, and form and pattern of swimming. Also, contaminant-induced effects such as hypoactivity and hyperactivity have been measured as the frequency of entries and exits through a dividing partition within laboratory exposure chambers[84] or number of grid lines crossed. To survive and grow in their environments, fish must develop normal behavioral responses that enable them to perform critical life functions such as habitat selection, competition, predator avoidance, prey selection, and reproduction.[72,85] Environmental pollutants which alter water quality can elicit a range of untoward behavioral responses of fish and the magnitude of these responses are dependant on the type and concentration of the pollutant, and the duration of exposure.[85]

Behavioral responses of fish are clearly excellent indicators of contaminant exposure and resultant injuries. The majority of behavioral studies with fish have been conducted in controlled laboratory settings; however, a few investigations have determined contaminant-induced behavioral effects in the field. The impairment of behavioral function can have immediate ecological consequences including the limitation or elimination of natural populations through reproductive failure or mortality. Several field studies have confirmed that fish avoid contaminants such as copper and zinc.[86,87] In such cases the avoidance occurred at ambient but sublethal concentrations in stream areas that supported robust populations prior to the mining activities. The ecological consequences of this behavior was the loss of the fish population from the area. The consequences for the species would be the loss of suitable habitat for reproduction. Migratory behavior of salmon smolts was disrupted as a result of exposure to copper and zinc,[88] arsenic,[89] and selenium[90] and spawning migrations were disrupted by paper mill effluents[91] and metals,[92] which would result in impaired reproduction. The abandonment of nests and loss of parental behavior have been studied in the field with centrarchids exposed to agricultural chemicals.[93] Recent advances in telemetry technology will enable verification of contaminant effects on locomotory responses, competition, and predator-prey interactions and should further our understanding of the causal relationship between observed behavioral changes and the impacts of contaminants on natural populations. As the use of fish behavioral responses in ecotoxicology evolves, verification of contaminant-induced behavioral effects in the field will be necessary to understand the causal relationship between observed behavioral changes and the impacts of contaminants on natural populations and communities.[80,85]

Biochemical Parameters. A tremendous need exist to develop and apply clinical and diagnostic tests in assessments of pollutant hazards to aquatic organisms in their natural environments. These tests would decrease the time required for safety evaluation of contaminants, define more adequate no-effect exposure concentrations, and provide a better understanding of the mode of action of chemicals.[94] Numerous biochemical responses of fish have been linked to exposure to specific classes of chemicals. For example, increased levels of the metal binding protein metallothionein in fish is indicative of exposure to metals.[95] The physiological and biochemical effects of organochlorine chemicals on the skeletal system of fishes were first demonstrated over 20 years ago.[96,97,98] These early laboratories studies demonstrated the relationship between biochemical and physiological responses and whole animal effects, which lead to the examination of fish from natural systems.[100,101] Exposure of fish to organophosphate compounds has been shown to inhibit their cholinesterase [101,102] and this inhibition on cholinesterase has been linked to reduced growth of a natural population of fish.[103] In another approach significant correlations were seen between brain cholinesterase inhibition, and several locomotory responses including speed, distance and tortuosity of movement following pesticide exposure.[104] In other tests, brain cholinesterase activity was found to correlate with swimming activity and with increased vulnerability to predation among early lifestage rainbow trout exposed to an organophosphate insecticide[105]. More recently, aryl hydrocarbon receptor (AHR) function, which controls levels of cytochrome P450 1A in fish and the activity of hepatic ethoxyresorufin O-deethylase (EROD), has been evaluated and used as a means of detecting exposure of fish to environmental contaminants such as PCBs, PAHs, and others in both laboratory and field studies.[106,107,108,109,110] In addition to those mentioned above, numerous other biochemical endpoints in fish are used as diagnostic indicators of contaminant exposure as well as effects.[100] The challenge for the future is to amass more data on the biochemical responses of fishes to pollutants and to determine how these organismal responses translate into population and community levels impacts.

FIELD APPROACHES

As previously discussed there are many ecological and economic reasons for using fish in biomonitoring studies. Likewise, there is a broad continuum of test systems that have been successfully used to monitor the effects of contaminants in fish, including in-situ cages, microcosms, mesocosms, and in natural systems with wild populations. There are tradeoffs associated with the use of each of these test systems. For example, the small-scale systems (i.e., in-situ cages, microcosms, mesocosms) increase experimental control, lower the variability of data, but decreases the accuracy of ecological assessments.[111] On the other hand, larger-scale natural systems with wild populations decrease experimental control and increase data variability, but environmental assessments utilizing larger-scale systems are more accurate compared to ones conducted with small-scale or laboratory systems. A thorough understanding of the uses and limitations of each experimental system can allow one to gain maximum inference from the information obtained.

Cage Studies. *In-situ* cage studies are used to isolate individual fish or fish populations within a desired exposure plume (i.e. treatment) or reference area. The use of cages insures that the fish remain in one location for maximum exposure or non-exposure to a contaminant or contaminant mixture. These studies are frequently used to measure acute mortality or bioconcentration of a chemical.[101,102] In some instances they can be used to measure growth as long as food is of sufficient supply or supplemental feeding is used. However, care must always be used in caged studies to insure that factors such as high temperatures or low dissolved oxygen do not influence the test results. Temperature and dissolved oxygen may not be a problem in streams or the epilimnion of lakes, but in many shallow, poorly mixed systems temperatures can fluctuate over 10°C and range from anoxia to supersaturation of oxygen if dense beds of macrophytes are present. Free-ranging fish have great abilities to find refuge from adverse physical or chemical conditions and in the absence of movement effects such as increased mortality may erroneously be attributed to a toxicant. Temperature and dissolved oxygen can be monitored continuously using continuous in-situ monitors. Alternatively, the use of max/min thermometers and measurement of dissolved oxygen in early morning (following the dark respiration period) can be used to discount these water quality factors.

Some fish are more amenable to in-situ caging than others. Fish that normally aggregate, e.g., fathead minnows, are easily studied in cages.[112] Others, such as channel catfish, can be used but should be stocked at optimum densities to over-ride effects such as social aggression.[113] Optimum numbers of fish for stocking are often available in aqua-culture publications along with recommendations for feeding, disease prevention, etc. Some fish species, particularly those that exhibit territorial reproductive behavior or develop social hierarchies, should be individually isolated or avoided due to density-dependent effects on survival or growth which may decrease the statistical sensitivity of responses.[114]

Microcosms and Mesocosms Studies. Microcosms and mesocosms are self-contained physical models which simulate the structural and functional aspects of natural ecosystems. Size is the primary distinguishing factor between the two types of experimental systems with microcosms ranging from 0.001 to 10 m^3 and mesocosms ranging from 10 to 1000 m^3 in size. Fish have been successfully used in all types of experimental ecosystems to answer a variety of ecotoxicological questions. Reviews of the types, sizes, and applications of microcosms and mesocosms are numerous.[111, 115,166] Fish are ideal ecological indicators for use in complex experimental systems because they serve as top predators and, thus, integrate a variety of effects that can occur at lower trophic levels. In addition, fish are known to forage optimally to gain the maximum physiological benefit at the least expenditure of energy; thus, they provide a direct biological interpretation of lower level ecosystem effects that are easily interpreted and tangible.[177] Furthermore, because fish frequently feed over a wide spatial area they can integrate the heterogenous distribution of invertebrate food organisms in aquatic systems.

The study of the effects of contaminants on fish in microcosms or mesocosms has numerous advantages of experimental control and realism. These types of systems are accurately dosed because of their known dimensionality (e.g. volume, surface area, and sediment/substrate composition). This allows the researcher control and added insight into the duration, intensity, and frequency of contaminant exposure which frequently are unknown variables in studies of natural systems. In addition, the species composition, biomass, and age distribution of fishes are controlled which allows the study of the effects of contaminants in relation to the biological effects of predation or competition (e.g., density-dependance) which can obfuscate the results of contaminant studies in natural systems.

Microcosms and mesocosms are ideal systems for comparing the results of laboratory and field effects on fishes. For example, laboratory studies are usually conducted in the presence of optimum temperature, water quality, and feeding conditions which are used to isolate the direct effects of contaminants on the behavior, physiology, survival, growth, and reproduction of fishes. However, the effects of contaminants in systems simulating more natural conditions of variations in exposure, food supply, and thermal stress can either increase or decrease the sensitivity of fish populations to contaminants.[33,101,108] Fish are generally more sensitive to pyrethroid insecticides in the laboratory than in the field. Integrated laboratory/field studies with bluegill have shown that standard laboratory studies overestimate the risk of the pyrethroid insecticide esfenvalerate, due to unrealistic laboratory exposures.[112,119] The high octanol:water partition coefficient of esfenvalerate led to its rapid dissipation in 0.1 ha mesocosms, which reduced its bioavailability.[119] However, the use of non-standard pulsed laboratory exposures resulted in more accurate predictions of fate and effects of esfenvalerate in mesocosms.[119] Further, the rapid dissipation of esfenvalerate (half-life of 10 h) resulted in only short-term reductions in zooplankton populations, which precluded secondary food-chain-related effects on young bluegill; in fact, growth rates of juvenile bluegill increased with increasing esfenvalerate concentrations due to decreases in intra-specific food competition as a result of increased bluegill mortality at increasing chemical concentrations.[119] In contrast, exposure of fathead minnows to the organophosphate insecticide chlorpyrifos in mesocosms elicited significant decreases in their growth rates.[120] Chlorpyrifos dissipated rapidly from mesocosms (half life of 10 h) and zooplankton populations were significantly reduced; fathead minnows, as predicted from laboratory data, were not directly affected by chlorpyrifos; however, their growth rates decreased in the highest treatment due to decreases in food availability. Similar effects on fish exposed in mesocosms to the polynuclear aromatic hydrocarbon, fluorene, have been documented.[121,122] Although fluorene reduced growth of bluegill under laboratory conditions at 0.5 mg/L, growth decreased at 4-fold lower levels (0.12 mg/L) in lentic mesocosms due to decreases in food supply. Although the standard laboratory growth test was not predictive of growth effects in the mesocosm, significantly reduced feeding which was measured in the laboratory at 0.06 mg/L fluorine, was predictive of effects of fluorene under field conditions.[122]

Mesocosms have also been used to isolate and identify biosynthetic processes which can increase the toxicity of chemicals to fish. For example, selenium is relatively non-toxic in the inorganic forms of selenate and selenite. However, inorganic selenium can be microbially transformed to various forms of organic selenium which can be extremely toxic to fish.[118] Such processes have been demonstrated in microcosms[123] and outdoor mesocosms[124] and have been shown to result in reproductive failure in fish.[125,126]

Natural Systems

Perturbations that may alter the survival, growth, and reproduction of fishes is a primary concern in protection of fish populations. Adverse changes in these parameters, which can result from a multitude of both contaminant and non-contaminant-related factors, can cause changes in demographic parameters such as numbers, age distribution, population growth, and persistence of populations. The ultimate goal of laboratory, in-situ cage, microcosm, and mesocosm studies with fish is to provide estimates of safe contaminant concentrations which are protective of fish populations or communities in natural ecosystems.

Many different approaches have been used to study fish populations, including long-term population dynamics studies and population modeling. For example, the study of fish population dynamics in the field over a number of years can be accomplished by measuring population size in relation to actual rates of reproduction and survival. These types of studies are very resource-intensive, and must be conducted for long periods of time (years) to establish population trajectories (e.g., increasing, decreasing, or stable numbers). These studies can be utilized for both research and management purposes, but are most useful under conditions where a single, manageable factor (e.g., the relationship between fishing mortality and population characteristics of largemouth bass) affects a species.

Fish population models incorporate actual population data along with available data from the literature to estimate the potential response of populations in relation to factors that may effect demographic parameters. Population modeling has evolved over the past 50 years and has been shown to be a very valuable tool for estimating the status of populations and exploring their vulnerability to various environmental stresses.[127] However, there may be many factors acting singly or in concert to affect population size, including fishing mortality, natural mortality, contaminants, food limitation, or weather conditions.[128] Thus, determining the contribution of any particular factor to overall impacts on population status is difficult due to the high cost associated with field investigations. Recently, applications of fish population modeling have increased due to the use of emerging computer technologies that facilitates the large number of mathematical calculations and iterations used in modeling procedures. Individual-based models, evaluated under spatially explicit conditions, are now utilized, and they greatly facilitate the assessment of the effects of contaminants on fish in relation to the multitude of other factors that may be involved.[129,130] In most cases, however, the effects of contaminants on fish populations are determined by use of a combination of approaches that develop multiple lines of evidence of impact. Many of these approaches combine laboratory and field techniques that can be applied at multiple levels of biological organization, including the cell, tissue, organ, individual, and population levels. In some cases, even community based approaches are being applied.

Population-Level Indicators. Assessment of population size is one approach used to monitor the effects of contaminants on fish populations. The assumptions, needs, and approaches to monitoring fish population size are discussed in most fisheries textbooks.[131,132] The procedures used to assess the size of fish populations differ depending on the size of systems being studied and the relative potential for capturing fish. For example, estimation of fish population size in ponds, lakes, estuaries, and in many streams and rivers is based on a number of sub-sampling and estimation techniques such as the mark:recapture method. In these systems it is impossible to make absolute estimates of the number of fish due to the size of the system and difficulty in sampling the entire population. Thus, in a mark recapture approach a sub-sample of the population is obtained using robust sampling methods such as electrofishing or seining. The individuals collected are usually measured, marked, aged (discussed later), and returned to the system unharmed. At a later date, usually at least a week or so later, the area is re-sampled using the same technique and level of effort. The proportion of marked to total individuals in this sample is then determined and used to estimate the number of total fish in the population. In most cases, a multiple mark-recapture study is conducted in which multiple samples are collected on different days. In each case, a unique mark is given to each day's subsampling. By doing multiple iterations of the process, one essentially obtains replicate estimates over time and simultaneously increases the total number of fish marked in the sample. Various mathematical approaches can be used to estimate population size in relation to the estimates of statistical reliability.

In small ponds or streams that are physically isolated or that can be mechanically (e.g. block nets) isolated, other approaches such as depletion sampling methods are used to estimate population size. In these approaches multiple samples are taken with the objective of catching as large a proportion of the population as possible. In each subsequent sample, the number of individuals declines until only small numbers or none are captured. One can then statistically interpolate to determine the total number of fish present along with the statistical confidence of that estimate.

Determination of size and age structure is a common technique used to study populations of fish in the field. The size and age structure of fish populations can be calculated from a single sample. Size and age structure is also frequently determined along with population size as described above. Size structure can be determined with length frequency histograms which provide estimates of relative numbers and sizes of various length groups in the population. In some cases these size groups can be assumed to be equivalent to age groups. However, this approach should be verified or augmented using actual aging of individual fish based on annual growth rings that are incorporated into scales, otoliths, or other bony structures due to temperature, light, diet, or other environmental factors that alter growth rates on a regular schedule.

Growth rates of fishes also are used to assess the status and health of fish populations and can be determined using several techniques including length at age calculations and mark:recapture techniques of known-age fish. Growth rates can be back-calculated using scales or other bony stuctures in which the distance between annular rings is used to back-calculate lengths at various ages. This technique is based on the assumption that good growing conditions (e.g. food supply, temperature) result in increased somatic growth rates that are reflected in increasing distances between annular increments on the scale, otolith, spines, or other bony structures. Growth rates can also be determined from recaptures of fish previously tagged or marked within a given year. Tagging can be done via various methods including physical (e.g. fin marks; numbered, bar-coded, or magnetic tags; dye injection; etc.), chemical (e.g. staining of bone with tetracylcline or other substance), or genetic marks. Physical marks can be used to mark individual fish, whereas the latter approaches are cost-effective approaches for group marks.

Genetic marks, such as the presence and location of specific genotypes or alleles, can also be used to mark, identify, and interpret environmental effects on fish populations.[133] For example, the genetic composition of stocked fish (narrowed diversity in this case) can be used to determine approximate population size and genetic dilution of the newly stocked individuals in relation to the genetic composition of the natural population. Alternatively, assessment of the genetic structure of natural populations can be used to infer the presence of contaminants or other factors (e.g. physical isolation) that may decrease the genetic diversity of populations. Fore et al.[133] determined that decreasing water quality led to the narrowing of genetic diversity of populations of *Pimephales promelas* at fifteen sites in Ohio. The authors suggested that genetically sensitive organisms (e.g. low contaminant metabolic capacity) are removed from the population leaving only genotypes that are less sensitive to the contaminant. Of particular concern is that these contaminant-resistant fish may be genetically restricted for other traits (e.g. temperature tolerance) that are necessary for them to survive other common stressors, thus resulting in less adaptable or fit populations.

Reproductive success is essential for the persistence and growth of fish populations. Thus, reproductive indicators or reproductive biomarkers are frequently used as an index of contaminant effects on field populations.   Endpoints of interest usually include parameters such as gonadal/somatic index, egg size, fecundity, and age at first reproduction. Munkittrick and Dixon[134] used white suckers as an indicator species to developed a framework for utilizing reproductive biomarkers in conjunction with other components of population structure in the assessment of population-level indicators of contaminant stress. White suckers were chosen because they are insectivores that are likely to respond to both direct (contaminant) and indirect (contaminant effects on the resource base) effects that might be dampened at higher trophic levels impacted by fishing.

Community-Level Indicators. Community level indicators that examine species number, numbers within species, and the types of species present have also been used to evaluate the response of aquatic environments to contaminants and other stressors. The basic approach is to sample the fish community and compare its composition to that expected from reference conditions. The data can be evaluated from several different approaches including: diversity, similarity, and biotic indices.[135] Diversity indices generally provide an index of community structure based on the number of species, total numbers, and distribution of numbers across species within an individual sample. Similarity indices, however, actually compare two different samples relative to the numbers of unique and common species that appear within the two samples. Biotic indices, on the other hand, are used to assess community health based on the relative tolerance of environmental stresses of each species found in a sample: the presence of intolerant species is indicator of good water quality conditions.

Karr et al.[136] developed the Index of Biotic Integrity, which is basically a multi-metric assessment that combines several population and community metrics including total numbers; species richness; presence or absence of sensitive/insensitive species; incidence of disease; and presence of physical deformities. Although this approach is not well based in mathematical or statistical theory it does combine the attributes of fish communities known to be reflective of good community health. Metrics are scored and ranked based on regional comparisons to a range of impacted to pristine conditions as assessed by regional fishery professionals. The use of multiple metrics provides the redundancy necessary to compensate for abnormal conditions or aberrant responses. In practice the IBI has been very successful and has been shown to be sensitive to chemical impacts by chlorine, ammonia, metals, etc., in addition to physical factors such as dredging and sedimentation. The IBI and similar approaches are used for both fish and invertebrate community assessments by numerous Federal (e.g. USEPA) and state (e.g. Ohio; North Carolina; Arkansas, etc.) agencies and provide a useful biological component to be paired with toxicity testing and residue analysis in comprehensive aquatic environmental assessments.

APPLICATIONS AND CASE STUDIES

Fish population assessments have been one of the most effective approaches to documenting and monitoring the effects of environmental contaminants in the environment. Fish population assessment approaches were originally developed because of the economic and recreational values of fisheries. However, fish are also recognized as having key functional roles in aquatic ecosystems and are known to be sensitive to the effects of contaminants. Thus, fish population assessments are one of the most fundamental approaches used to monitor environmental quality. Assessments of the response of fish populations can be conducted by measuring either exposure (e.g. chemical residues; biomarkers; etc.) or effects (e.g. pathologies; reductions in survival and growth resulting in altered population structure). Effects can occur due to many factors, however, including over-harvest, winter-kill, climatic fluctions, or biological interactions (e.g. competition or predation) . Thus, determining the effects of contaminants on fish populations requires the assessment of both exposure and effects.

Adams et al.[137] conducted a comprehensive assessment of the exposure and effects of PCBs and mercury on redbreast sunfish populations in East Fork Poplar Creek, a third order stream receiving point sources of contaminants.   At each of five sites (4 impact sites and 1 reference) 15 adult male sunfish were collected using electrofishing during four sampling intervals conducted over two years. In this study the exposure and effects indicators that were measured ranged from the molecular to the organismal level of biological organization. Exposure indicators included DNA damage (strand breaks) and liver detoxification enzymes (7-ethoxyresorufin O-deethylase, i.e. EROD; cytochrome P-450; cytochrome b_5; and NADH reductase). Effects indicators included measures of carbohydrate-protein metabolism (glutamate oxyaloacetate transaminase, i.e. SGOT; serum protein; serum glucose), lipid metabolism (total triglicerides; serum cholesterol; serum triglycerides; total triglycerides; phospholipids), histopathological condition (liver parasites; liver macrophage aggregates), and general health indicators (condition factor; total lipids; RNA/DNA ratio, and liver somatic index). Canonical discriminate analysis using all of the assessment factors led to clear separation of sites (Figure 1). Site 1 was in immediate vicinity of the contaminant input and 2, 3, and 4 were arranged in a downstream gradient. All sites were significantly different with the exception of sites 3 and 4 which had overlapping confidence intervals. Liver parasites, liver somatic index, CB5, total body lipids, serum protein, serum triglicerides, RNA/DNA ratio, and serum cholesterol were the primary variables responsible for individual site discrimination.   Actual ranks of these individual variables by site are presented in Table 1 along with several other variables that appeared to be good indicators yet were not statistically significant.   These results indicate that biochemical and organismal level indicators of fish are quite useful in the biomonitoring of contaminant exposure and effects, but that assessment of a suite of variables is often necessary to adequately discriminate among sites.

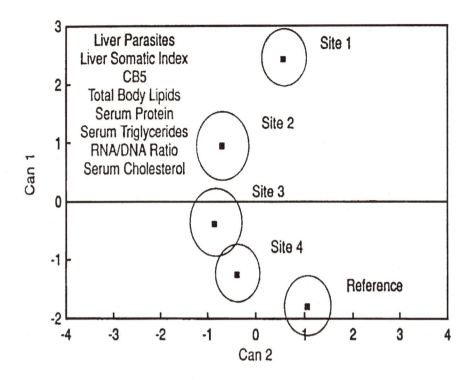

Figure 1. Segregation of integrated health responses for redbreast sunfish from four PCB/mercury contaminated sites and a reference site using multiple bioindicators analyzed via canonical discriminate analysis. The points and circles represent the means and 95% confidence radii of the site means. The statistically significant variables are listed in the upper left-hand corner. Figure from Adams et al. (1990)

Munkittrick and Dixon[134] proposed a useful framework for evaluating the effects of contaminants on population structure of white sucker populations. This framework identified key population-level parameters that could be used as indices of effects based on differential rates of reproduction, growth, and survival. Population-level parameters examined in the framework included mean age, maximum age, growth rate, condition factor, age at maturation, reproductive life span, fecundity, egg size, catch per unit effort, and population size. Effects were categorized relative to five response patterns that varied according to target source (e.g. direct on adults; indirect food chain, etc.), contaminant event (e.g. acutely lethal spill; chronic effects on reproduction, etc.); comparable non-contaminated anthropgenic event (e.g. fisheries exploitation, habitat destruction, etc.), effect on population (e.g. reduced abundance of adults; increase food competition, etc.), effect on resource base (e.g. increases or decreases in food availability), and initial population response (e.g.

Table 1. Selected indicators of exposure and effects by sampling site as measured by Adams et al. (1990) in redbreast sunfish exposed to point sources of mercury and PCBs. Numbers represent percentage of fish appearing in the lowest quartile of ranking (i.e. high ranks in the lowest quartile implies poorest health status).

Response Variable	Sampling Site				
Exposure Category	**1**	**2**	**3**	**4**	**Reference**
EROD	22	41	16	22	0
NADPH Reductase	44	34	11	14	9
Cytochrome P450	37	26	8	25	19
Effects Category	**1**	**2**	**3**	**4**	**Reference**
Liver Somatic Index	22	7	35	32	28
RNA/DNA Ratio	11	3	11	15	11
Total Body Lipids	32	20	16	10	12
Serum chlolesterol	22	28	27	20	26
Serum protein	25	30	16	24	19
Liver macrophage aggregates	24	10	5	2	4

increased growth rates of survivors; decreased growth rates, etc.). The five response patterns can result, as the authors indicated, by either a contaminant or numerous non-contaminated causes. However, the combination of this conceptual model with the assessment of exposure can often unequivocally document the contribution of contaminants to alterations in population structure.

Munkittrick and Dixon [134] applied this conceptual model to several case studies of contaminant exposures including mixed metal mining waste, atmospheric exposure to metals, [138,139], and acidification.[140,141] In all cases there were distinct changes in selected parameters that were revealed in altered population structure. For example, in the studies by Trippel and Harvey [140,141] five lakes were examined for the effects of acidification on population structure of white sucker populations. Previous information had revealed that acidification of Lake George had resulted in decreases in pH and increased aqueous concentrations of metals including aluminum and manganese; fish sampled from the lake exhibited abnormally low concentrations of plasma calcium and increased incidence of ovarian absorption[142] and 3-fold increases in cadmium in livers.[143] Munkittrick and Dixon[134] analyzed the population-level parameters of these studies and classified the results as a type I population response as a result of increased adult mortality (decrease in overall population size) which led to a decrease in food limitation. Subsequently, average growth rates, condition, and age at maturity increased. Predictions of increased fecundity and egg size did not occur which the authors attributed to prolonged physiological stress of exposure to low pH conditions.

Subsequent application of the above approach has proposed using smaller sentinel fish species that have a smaller home range and increased site fidelity for increased confidence in assessment of exposure and response. For example, Gibbons et al.[144] evaluated the response of spoonhead sculpoin (*Cottus ricei*) to exposure to a bleached-kraft pulp mill effluent in Canada. Hepatic EROD activity, decreased length, weight, condition, age, and liver somatic index indicated that sculpin were exposed to effluent. Moreover, fish from the side of the river opposite the effluent were similar to reference fish, and the effects of the effluent plume deceased downstream with distance. Thus, integrated exposure and effects assessments will likely become highly sensitive and broadly applied as techniques based on life history and behavior are otimized.

CONCLUSIONS

In conclusion, fish have been used extensively in biomonitoring and ecotoxicology. The requirement of fish data is legally mandated in the United States in three general areas: chemical registration under TSCA, FIFRA, and FDA regulations; compliance monitoring under the Clean

Water Act and FIFRA; and damage assessment investigations under CERCLA and OPA. In Europe the use of fish is mandated in environmental regulations coordinated by the Organization For Economic Cooperation and Development. Environmental legislation in conjunction with international environmental awareness has resulted in tremendous improvements in environmental quality over the last three decades. The development and standardization of toxicity and bio-accumulation tests with fish, chemical fate studies, laboratory assays, and controlled field tests have played a crucial role in both maintaining and improving environmental quality. Data collected from studies with fish support drinking water standards; establish tolerance limits for chemical substances; effluent monitoring, treatment and mitigation; and provide information for the protection of aquatic ecosystems.

There are numerous benefits in using fish in ecotoxicological monitoring. As sentinel monitoring organisms, fish are sensitive to changes in water quality and the toxicological responses of fish are effective in detecting the presence of hazardous conditions. There are numerous standardized assays and the tests are rapid and cost effective. Fish toxicity tests vary from short-term, acute exposures emphasizing mortality to chronic exposures of a duration that include a full life cycle or portion thereof which emphasize the consequences of sublethal exposure relative to health, long-term viability and reproduction. Biochemical, histological and behavioral methods have been developed for assessing sublethal exposure, as well as providing predictive diagnostic biomarkers for exposure and effect. Although laboratory studies provide precision, accuracy, and reproducibility, they lack predictability and realism because they fail to account for many biotic and abiotic factors that modify responses to chemical stressors in natural systems. This short coming is addressed by a range of field approaches which range in complexity from caged fish studies to controlled microcosm or mesocosm studies to assessments of natural systems. A range of population-level endpoints and statistical methods have been developed to assess the health and status of free ranging fish populations, including tag and recapture, genetic markers, growth, reproductive and health endpoints. Community-level responses are also investigated and include methods such as species diversity, species similarity and biotic indices of community health. Fish population assessments of exposure and effects have proven to be highly effective in documenting and monitoring the impact of environmental contaminants in the environment, and are fundamental to monitoring environmental quality. The use of fish in ecotoxicity testing has become a multi-disciplinary effort and the integration of biochemical, organismal, population and community approaches with statistical risk assessment protocols promises to increase the accuracy and predictive capabilities of biomonitoring and ecotoxicological testing in the future.

REFERENCES

[1] J.H. Ausubel, D.G. Victor, and I. K. Wernick. 1995. The environment since 1970. Consequences: The Nature and Implications of Environmental Change. 1:2-15.

[2] *Our Living Resources: A Report to the Nation on the Distribution, Abundance, and Health of U.S. Plants, Animals, and Ecosystems.* E.T. LaRoe, G.S. Farris, C. E. Puckett, P.D. Doran, and M.J. Mac, editors. U.S. Department of the Interior, National Biological Service. U.S. Government Printing Office, Washington, D.C. 1995. 530 pp.

[3] *The Potential Effects of Global Climate Change on the United States.* J.B. Smith and D.A. Tirpak, editors. United States Environmental Protection Agency, Office of Policy, Planning, and Evaluation, Office of Research and Development, Washington, D.C. Hemisphere Publishing Corporation, NY, 1990. 689 pp.

[4] S. Rowland. 1991. Stratospheric ozone in the 21st century: the chlorofluorocarbon problem. Environ. Sci. Technol. 25:622-628.

[5] R.C. Smith, B.B. Pcezelin, K.S. Baker, R.R. Bidigare, N.P. Boucher, T. Coley, D. Karentz, S. MacIntyre, H.A. Matlick, D. Menzies, M. Ondrusek, Z. Wan, K.J. Waters. 1992. Ozone depletion: Ultra-violet radiation and photo plankton biology in Antarctic waters. Sci, 255, 952-958.

[6] J.B. Kerr, and C.T. McElroy. 1993. Evidence of large upward trends of ultraviolet-B radiation linked to ozone depletion. Science 262:1031-1034.

[7] M. Tevini. 1993. Molecular Biological Effects of Ultraviolet Radiation. In: UV Radiation and Ozone Depletion. Lewis Publishers, Boca Raton, FL.

[8] SCOPE (Scientific Committee on Problems of the Environment). 1993. Effects of increased ultraviolet radiation on global ecosystems. Proceedings of a Workshop, October, 1992, Tramariglio (Sassari), Sardinia. 47 pp.

[9] OTA (Office of Technology Assessment). 1987. Technologies to maintain biological diversity. U.S. Government Printing Office, Washington, DC.

[10] M.A. Cairns and R.T. Lackey. 1992. Biodiversity and management of natural resources: The issues. Fisheries. 17:6-10.

[11] J.F. Franklin. 1993. Preserving Biodiversity: Species, ecosystems, or landscapes. Ecological Applications. 3:202-205.

[12] NAPAP (National Acid Precipitation Assessment Program). Report 13. In: National Acid Precipitation Assessment Program, Acidic Deposition; State of the Science and Technology.

Volume II, 1990.

[13] NAPAP (National Acid Precipitation Assessment Program). 1993. 1992 Report to Congress. National Acid Precipitation Assessment Program, Office of the Director, Washington, D.C., 130 pp.

[14] M.K. Saiki, M.R. Jennings, and T.W. May. 1992. Selenium and other elements in freshwater fishes from the irrigated San Joaquin Valley, California. The Science of the Total Environment 126:109-137.

[15] A.D. Lemly, S.E. Finger, and M.K. Nelson. 1993. Sources and impacts of irrigation drainwater contaminants in arid wetlands. Environmental Toxicology and Chemistry. 12:2265-2279.

[16] H.L. Garie, and A. Mcintosh. 1986. Distribution of benthic macroinvertebrates in a stream exposed to urban runoff. Water Resources Bulletin 22: 447-455.

[17] R.B.E. Shutes, D.M. Revitt, A..S. Mungur, and L.N.L. Scholes. 1997. The design of wetland systems for the treatment of urban run off. Water Science and Technology 35:19-25.

[18] J.R. Davis. 1997. Revitalization of a northcentral Texas river, as indicated by benthic macroinvertebrate communities. Hydrobiologia. 346: 95-117.

[19] J.G. Galindo-Reyes, M.A. Guerrero-Ibarra, C. Villagrana-Lizarraga, L.G. Quezada-Urenda, S.A. Escalante. 1992. Pesticide contamination in water, sediments, shrimps and clams in two coastal ecosystems in Sinaloa, Mexico. Tropical Ecology. 33:172-180.

[20] W.F. Ritter. 1990. Pesticide contamination of ground water in the United States: A review. Journal Environmental Science Health. B25:1-29

[21] S.J. Larson, P.D. Capel, and M.S. Majewski. 1997. Pesticides in surface waters: Distribution, trends, and governing factors. Ann Arbor Press, Inc. Chelsea, MI, USA, 373 pp.

[22] E.B. Welch. 1992. *Ecological effects of wastewater: Applied limnology and pollutant effects*, Second edition. Chapman and Hall. New York, NY USA. 425 pp.

[23] N. Pirrone, G.J. Keeler, and J.O. Nriagu. 1996. Regional differences in worldwide emissions of mercury to the atmosphere. Atmosphere and Environment. 30:2981-2987.

[24]. The Council on Environmental Quality. 1992. *Environmental Quality, 23rd Annual Report*. U.S. Government Printing Office. Washington, D.C. 451 pp.

[25] U.S. Environmental Protection Agency. 1986. Quality Criteria for Water 1986. USEPA 440/5-86-001. U.S. USEPA Office of Water Regulations and Standards, Washington, DC.

[26] J. Cairns, Jr. 1996. Determining the balance between technological and ecosystem services. Pages 13-30, In: *Engineering Within Ecological Constraints*, edited by P.C. Schulze.

Washington Academy Press. Washington, D.C.

[27] (USEPA) U.S. Environmental Protection Agency. 1997. Rapid Bioassessment Protocols for use in Streams and Rivers: Periphyton, Benthic, Macroinvertebrates, and Fish. EPA 841-D-97-002.

[28] D.J. Hoffman, B.A. Rattner, G.A. Burton, Jr., and J. Cairns, Jr. 1995. Introduction. In: *Handbook of Ecotoxicology*, D.J. Hoffman, B.A. Rattner, G.A. Burton, Jr., and J. Cairns, Jr., editors. Lewis Publishers, CRC Press, Inc., Boca Raton, Florida. 755 pp.

[29] W.H. Konemann. Ecotoxicology and environmental quality. Pages 94-103, In: *Environmental Protection: Standards, Compliance And Costs*, edited by T.J. Lack, Ellis Horwood Limited Publishers, Chichester.

[30] Annual Book of ASTM Standards. 1993. Volume 11.04. ASTM Designations E 729-88a, E 1023 - 84, and E 1241 - 92. American Society for Testing and Materials. Philadelphia, PA. 1622 pp.

[31] G. B. Wiersma, R.C. Rogers, J.C. McFarlane, and D.V. Bradley, Jr. 1980. Biological monitoring techniques for assessing exposure. Pages 123-133. In : *Biological Monitoring of Environmental Effects*, D.L. Worf, editor. Lexington Books. D.C. Heath Company. Lexington, MA. 227 pp.

[32] J.D. Walker. 1985. Pp. 669-702, in G.M. Rand, ed., Fundamentals of Aquatic Toxicology: Effects, Environmental Fate, and Risk Assessment. Taylor and Francis Publishers, Washington, DC. 1125 pp.

[33] M.G. Zeeman. 1995. Ecotoxicity Testing and Estimation Methods Developed Under Section 5 of the Toxic Substances Control Act. Pp. 703-716, in G.M. Rand, ed., Fundamentals of Aquatic Toxicology: Effects, Environmental Fate, and Risk Assessment. Taylor and Francis Publishers, Washington, DC. 1125 pp.

[34] L.W. Touart. 1995. The Federal Insecticide, Fungicide, and Rodenticide Act. Pp. 657-668, in G.M. Rand, ed., Fundamentals of Aquatic Toxicology: Effects, Environmental Fate, and Risk Assessment. Taylor and Francis Publishers, Washington, DC. 1125 pp.

[35] K.R. Solomon, D.B. Baker, R.P. Richards, K.R. Dixon, S.J. Klaine, T.W. LaPoint, R.J. Kendall, J.M. Giddings, J.P. Giesy, L.W. Hall, Jr. and W.M. Williams. 1996. Ecological risk assessment of atrazine in North American surface waters. Environ. Toxicol. Chem. 15:31-76.

[36] S.J. Klaine, G.P. Cobb, R.L. Dickerson, K.R. Kendall, E.E. Smith, and K.R. Solomon. 1996. An ecological risk assessment for the use of the biocide, dibromonitriliopropionamide (DBNPA),

in industrial cooling systems. Environ. Toxicol. Chem. 15:21-30.

[37] S.J. Klaine, G.P. Cobb, R.L. Dickerson, K.R. Kendall, E.E. Smith, and K.R. Solomon. 1996. An ecological risk assessment for the use of the biocide, dibromonitriliopropionamide (DBNPA), in industrial cooling systems. Environ. Toxicol. Chem. 15:21-30.

[38] D.J. Urban, and N. J. Cook. 1986. Hazard Evaluation Division, Standard Evaluation Procedure, Ecological Risk Assessment. U.S. Environmental Protection Agency, EPA 540/9-85-001.

[39] P.G. Vincent. 1995. FDA's Implementation of the National Environmental Policy Act. Pp. 735-762, in G.M. Rand, ed., Fundamentals of Aquatic Toxicology: Effects, Environmental Fate, and Risk Assessment. Taylor and Francis Publishers, Washington, DC. 1125 pp.

[40] G.W. Hudiburgh, Jr. 1995. The Clean Water Act. Pp. 717-734, in G.M. Rand, ed., Fundamentals of Aquatic Toxicology: Effects, Environmental Fate, and Risk Assessment. Taylor and Francis Publishers, Washington, DC. 1125 pp.

[41] USEPA. 1994. Short-term methods for estimating the chronic toxicity of effluents and receiving waters to freshwater organisms. EPA/600/4-91-002, Cincinnati, OH. 246 pp.

[42] USDI. 1987. Type B Technical Information Document: Injury to Fish and Wildlife Species. U.S. Dept. of the Interior, CERCLA 301 Project, Washington DC.

[43] N.J. Grandy. 1985. Role of the OECD in Chemicals Control and International Harmonization of Testing Methods. Pp 763-774 in G.M. Rand, ed., Fundamentals of Aquatic Toxicology: Effects, Environmental Fate, and Risk Assessment. Taylor and Francis Publishers, Washington, DC. 1125 pp.

[44] E.E. Little and D.L. Fabacher. 1996. Exposure of freshwater fish to simulated solar UVB radiation. Chapter 8, In: Techniques in Aquatic Toxicology, G.K. Ostrander, editor. Lewis Publishers. Boca Raton, Florida.

[45] T.W. LaPoint, J.F. Fairchild, E.E. Little, and S.E. Finger. 1989. Laboratory and field techniques in ecotoxicological research: Strengths and limitations. Pages 239-255. In: *Aquatic Ecotoxicology: Fundamental Concepts and Methodologies*, Volume II, A. Boudou and F. Ribeyre, editors. CRC Press, Inc. Boca Raton, Florida.

[46] J. Cairns, Jr. 1992. The threshold problem in ecotoxicology. Ecotoxicology. 1:3-16.

[47] T.W. LaPoint and J.F. Fairchild. 1994. Use of mesocosm data to predict effects in aquatic ecosystems: Limits to interpretations. Pages 241 - 255, In: *Aquatic Mesocosm Studies in Ecological Risk Assessments*. C.H. Ward. B.T. Walton, and T.W. LaPoint, editors. Lewis Publishers, Boca Raton, Florida.

[48] M. Waldichuk. 1989. Aquatic toxicology in management of marine environmental quality: Present trends and future prospects. Pages 8-22, In: *Aquatic Toxicology and Water Quality Management.* J.O. Nriagu and J.S.S. Lakshminarayana, editors. John Wiley & Sons, New York, New York.

[49] American Institute of Biological Sciences. 1978. Criteria and rationale for decision making in aquatic hazard evaluation. AIBS report to the Office of Pesticide Programs, USEPA, Washington, DC. 79pp.

[50] A.S. Murty. 1986a. Toxicity of pesticides to fish. Volume I. CRC Press, Inc. Boca Raton, Florida. 178pp.

[51] A.S. Murty. 1986b. Toxicity of pesticides to fish. Volume II. CRC Press, Inc. Boca Raton, Florida. 143pp.

[52] P.R. Parrish. 1985. Acute toxicity tests. Pages 31-57, In: G.M. Rand and S.R. Petrocelli, editors. *Fundamentals of Aquatic Toxicology.* Hemisphere Publishing Corporation. Washington, D.C.

[53) J.B. Sprague. 1969. Measurement of pollutant toxicity to fish. I. Bioassay methods for acute toxicity. Water Research. 3:793-821.

[54] L. Cleveland. 1992. Riesgos de algunos quimicos agricolas en las pesquerias (Hazards of select agricultural chemicals to fisheries). Ingenieria Hidraulica en Mexico Vol. VII Nums. 2/3 II Epoca. Mayo - Diciembre, 1992.

[55] C.E. Stephan. 1977. Methods for calculating an LC50. Aquatic Toxicology and Hazard Evaluation, ASTM STP 634:65-48.

[56] D.I. Mount and W.A. Brungs, 1967. A simplified dosing apparatus for fish toxicology studies. Water Res. 1, 21-29.

[57] P.K. Dayton. 1998. Reversal of the burden of proof in fisheries management. Science. 279:821-822.

[58] American Public Health Association, American Water Works Association and Water Pollution Control Federation. 1980. Standard Methods for the Examination of Water and Wastewater, 15th edition. American Public Health Association, Washington, DC.

[59] P.R. Parrish 1985. Acute toxicity tests. Pages 31-57 In: G.M. Rand and S.R. Petrocelli, editors. *Fundamentals of Aquatic Toxicology.* Hemisphere Publishing Co., Washington, DC.

[60] W.J. Birge, J.A. Black and A.G. Westerman. 1985. Short-term fish and amphibian embro-larval tests for determining the effects of toxicant stress on early life stages and estimating chronic

values for single compounds and complex effluents. Environmental Toxicology and Chemistry. 4:807-822.

[61] J.M. McKim. 1985. Early life stage toxicity tests. Pages 58-95 In: G.M. Rand and S.R. Petrocelli, editors. *Fundamentals of Aquatic Toxicology*. Hemisphere Publishing Co., Washington, DC.

[62] J.Cairns, Jr., P.V. McCormick and B.R. Niederlehner. 1993. A proposed framework for developing indicators of ecosystem health. Hydrobiologia. 263:1-44.

[63] C.E. Warren. 1971. Biology and Water Pollution Control. W.B. Saunders Co., Philadelphia, PA. 434 pp.

[64] M.M. Ellis. 1937. Detection and measurement of stream pollution. Bulletin of the U.S. Bureau of Fisheries 48(22):365-437.

[65] F. L. Mayer, Jr. and M.R. Ellersieck. 1986. Manual of acute toxicity: interpretation and data base for 410 chemicals and 66 species of freshwater animals. U.S. Fish and Wildlife Service Research Publlication 160. 574p.

[66] F. L. Mayer, Jr.n K.S. Mayer, and M.R. Ellersieck. Relation of survival to other endpoints in chronic toxicity tests with fish. Environmental Toxicology and Chemistry 5(6):737-748.

[67] D.M. Woltering. 1984. The growth response in fish chronic and early life stage toxicity test: A critical review. Aquatic Toxicology 5:1-21.

[68] D.F. Woodward, W.G. Brumbaugh, A.J. DeLonay, and E.E. Little. 1994. Effects on rainbow trout fry of a metal-contaminated diet of benthic invertebrates from the Clark Fork river, Montana. Transactions of the American Fisheries Society 123:51-62.

[69] L. Cleveland, D.R. Buckler, F.L. Mayer, Jr, and D.R. Branson, 1982. Toxicity of three preparations of pentachlorophenol to fathead minnows - a comparative study. Environmental Toxicology and Chemistry 1:205-212.

[70] L. Cleveland, E.E. Little, S.J. Hamilton, D.R. Buckler, and J.B. Hunn, 1986. Interactive toxicity of aluminum and acidity to early life stages of brook trout. Transactions of the American Fisheries Society 115(4):610-620.

[71] L. Cleveland, E.E. Little, D.R. Buckler, and R.H. Wiedmeyer, 1993. Toxicity and bioaccumulation of waterborne and dietary selenium in juvenile bluegill (Lepomis macrochirus). Aquatic Toxicology 27(3/4):265-279.

[72] E.E. Little, B.A. Flerov, and N.N. Ruzhinskaya. 1985. Behavioral approaches in aquatic toxicity: a review. In. P.M. Mehrle ,Jr., R.H. Gray, and R.L. Kendall, eds., *Toxic*

Substances in the Aquatic Environment: An International Aspect. American Fisheries Society, Water Quality Section, Bethesda, Maryland, pp. 72-98.

[73] E.E. Little, and S.E. Finger. 1990. Swimming behavior as an indicator of sublethal toxicity in fish. Environmental Toxicology and Chemistry 9(1):13-19.

[74] E.E. Little, A.J. DeLonay. 1996. Measures of fish behavior as indicators of sublethal toxicosis during standard toxicity tests. In: La Point TW, Price FT, Little EE, editors. Environmental Toxicology and Risk Assessment: Fourth Volume. W. Conshohocken, PA : American Society for Testing and Materials. (STP 1262) p 216-33.

[75] G.J. Atchison, M.G. Henry, and M.B. Sandheinrich. 1987. Effects of metals on fish behavior: a review. Environmental Biology of Fishes 18:11-25.

[76] C.J. Bull and J.E. McInerney. 1974. Behavior of juvenile coho salmon (Oncorhynchus kisutch) exposed to Sumithion (fenitrothion), an organophosphate insecticide. Journal of the Fisheries Research Board of Canada 31:1867-1872.

[77] P.M. Mehrle, D.R. Buckler, E.E. Little, L.M. Smith, J.D. Petty, P.H. Peterman, D.J. Stalling, G.M. DeGraeve, J.J. Coyle, and W.J. Adams. 1988. Toxicity and bioconcentration of 2,3,7,8-Tetrachlorodibenzodioxin and Tetrachlorodibenzofuran in rainbow trout. Environmental Toxicology and Chemistry 7: 47-62.

[78] D.F. Woodward, E.E. Little, and L.M. Smith. 1987. Toxicity of five shale oils to fish and aquatic invertebrates. Archives of Environmental Contamination and Toxicology 16: 239-246.

[79] J.A. Brown, P.H. Johansen, P.W. Colgan, and R.A. Mathers. 1987. Impairment of early feeding behavior of largemouth bass by pentachlorophenol exposure: A preliminary assessment. Transactions of the American Fisheries Society 116:71-78.

[80] M.B. Sandheinrich and G.J. Atchison. 1990. Sublethal toxicant effects on fish foraging behavior: empirical vs mechanistic approaches. Environmental Toxicology and Chemistry 9:107-120.

[81] T.L. Beitinger. 1990. Behavioral reactions for the assessment of stress in fishes. Journal Great Lakes Research 16: 495-528.

[82] D.C. Miller. 1980. Some applications of locomotor response in pollution effects monitoring. Rap. P.-V Reun. Cons. Int. Explor. Mer 179:154-161.

[83] G.M. Rand. 1977. The effect of exposure to a subacute concentration of parathion on the general locomotor behavior of the goldfish. Bulletin of Environmental Contamination and

Toxicology 18:259-266.

[84] E.G. Ellgaard, J.C. Ochsner, and J.K. Cox. 1977. Locomotor hyperactivity induced in bluegill sunfish, Lepomis macrochirus, by sublethal concentrations of DDT. Canadian Journal of Zoology 55:1077-1081.

[85] E.E. Little, J.F. Fairchild, and A.J. DeLonay. 1993. Behavioral methods for assessing impacts of contaminants on early life stage fishes. American Fisheries Society Symposium 14:67-76.

[86] J. Lipton, E.E. Little, J.C.A. Marr, and A.J. DeLonay. Use of behavioral avoidance in Natural Resource Damage Assessment," Environmental Toxicology and Risk Assessment: Biomarkers and Risk Assessment Fifth Volume, ASTM STP 1306, D.A Bengston and D.S. Henshel, Eds., American Society for Testing and Materials, Philadelphia, 1997.

[87] A.J. DeLonay, E.E. Little, J. Lipton, And J.A. Hansen. "Behavioral avoidance as evidence of injury to fishery resources: Applications to natural resource damage assessments," Environmental toxicology and Risk Assessment: Biomarkers and Risk Assessment Fourth Volume, ASTM STP 1262, T.W. LaPoint, F.T. Price, and E.E. Little, Eds., American Society for Testing and Materials, Philadelphia, 1997.

[88] H.W. Lorz and B.P. McPherson. 1976. Effects of copper or zonc in freshwater on the adaptation to seawater and ATPase activity and the effectsof copper on migratory dispositions of coho salmon (*Oncorhynchus kisutch*), J. Fish. Research Bd. Can. 33: 2023-2030.

[89] J.W. Nichols, G.A. Wedemeyer, G.A., F.L. Mayer, W.W. Dickhoff, S.V. Gregory, W.T. Yatsutake, and S.D. Smith. 1984. Effects of freshwater exposure to arsenic trioxide on the parr-smolt transformation of coho salmon (*Oncorhynchus kisutch*). 3: 143-149

[90] S. Hamilton, A. Palmisano, G.A. Wedemeyer, and W.T. Yatsutake. 1986. Impacts of selenium on early lifestages and smoltification of fall chinook salmon. Transactions of the 51st Wildlife and Natural Resources Conference, pp 343-356.

[91] J.R. Brett and D. MacKinnon. 1954. Some aspects of olfatory perception in migrating adult coho and spring salmon. J. Fish Research Board Can. 11:310-318.

[92] R.L. Saunders and J.P. Sprague. 1976. Effects of copper-zinc mining pollution on a spawning migration of Atlantic salmon. Water Research 1: 419-432.

[93] P.W. Bettolli, P.W. Clark. 1992. Behavior of sunfish exposed to herbicides: a field study. Envrionmental Toxicology and Chemistry 11:1461-1467.

[94] P.M. and F.L. Mayer. 1980. Clinical tests in aquatic toxicology: state of the art. Environmental Health Perspectives 34:139-143.

[95] J.F. Klaverkamp, W.A. Macdonald, D.A. Duncan, and R. Wagemann. 1984. Metallothionein and acclimation to heavy metals in fish. A review. In: *Contaminant Effects on Fisheries.* Edited by V.W. Cairns, P.V. Hodson and J.O. Nriagu. John Wily and Sons, Toronto. Pp 99-113.

[96] P.M. Mehrle, Jr. and F.L. Mayer, Jr. 1975. Toxaphene effects on growth and bone composition of fathead minnows, Pimephales promelas. Journal of the Fisheries Research Board of Canada 32:593-598.

[97] P.M. Mehrle, Jr. and F.L. Mayer. 1977. Bone development and growth of fish as   affected by toxaphene. Fate of Pollutants in Air and Water Environments. Part 2. Wiley Interscience Publishers, New York, NY.:301-314.

[98] S.J. Hamilton, P.M. Mehrle, Jr. F.L. Mayer, J.R. Jones. 1981. Method to evaluate mechanical properties of bone in fish. Transactions of the American Fisheries Society 110:708-717.

[99] P.M. Mehrle, Jr. T.A. Haines, S.J. Hamilton, J.L. Ludke, F.L. Mayer, and M.A. Ribick. 1982. Relationship between body contaminants and bone development in east-coast striped bass. Transactions of the American Fisheries Society 111:231-241.

[100] *The Role of Biochemical Indicators in the Assessment of Ecosystem Health - Their Development and Validation.* National Research Council Canada, Associate Committee on Scientific Criteria for Environmental Quality. NRCC No. 24371. 1985.

[101] B.R. Hobden and J.F. Klaverkamp. 1977. A pharmacological characterization   of acetylcholinesterase from rainbow trout [salmo gairdneri] brain. Comparative Biochemistry Physiology 57:131-133.

[102] J.F. Klaverkamp,and B.R. Hobden. 1980. Brain acetylcholinesterase inhibition and hepatic activation of acephate and fenitrothion in rainbow trout (salmo gairdneri). Canadian Journal Fisheries Aquatic Science 37:1450-1453.

[103] W.L Lockhart, D.A. Metner, F.J. Ward, and G.M. Swanson. 1985. Population and cholinesterase responses in fish exposed to malathion sprays. Pesticide Biochemistry And Physiology 24:12-18.

[104] S.K. Brewer, A.J. DeLonay, S.L. Beauvais, E.E. Little, E.E., S.B. Jones. The use of Autoated monitoring to assess behavioral toxicology in fish:Linking bheavior and physiology, Environmental toxicology and Risk Assessment: Biomarkers and Risk Assessment Eightth Volume, ASTM STP 1364, D.S. Henshel, M.C. Black, and M.C. Harrass, Eds., American Society for Testing and Materials, Philadelphia, 1997.

[105] E.E. Little. Personal communication.

[106] M.E. Hahn, B.R. Woodin, J.J. Stegeman, and D.E. Tillitt. 1998. Aryl hydrocarbon receptor function in early vertebrates: inducibility of cytochrome P450 1A in agnathan and elasmobranch fish. Comparative Biochemistry and Physiology (in press, 1998).

[107] S.A. Ploch, L.C. King, M.J. Kohan, and R.T. Di-Giulio. 1998. Comparative in vitro and in vivo benzo[a]pyrene-DNA adduct formation and its relationship to CYP1A activity in two species of ictalurid catfish. Toxicology and Applied Pharmacology 149:90-98.

[108] L. Taysse, C. Chambras, D. Marionnet, C. Bosgiraud, and P. Deschaux. 1998. Basal level and induction of cytochrome P450, EROD, UDPGT, and GST activities in carp (Cyprinus carpio) immune organs (spleen and head kidney). Bulletin Environmental Contamination and Toxicology 60:300-305.

[109] M.R. Soimasuo, A.E. Karels, H. Leppanen, R. Santti, and A.O. Oikari. 1998. Biomarker responses in whitefish (Coregonus lavaretus L. s.l.) experimentally exposed in a large lake receiving effluents from pulp and paper industry. Archives Environmental Contamination and Toxicology 34:69-80.

[110] P. Flammarion and J. Garric. 1997. Cyprinids EROD activities in low contaminated rivers: a relevant statistical approach to estimate reference levels for EROD biomarker? Chemosphere 35:2375-2388

[111] Fairchild, J.F. and E.E. Little. 1993. Use of mesocosm studies to examine the direct and indirect impacts of water quality on early lifestages of fishes. Transactions of the Amer. Fisheries Society Symposium 14:95-103.

[112] J.F. Fairchild, F.J. Dwyer, T.W. LaPoint, S.A. Burch, and C.G. Ingersoll. 1993. Evaluation of a laboratory-generated NOEC for linear alkylbenzene sulfonate in outdoor experimental streams. Environmental Toxicology and Chemistry 12:1763-1775.

[113] S.A. Davis, T.E. Schwedler, J.R. Tomasso, and J.A. Collier. 1991. Production characteristics of pan-size channel catfish in cages and open ponds. Journal of the World Aquacultural Society 22:183-186.

[114] M.G. Henry and G.S. Atchison. 1984. Behavioral effects of methyl parathion on social groups of bluegill (*Lepomis macrochirus*). Environmental Toxicology and Chemistry 3:399-408.

[115] J.R. Voshell, ed. 1989. Using Mesocosms to Assess the Aquatic Ecological Risk of Pesticides: Theory and Practice. Misc. Publ. Ent. Soc. Amer. 75, Lanham, MD.

[116] R.L. Graney, J.P. Giesy, and J.R. Clark. 1995. Field Studies. Pp 257-306 in G.M. Rand, ed.,

Fundamentals of Aquatic Toxicology: Effects, Environmental Fate, and Risk Assessment. Taylor and Francis Publishers, Washington, DC. 1125 pp.

[117] E.E. Werner, G.G. Mittelbach, D.J. Hall, and J.F. Gilliam. 1983. Experimental tests of optimal habitat use in fish: the role of relative habitat profitability. Ecology 64:1525-1539.

[118] A.D. Lemly. 1996. Assessing the toxic threat of selenium to fish and aquatic birds. Environmental Monitoring and Assessment 43:19-35.

[119] E.E. Little, F.J. Dwyer, J.F. Fairchild, A.J. DeLonay, and J.L. Zajicek. 1993. Survival of bluegill and their behavioral responses during continuous and pulsed exposures to esfenvalerate, a pyrethroid insecticide. Environmental Toxicology and Chemistry 12:871-878.

[120] R.E. Siefert, S.J. Lozano, J.C. Brazner, and M.L. Knuth. 1989. Littoral enclosures for aquatic field testing of pesticides: effect of chlorpyrifos in a natural system. Pp. 57-73, in J.R. Roshell, ed., Using Mesocosms to Assess the Aquatic Ecological Risk of Pesticides: Theory and Practice. Miscellaneous Publications of the Entomological Society of America 75, Lanham, MD.

[121] T.P. Boyle, S.E. Finger, R.L. Paulsen, and C.F. Rabeni. 1985. Comparison of laboratory and field assessment of fluorene-part II: effects on the ecological structure and function of experimental pond ecosystems. Pages 134-151 in T.P. Boyle, editor, Validation and Predictability of Laboratory Methods for Assessing the Fate and Effects of Contaminants in Aquatic Ecosystems, ASTM STP 865, American Society for Testing and Materials, Philadelphia, PA.

[122] S.E. Finger, E.E. Little, M.G. Henry, J.F. Fairchild, and T.P. Boyle. 1985. Comparison of laboratory and field assessment of fluorene-part I: effects of fluorene on the survival, growth, reproduction, and behavior of aquatic organisms in laboratory tests. Pages 120-133 in T.P. Boyle, editor, Validation and Predictability of Laboratory Methods for Assessing the Fate and Effects of Contaminants in Aquatic Ecosystems, ASTM STP 865, American Society for Testing and Materials, Philadelphia, PA.

[123] J.M. Besser, T.J. Canfield, and T.W. LaPoint. 1993. Bioaccumulation of organic and inorganic selenium in a laboratory food chain. Environmental Toxicology And Chemistry 12:57-72.

[124] R.O. Hermanutz, K.N. Allen, T.H. Roush, and S.F. Hedtke. 1992. Effects of elevated selenium concentrations on bluegills (*Lepomis macrochirus*) in outdoor experimental streams. Environmental Toxicology and Chemistry 11:217-224.

[125] R.B. Gillespie and P.C. Baumann. 1986. Effects of high tissue concentrations of selenium on

reproduction by bluegills. Transactions of the American Fishery Society. 115:208-213.

[126] J.J. Coyle, D.R. Buckler, C.G. Ingersoll, J.F. Fairchild, and T.W. May. 1993. Effect of dietary selenium on the reproductive success of bluegills (*Lepomis macrochirus*). Environmental Toxicology and Chemistry 12: 551-565.

[127] L.W. Barnthouse. 1992. The role of models in ecological risk assessment: A 1990's perspective. Environmental Toxicology and Chemistry 11:1751-1760.

[128] L.W. Barnthouse, G.W. Suter, II, A.E. Rosen. 1990. Risks of toxic Contaminants to exploited fish populations. Environmental Toxicology and Chemistry. 9:297-311.

[129] D.L. DeAngelis, L.W. Barnthouse, W. Van Winkle, R.G. Otto. 1990. A critical appraisal of population approaches in assessing fish community health. Journal Great Lakes Research. 16:576-590.

[130] J.S. Jaworska, K.A. Rose, L.W. Barnthouse. 1997. General response patterns of fish populations to stress: An evaluation using an individual-based simulation. Journal of Aquatic Ecosystem Stress Recovery. 6:15-31.

[131] L.A. Nielsen and D.L. Johnson (eds.) 1983. Fisheries Techniques. American Fisheries Society. Bethesda, MD.

[132] R.C. Summerfelt and G.E. Hall (eds). 1987. *Age and Growth of Fish.* Iowa State University Press. 544 pp.

[133] S.A. Fore, S.I. Guttman, A.J. Bailer, D.J. Altfater and B.V. Counts. 1995. Exploratory analysis of population genetic assessment as a water quality indicator:I. Pimephales notatus. Ecotoxicology and Environmental Safety 30:24-35.

[134] K.R. Munkittrick and D.G. Dixon. 1989. Use of white sucker (*Catostomus commersoni*) populations to assess the health of aquatic ecosystems exposed to low-level contaminant stress. Canadian Journal of Fisheries and Aquatic Sciences 46:1455-1462.

[135] Washington, H.G. 1984. Diversity, biotic, and similarity indices: A review with special relevance to aquatic ecosystems. Water Research. 18:653-694.

[136] J.R. Karr. 1987. Biological monitoring and environmental assessment. Environmental Management 11:249-256.

[137] S.M Adams, L.R. Shurgart, G.R. Southworth, and D.E. Hinton. 1990. Application of bioindicators in assessing the health of fish populations experiencing contaminant stresss. Pp. 333-353, in J.F. McCarthy and L.R. Shugart, eds., Biomarkers of Environmental Contamination, Lewis Publishers, New York, NY.

[138] G.A. McFarlane and W.G. Franzin. 1978. Elevated heavy metals: a stress on populations of white sucker (*Catostomus commersoni*), in Hamill Lake, Saskatchewan. Journal of the Fishery Research Board of Canada. 35:963-970.

[139] G.A. McFarlane and W.G. Franzin. 1980. An examination of Cd, Cu and Hg. Concentrations in livers of Northern pike, (*Esox lucius*) and white sucker (*Catostomus commersoni*) from five lakes near a base metal smelter in Flin Flon, Manitoba. Canadian Journal of Fisheries and Aquatic Sciences. 37:1573-1578.

[140] E.A. Trippel and H.H. Harvey. 1987a. Abundance, growth and food supply of white sucker (*Catostomus commersoni*) in relation to lake morphomety and pH. Canadian Journal of Zoology. 65:558-564.

[141] E.A. Trippel and H.H. Harvey. 1987b. Reproductive responses of five white sucker (*Catostomus commersoni*) populations in relation to lake acidity. Canadian Journal of Fisheries and Aquatic Sciences. 44:1018-1023.

[142] R.J. Beamish, W.L. Lockhart, J.C. Vanloon, and H.H. Harvey. 1975. Long-term acidification of a lake and resulting effects on fishes. Ambio 4:98-102.

[143] L.I. Bendall-Young, H.H. Harvey, and J.F. Young. 1986. Accumulation of cadmium by white suckers (Catostomus commersoni) in relation to fish growth and lake acidification. Canadian Journal of Fisheries and Aquatic Sciences. 43:806-811.

[144] W.N. Gibbons, K.R. Munkittrick, and W.D. Taylor. 1998. Monitoring of aquatic environments receiving industrial effluents using small fish species 1: Response of spoonhead sculpin (Cottus ricei) downstream of a bleached-kraft pulp mill. Environmental Toxicology and Chemistry 17:2227-2237.

Environmental Science Forum Vol.96 (1999) pp. 233-242
© 1999 Trans Tech Publications, Switzerland

Submerged Bryophytes in Running Waters, Ecological Characteristics and their Use in Biomonitoring

H. Tremp

University of Hohenheim, Institute of Landscape- and Plant Ecology 320,
DE-70593 Stuttgart, Germany

Keywords: Submerged Bryophytes, Bioindicator, Ecology, Running Waters, Acidification, Waste Heat, Eutrophication

Abstract

The article tries to give a short review on bioindication-properties of submerged bryophytes. The main focus lies on their possible use as reaction-indicators under field conditions at the population and community level. Furthermore a few remarks are given on the accumulation capacities of submerged bryophytes, a feature typically considered at the individual-species level.

One general problem encountered when working with bryophytes in running waters, much more so than in standing waters, is the lack of ecologically rigorous definitions of the terms watermosses, stream bryophytes or aquatic mosses.

Despite the fact that submerged bryophytes usually show higher accumulation capacities in comparison to abiotic compartments and other stream organisms, often the ecological context of the observed accumulation patterns and, thus, its significance for the stream ecosystem is little understood. What is clearly needed, is a better understanding of the ecology of submerged bryophytes and how it is affected by the interaction of stream type and moss species.

In this paper I will present three examples on how aquatic bryophytes might be used as sensitive indicator species. First the reaction of some submerged bryophytes towards heated effluents is given. Secondly, I will summarize how these bryophytes react towards anthropogenic acidification of their habitats, and their use for the establishment of bryophyte based bioindicator keys for acidification. Finally, most aquatic bryophyte species are usually restricted to smaller streams of relatively low trophy. Under these conditions aquatic mosses should be included in multispecies-trophy-indicator systems. Some thoughts about possible advantages and disadvantages of this approach are presented here.

1 Introduction

Bryophytes lack vascular systems and most species are entirely dependent on ectohydric water transport. Therefore bryophyte growth in terrestrial systems is largely restricted because of insufficient humidity. On the other hand, only some central European bryophyte species are able to grow over prolonged periods while being completely submersed. Among the factors thought responsible for the obvious limitations of bryophyte growth in aquatic environments is the lack of aerenchyma and the low CO_2 diffusion coefficient in water which is about four orders of magnitude lesser than in air [1]. Thus in contrast to vascular plants and limnic algae, bryophytes can be considered as 'wanderers' between two worlds. They are viable in both air and in water, but nowhere completely at home [2]. The term submerged macrophytes is often used for practical reasons, and encompasses all submerse aquatic plants visible to the naked eye, including bryophytes. However, because of their different physiology and ecology it is hazardous to generalize uncritically amongst bryophytes and vascular plants [3].

Even within the bryophytes, the terms 'watermosses', 'stream bryophytes' or 'aquatic mosses' are difficult to define ecologically in a rigorous way. All three assume that the aquatic medium is the most favoured site, where these bryophytes show maximum growth and complete their life cycle, including spore-germination, protonema- and bud formation, establishment of the gametophyt, sporophyt-growth and spore dispersal. The last two stages of development are only found in a few species of the ecologically heterogeneous group of hydrophilous bryophytes. Even the capsules of *Fontinalis antipyretica* may be found more frequently on plants exposed to the air than those completely submersed [4]. Most of the so called watermosses prefer the spray zone and bank sites along running waters (Fig. 1), where they are able to complete their entire life-cycle.

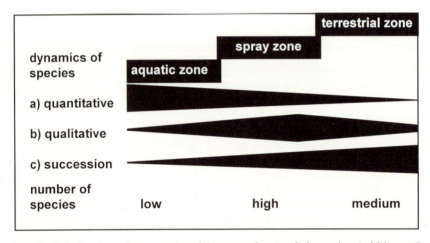

Fig. 1: Relative bryophyte species richness and natural dynamics (within a 10 year period) for three habitats of highly structured dynamic mountain streams. Species dynamics refers to both percentage coverage (a) and species composition (b).

Bryophyte species composition in the spray zone (amphibious habitat) oscillates between more terrestrial or aquatic conditions depending on the changing hydrological regime. Annual quantitative dynamics are most pronounced in the aquatic zone (Fig. 1).

With lower water-levels in winter, even typical watermosses like *Fontinalis antipyretica, Fontinalis squamosa* or *Fissidens fontanus* may survive for several months out of water, just as they survive for several years in amphibious sites. For practical use we should consider all bryophytes as aquatic that are able to tolerate total submersion for prolonged periods of time [5].

In moderate dynamic brooks and streams, amphibious sites are most important for the submersed establishment of hydrophilous bryophytes, because from these 'safe' sites they are more easily able to enter and reenter the submerged zone should their underwater habitats became destroyed. The direct establishment of vegetative fragments [6] and spores [7] at submersed sites is also possible, but comparatively more improbable. That submersed sites are generally less favourable for bryophyte establishment is indicated by increasing species-richness, cover, and diversity within a gradient from submerged to amphibious to terrestrial sites [8, 9].

From accumulation studies in which submerged bryophytes are often viewed as 'living ion exchangers' it is obvious that aquatic bryophytes are able to live in a broad range of water chemistries and are tolerant of various types of pollution [10]. In contrast, in vegetational studies, aquatic bryophytes are considered as living organisms with specific site demands and species distribution patterns that are the result of the specific stream type and, for example, the level of pollution. In the future it should be desirable to achieve a broader overlap between these contrasting approaches.

This article refers mainly to aspects of the possible use of aquatic bryophytes as reaction-indicators. In this context the general ecology, particulary the habitat component, and, thus, the growth patterns of the bryophytes might provide significant insights on the interpretation of accumulation data. Apart from screening indicators like productivity or nutrient status which provide an overview about general stream condition, sensitive indicator species are also needed to assess stream integrity in a more specific way. In this context bryophytes should also be considered.

2 Submerged bryophytes as accumulation indicators

In addition direct measurements of physical and chemical parameters, biological indicators are increasingly used for determining and monitoring water quality in lotic environments [11]. A special property of submerged bryophytes is their high capacity for accumulating heavy metals and chlorinated hydrocarbons [12]. This capacity has also been used for the monitoring of radionuclids [13, 14]. An important ecological property of submerged bryophytes is the trapping of suspended particles which may serve other organisms as a food supply. In anthropogenically disturbed systems, suspended coal particles, for example, lead to severe damage on submerged bryophytes [15]. Many investigations have shown that aquatic bryophytes have a higher accumulation-capacity for heavy metals than have sediments or vascular macrophytes. These plants have the disadvantage of a more pronounced seasonality. Furthermore haptophytic bryophytes are nutritionally independend from their growing substrate, and thus their accumulation in phylloids and cauloids reflects the contamination of the water-body alone. Mainly because of these special features, standardized sampling and analyses-methods for different bryophyte-species have been developed for this use as accumulation-indicators [16, 17]. Finally, as a further advantage compared to many vascular macrophytes, aquatic bryophytes or their parts seem to be no major food of herbivores.

Despite the increasing use of aquatic bryophytes in heavy-metal-accumulation studies [18], nearly nothing is known about the morphological, metabolic, physiological, and genetical differences that determine or modify the accumulation-capacity of different species [19]. Furthermore, hardly any information exists about found accumulation-pattern particularly of those mosses transplanted for active monitoring. Some considerations about potential ecosystem consequences (e.g. extinction of fishes) of accumulated substances have been made by Mouvet et al. [12].

Some authors have found that in acidified streams the hydrogen ions make a more significant contribution to the 'loading' of the bryophytes than the concentration of metal cations in the water. Therefore accumulation patterns by themselves are of little value [20] unless they are interpretated in an ecological context [7]. Also the influence of temperature for adsorption processes is viewed differently by several authors [e.g. 21, 22]. Often a certain experimental temperature is interpreted as an adequate representation of seasonal effects, however temperature by itself is only one component of seasonality. So it can be expected that exchange surface, e.g. phylloid/cauloid-ratio, may differ seasonally and in dependence of individual site factors. In general, the longstanding discussion about the physiological and morphological adaptions of submerged bryophytes [23] still seems to suffer from lack of good data.

3 The effect of environmental factors on submerged bryophytes

Stream ecosystems usually provide a set of different aquatic habitats with their own selective forces, and thus provide habitats for bryophyte species with different adaptions [5]. Especially in montaneous streams, physical disturbance due to highly dynamic discharge patterns and frequent substrate shifting prevent the development of a vascular macrophyte vegetation. In contrast, crypto-gams can better withstand the grinding effect of particles by sprouting of their remaining cells, stalks, and rhizoid parts [7]. It is because of these adaptions that bryophytes can exist in highly dy-namic permanent changing aquatic environments. Single bryophyte species differ not markedly in these adaptions. The most striking interspecific difference is their apparent growth-form. Fastgro-wing, mat-forming species like *Rhynchostegium riparioides* normally overgrow turfs (e.g. *Scapania undulata*) and cushions. Furthermore in physical highly disturbed underwater stream environments, intraspecific competition between bryophytes plays only a subordinate role. In such streams, bryo-phyte cover rarely exceeds 10 %. Also, at least under oligotrophic to mesotrophic conditions, the regular disturbances contribute to the fact that biodiversity of submerged bryopytes in mountain-streams generally is maximized.

On the other hand, streams with lower discharge dynamics are usually dominated by vascular macrophytes. Under those conditions these higher plants have an advantage because of their higher productivity. That's why vascular macrophytes may be better used as indicators for eutrophication, whereas because of their generally low productivity, submerged bryophytes do not reflect the nutri-ent status of a stream over a broad scale.

Another factor that apparently determines the luxuriance of bryophyte growth [24], is the 'aeration of the water'. Bryophytes are unable to use HCO_3^- as a carbon source and are thus dependent on the amount of CO_2 getting into solution, which is higher under turbulent flow conditions. In nutrient enriched streams, bryophyte production is often reduced by epiphytic algae and sedimentation of fine particulate matter.

The fact that we can distinguish between a hardwater and softwater bryoflora is not well explained ecologically. As the pH rises, the carbon dioxide-bicarbonate equilibrium shifts to bicarbonate, which bryophytes can not use as a carbon source. On the other hand, bryophytes also occur in lakes, where current velocities are low even under the influence of wave action. Thus, also under very low CO_2 diffusion gradients, many bryophytes are able to balance their C-budget, although there are distinctive differences in the ability to 'capture' the available CO_2 between mat/turf-forming species like *Scapania undulata* and streamer-forming species like *Fontinalis antipyretica* [1].

Apparently there are a number of environmental factors including streambed structure and flow dy-namics, that limit and thus restrict the use of submerged bryophytes as biomonitors in streams.

4 Anthropogenic environmental disturbances and their effect on submerged bryophytes

Environmental factors may exist as physical and chemical gradients. Some of these factors influence the distribution pattern of submerged bryophytes and allow conclusions about the function of certain morphology features of these mosses and/or their physiological properties. Most of these species specific properties can only be specified if the gradient in a particular environmental factor is strong enough to become a dominant structuring force at a site.

Nevertheless unidirectional environmental factors must be considered in special cases in multifacto-rially influenced running waters and can only provide simplified models. In the following three an-thropogenic environmental disturbances and their effect on aquatic bryophytes will be considered: waste heat, acidification and eutrophication.

4.1 Temperature - waste heat

Compared to terrestrial bryophytes, species growing under water are exposed to less extreme tempe-ratures. Nevertheless, because of their persistence throughout the year, submerged bryophytes are subject to a wide range of seasonal temperatures [25].

Therefore, it can be assumed that only the substantially elevated maximum temperatures caused by heated effluents directly influence submerged bryophytes. Under natural conditions, the effect of temperature on species composition is more indirect, probably because of the higher competition of epiphytic/filamentous algae.

A possible effect of temperature on species distribution pattern in the Kocher River (South-West Germany) is shown in Fig. 2. Throughout the year cooling water discharge from an industrial complex heats up the river over an approximately 300 m long section. This effect is mainly restricted to the orographic right side of the channalized river.

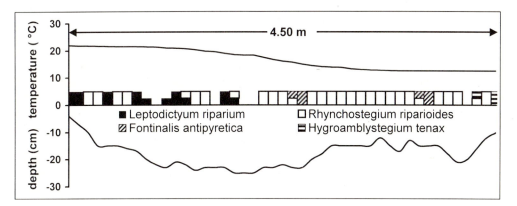

Fig. 2: Cross-transect of the Kocher River 10 m downstream of a cooling water discharge. Small rectangles indicate the presence of 4 bryophyte species at two levels of cover (half length ≤ 50%; full length > 50 %).

Leptodictyum riparium was only found in the area of higher temperatures, whereas *Hygroamblystegium tenax* and *Fontinalis antipyretica* show a tendency to be dominant under normal water temperature (Fig. 2). *Rhynchostegium riparioides* reacted indifferently towards temperature.

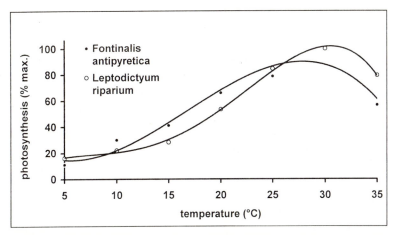

Fig. 3: Temperature dependence of photosynthetic capacity (in % of maximum) in short-term experiments with *Fontinalis antipyretica* and *Leptodictyum riparium.* In absolute values, *Fontinalis antipyretica* (max. 28 mg $O_2 \cdot l^{-1} \cdot h^{-1} \cdot dw(g)^{-1}$) show a one third higher photosynthetic rate as *Leptodictyum riparium.*

One interpretation of this species distribution pattern is that *Leptodictyum riparium*, has a photo-synthetic optimum at slightly higher temperatures (Fig. 3) and may be better able to resist competi-tion effects of the otherwise faster growing moss-species at the warmer locations.

Similarly, in laboratory experiments Sanford [26] found maximum growth-rates from *Amblystegium riparium* (syn. *Leptodictyum riparium*) at 23°C. CARBALLEIRA et al. [27] found that *Fontinalis antipyretica* is not able to acclimatisize on temperatures above 24,9°C. However physiological data (Fig. 3) provide only a vague idea, whether or not temperature may effect species distribution pat-terns, and other reasons like eutrophication (Tab. 1) must also be considered. *Fontinalis antipyre-tica* which grows under elevated temperatures throughout the year at such sites (Fig. 2) shows atypi-cal morphological features. Leaves are smaller and more slender whereas the same species else-where, growths more rigidly and has broader leaves.

Such atypically high temperatures are normally not tolerated by aquatic bryophytes of the temperate region without significant morphological or physiological modifications. Generally, ecomorphoses are often found among bryophytes growing along or inside streams. Apart from natural causes, these morphological changes might be caused by pollution effects.

4.2 pH - stream acidification

In Central Europe the problem of acidification of running waters is mainly restricted to brooks and smaller streams in silicaceous mountain areas. These waters usually have their source in acidified catchments of triassic red sandstone-, granite- and gneiss-formations. These catchments are only weakly buffered and the atmospheric acid load has lowerded alkalinity.

In the Black Forest where discharge of stream headwaters can rise several hundred fold after snow melt or heavy precipitation, the only vegetation present are bryophytes. In this stream systems these mosses play an important ecological role and show characteristic longitudinal zonation patterns along an acidification-gradient (Fig. 4).

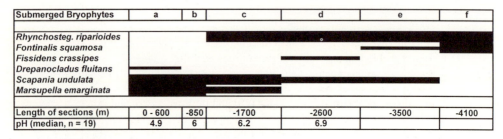

Submerged Bryophytes	a	b	c	d	e	f
Rhynchosteg. riparioides Fontinalis squamosa Fissidens crassipes Drepanocladus fluitans Scapania undulata Marsupella emarginata						
Length of sections (m)	0 - 600	-850	-1700	-2600	-3500	-4100
pH (median, n = 19)	4.9	6	6.2	6.9		

Fig. 4: Longitudinal zonation of submerged bryophytes along the acidification-gradient of a brook (Kaltenbach) in the Black Forest (Germany). The thickness of bars indicate dominant, frequent and rare occurence of submerged bryophytes.

Under normal run-off conditions for the most time of the year, the acidification-gradient shown in Fig. 4 is stable over several hundred meters as are the corresponding changes in the distribution of the bryophyte vegetation. In comparison, structurally similar headwater streams that do not show such a pH-gradient also have no pronounced pattern of their bryophyte vegetation, because Cal-cium- or Magnesiumcarbonate rich layers partially cover the main geological formation.

In order to explain the particular zonation patterns found, laboratory and field experiments were conducted [7], because in theory several gradients are imaginable along brooks over distances of several kilometers.

These authors found pronounced differences in the tolerance (in terms of stem growth) towards acid conditions within a pH range of 4.2 and 6.5 (Fig. 5). The exposure time was 40 days. Temperature and waterchemical parameters (K^+, Ca^{2+}, Al^{3+}, Fe^{3+}, NH_4^+, PO_4^{3-}, NO_3^-, alkalinity) were measured weekly. Only hydroniumion-concentration, just as autocorrelated parameters like aluminium, iron, calcium and potassium and acid binding capacity showed significant correlation to stem growth- of acid intolerant species *Rhynchostegium riparioides*.

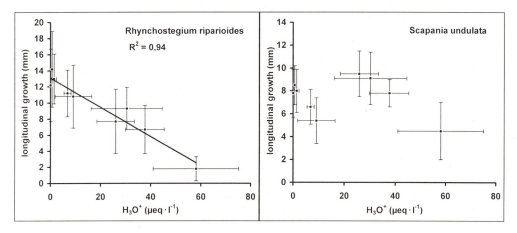

Fig. 5: Incremental growth of stems of of *Rhynchostegium riparioides* and *Scapania undulata* after 40-day exposure in 9 sections of small streams of different acidity. Means ± SD of longitudinal growth (n = 20) and hydroniumion-concentration (n = 4) is given. Temperature differences at exposion-sites differed up to 5 K, but had no significant influence on growing of bryophyte-tips. For details of method see [7].

Further enclosure-field-experiments showed that *Rhynchostegium riparioides* may survive in waters where acidity seldom exceeds pH 4.4. Under natural conditions, however, this species cannot be found at such sites probably due to lowered regeneration capability.

The zonation-patterns observed among submerged bryophyte communities (Fig. 4) have been used as a basis for a bioindication-key for easy biological evaluation of acidification gradients in small streams [11]. Investigations obtained from Wales [28] and theVosges Mountains [29] confirm these indicator properties of submerged bryophytes. This mapping-method which encompasses the whole stream, and not just some individual sites, should be used for a fast ecological screening in areas threatend by acidification.

4.3 Trophic level - eutrophication

In Germany two components of water quality have traditionally been used in assessment methods. One of which is the 'Saprobien-Index', based on the saprobic-system [30] the other is the determination of the trophic state. Trophy, in this context, represents the intensity of plant-growth. High trophy leads to eutrophication, a complex and characteristic property of waters. The typification of different classes of trophy traditionally consider the chlorophyll-content of the water. This measure, however, is not very useful in most streams where primary production within the water column is negliable and which is biomass dominated by benthic algae or macrophytes. Alternatively other criteria have been used to determine the trophic potential of running waters, as for example the concentration of total phosphorous, maximum pH or day/night differences in oxygen content. The rationale being, that the higher the autotrophic productivity of the water is, the higher are phosphate-concentration, pH, and day/night differences in oxygen [31].

In this context the question arises if submerged bryophytes can be good trophy-indicators considering their production-ecology. Theoretically they are not, because they are lacking productive properties like a fast succession of generations like algae, or the high productivity of submerged vascular macrophytes which profit from additional nutrients from sediment and possess a more effective gaseous metabolism.

Most aquatic bryophytes are comparatively low-productive, so even oligotrophic streams fulfill their low nutrient demands whereby this problem is closely related to stream velocity and the extent of the boundary layer on bryophytes surfaces. At higher current velocity as better supply of nutrients. Other previously mentioned factors like the availability of free CO_2 and the lack of competition seem to be probable factors which enables the bryophytes relative high production rates in running waters.

The conditions for the growth of most submerged bryophytes are fulfilled in oligotrophic to mesotrophic streams. As seen in Table 1 bryophytes are arranged according to their main distribution patterns in running waters of different trophic level. Typical species of large rivers [32], e.g. several more species of genus *Cinclidotus*, are not involved.

Tab. 1: Occurrence of selected bryophytes of brooks and smaller streams according to the trophic levels (<u>o</u>ligo-, <u>m</u>eso-, <u>e</u>u- and <u>p</u>olytrophic) of sites. The 20 point scale indicates possible presence (1) to exclusive occurrence. For hardness, s and h indicate that the main distribution of species is in soft- and/or hardwaters.

Species	trophic level							hardness
	o	o-m	m	m-e	e	e-p	p	
Marsupella emarginata	18	2						s
Scapania undulata	16	3	1					s
Cratoneuron commutatum	12	6	2					h
Riccardia chamaedryfolia	10	8	2					s
Hygrohypnum ochraceum	8	8	4					s
Cratoneuron filicinum	10	7	2	1				h
Chiloscyphus polyanthos	8	9	2	1				s
Fontinalis squamosa	6	9	4	1				s
Pellia endiviifolia	8	6	4	2				h
Amblystegium tenax	3	9	6	2				h
Cinclidotus aquaticus		15	3	2				h
Brachythecium rivulare	2	8	7	2	1			s/h
Fissidens crassipes	5	8	4	2	1			s/h
Rhynchostegium riparioides	1	5	6	5	2	1		s/h
Fontinalis antipyretica	2	3	4	4	4	3		s/h
Leptodictyum riparium	1	2	4	5	4	3	1	h

Although the theoretical background underlying the classification in Table 1 is not well understood, these empirically derived numbers may serve as a basis for discussion. Table 1 shows, for example, that only the three last species, which all grow plagiotrop, may be viable under eu-/polytrophic conditions. Polytrophic conditions are only tolerated by *Leptodictyum riparium* which can even be found on bacterial filter beds [4].

5 Conclusions about bioindicative properties of submerged bryophytes

In contrast to the more intensive use of terrestrial bryophytes as bioindicators of site conditions and in pollution monitoring, aquatic bryophyte species of running waters are seldom used as indicator species.

However, bryophytes as any other organism, can be useful indicator species under certain ecological conditions. To my opinion, to gain insight into the ecology of running waters and submerged bryophytes and the effects of human disturbances, it is important to work at the appropriate level of complexity. Together with other autotrophic species like vascular macrophytes and bentic algae, bryophytes may be used to indicate the pollution status of running waters integratively.

Despite of their current status, aquatic bryophytes have much to contribute to such a monitoring approach. In oligotrophic and mesotrophic streams, for example, bryophytes are well suited to indicate various forms of human impact and support and extend results obtained from other autotrophic species.

6 References

[1] J. T. Jenkins & C. F. Proctor. Water velocity, growth form and diffusion resistances to photosynthetic CO_2 uptake in aquatic bryophytes. Plant, Cell and Environment 8 (1985), pp. 317-323.

[2] F. Gessner. Hydrobotanik. Die physiologischen Grundlagen der Pflanzenverbreitung im Wasser, I. Energiehaushalt. VEB Deutscher Verlag der Wissenschaften, Berlin (1955).

[3] M. C. F. Proctor. The physiological basis of bryophyte production. Botanical Journal of the Linnean Society 104 (1990), pp. 61-77.

[4] E. W. Watson. British mosses and liverworts. Cambridge (1963).

[5] D. H. Vitt & J. M. Glime. The structural adaptions of aquatic musci. Lindbergia 10 (1984), pp. 95-110.

[6] J. Glime. Effects of temperature and flow on rhizoid production in *Fontinalis*. The Bryologist (1980), pp. 477-485.

[7] H. Tremp & A. Kohler. Wassermoose als Versauerungsindikatoren. Veröffentlichungen PAÖ (Projekt Angewandte Ökologie der Landesanstalt für Umweltschutz Baden-Württemberg) 6, Karlsruhe (1993).

[8] J. M. Glime & D. H. Vitt. A comparison of bryophyte species diversity and niche structure of montane streams and stream banks. Can. J. Bot. 65 (1987), pp. 1824-1837.

[9] D. H. Vitt, J. M. Glime & C. LaFarge-England. Bryophyte vegetation and habitat gradients of montane streams in western Canada. Hikobia 9 (1986), pp. 367-385.

[10] J. Mersch & M. Reichard. In situ investigation of trace metal availability in industrial effluents using transplanted aquatic mosses. Arch. Contam. Toxicol. 34 (1998), pp. 336-342.

[11] H. Tremp & A. Kohler. The usefulness of macrophyte monitoring-systems examplified on eutrophication and acidification of running waters. Acta bot. Gallica 142(6) (1995), pp. 541-550.

[12] C. Mouvet, E. Morhain, C. Sutter & N. Couturieux. Aquatic mosses for the detection and follow-up of accidental discharges in surface waters. Water, Air and Soil Pollution 66 (1993), pp. 333-348.

[13] R. Kirchmann & J. Lambinon. Bioindicateurs végétaux de la contamination d' un cours d' eau par des effluants d' une centrale nucléaire à eau pressurisée. Bulletin de la Société royale de Botanique de Belgique 106 (1973), pp. 187-201.

[14] H. T. Shacklette & J. A. Erdmann. Uranium in spring water and bryophytes at Basin Creek in Central Idaho. Journal of Geochemical Exploration 17 (1982). pp. 221-236.

[15] K. Lewis. The effect of suspended coal particles on the life forms of the aquatic moss *Eurhynchium riparioides* (Hedw.). I. The gametophyte plant. Freshwat. Biol. 3 (1973), pp. 251-257.

[16] B. A. Whitton, P. J. Say & B. P. Jupp. Accumulation of zinc, cadmium and lead by the aquatic liverwort *Scapania*. Environmental Pollution (Series B) 3 (1982), pp. 299-316.

[17] J. D. Wehr, A. Empain, C. Mouvet, P.J. Say & B.A. Whitton. Methods for processing aquatic mosses used as monitors of heavy metals. Water Res. 17 (9) (1983), pp. 985-992.

[18] J. M. Glime & R. E. Keen. The importance of bryophytes in a man-centered world. Journ. Hattori Bot. Lab. 55 (1984), pp. 133-146.

[19] J. López & A. Carballeira. Interspecific differences in metal bioaccumulation and plant-water concentration ratios in fife aquatic bryophytes. Hydrobiologia 263 (1993), pp. 95-107.

[20] L. A. Caines, A. W. Watt & D. E. Wells. The uptake and release of some trace metals by aquatic bryophytes in acidified waters in Scotland. Environmental Pollution (Series B) 10 (1985), pp. 1-18.

[21] F. Vray, J.-P. Baudin & M. Švadlenková. Effects of some factors on uptake and release of ^{106}Ru by a freshwater moss, *Platyhypnidium riparioides*. Arch. Environ. Contam. Toxicol. 23 (1992), pp. 190-197.

[22] B. Claveri & C. Mouvet. Temperature effects on copper uptake and CO_2 assimilation by the aquatic moss *Rhynchostegium riparioides*. Arch. Environ. Contam. Toxicol. 28 (1995), pp. 314-320.

[23] J. M. Glime & D. H. Vitt. The physiological adaptions of aquatic musci. Lindbergia 10. (1984), pp. 41-52.

[24] W. Tutin. The moss ecology of a Lakeland stream. Trans. British Bryol. Soc. 1(3) (1949), pp. 166-171.

[25] M. D. Fornwall & J. M. Glime. Cold and warm-adapted phases in *Fontinalis duriaei* Schimp. as evidenced by net assimilatory and respiratory responses to temperature. Aquatic Botany 13, (1982), pp. 165-177.

[26] G. R. Sanford. Temperature related growth patterns in *Amblystegium riparium*. The Bryologist 82(4) (1979), pp. 525-532.

[27] A. Carballeira, S. Díaz, M. D. Vázquez & J. López. Inertia and resilience in the responses of the aquatic bryophyte *Fontinalis antipyretica* Hedw. to thermal stress. Arch. Environ. Contam. Toxicol. 34 (1998), pp. 343-349.

[28] S. J. Ormerod & K. R. Wade. The role of acidity in the ecology of Welsh lakes and streams. In. R. W. Edwards, A. S. Gee & J. H. Stoner (ed.). Acid waters in Wales. Cluwer Academic Publishers (1990), pp. 93-119.

[29] J.-P. Frahm. Ein Beitrag zur Wassermoosvegetation der Vogesen. Herzogia 9 (1992), pp. 141-148.

[30] R. Kolkwitz & M. Marsson. Grundsätze für die biologische Beurteilung des Wassers nach seiner Fauna und Flora. Mitt. d. Kgl. Prüfungsanstalt f. Wasserversorgung und Abwässerbeseitigung Berlin-Dahlem 1 (1902), pp. 33-72.

[31] Bayerisches Landesamt für Wasserwirtschaft (ed.). Trophiekartierung von aufwuchs- und makrophytendominierten Fließgewässern. Bayer. Landesamt für Wasserwirtschaft, München. (1998).

[32] J.-P. Frahm. Moose als Bioindikatoren. Biologische Arbeitsbücher, Quelle & Meyer, Wiesbaden (1998).

Environmental Science Forum Vol.96 (1999) pp. 243-274
© 1999 Trans Tech Publications, Switzerland

Biomonitoring using Aquatic Vegetation

M.A. Lewis[1] and W. Wang[2]

[1] U.S. Environmental Protection Agency, National Health and Environmental Effects Research Laboratory, Gulf Ecology Division, 1 Sabine Island Drive, Gulf Breeze, FL 32561, USA

[2] U.S. Geological Survey, Water Resources Division, 720 Gracern Road, Columbia, South Carolina 29210, USA

Keywords: Biomonitoring, Aquatic Plants, Contaminants, Indicators

Chapter Outline

I. **Introduction**
    A. Plant Types
    B. Importance and State of Phytoassessment

II. **Toxicity tests**
    A. Standardization
    B. Duckweed
    C. Rooted Species
    D. Ecological Relevance

III. **Biomonitoring Techniques**
    A. Population Surveys
    B. Biometrics
        (i) Percent Cover, Shoots, Roots
        (ii) Biomass
        (iii) Productivity
        (iv) Physiological Effects
        (v) Pigment Content
    C. Indicator Species
    D. Numerical Indices

IV. **Bioaccumulation**

V. **Eutrophication and Phytoassessment**

VI. **Wetlands and Phytoassessment**

VII. **The Future/Recommendations**

VIII. **References**

I. Introduction

This Chapter provides an overview of the state-of-the-science as related to the phytoassessment techniques used in environmental biomonitoring and the hazard assessment process for single chemicals and complex mixtures. The emphasis is on freshwater angiosperms and bryophytes. Algal species, which are presented in-depth in Chapter [xxx], will be discussed in a few cases for comparison only. The discussion will include aquatic, wetland, and to a lesser extent, terrestrial plants, in the context of the laboratory and field assessment techniques used to determine their condition. In this Chapter, plants, macrophytes, and vascular plants are terms used interchangeably for angiosperms.

A variety of physical, chemical and biological factors in water and sediment control the distribution and condition of vascular aquatic vegetation. These include, among others, water level, light intensity, currents, salinity, substrate type and suspended solids. In addition, the diversity and density of vascular plants are also influenced by anthropogenic activities that affect water and sediment quality. The determination of their effects, primarily in rivers and lowland streams, has been an integral part of environmental monitoring and assessment activities in the United States and other countries. In this Chapter, the role of phytoassessment is discussed in the context of field biomonitoring and as it applies to the hazard assessment process for chemical substances such as pesticides, contaminated sediments and wastewaters. In addition, phytoremediation is briefly discussed as an important use of plant species for the environmental improvement of wastewaters and other sources of potential contaminants.

The use of vascular plants in field or *in-situ* biomonitoring activities utilizes techniques common to those for micro- and macro-algae and periphyton. In addition, some sampling methodologies have been adapted from techniques used for terrestrial species. Scientific understanding concerning the environmental requirements and sensitivities of vascular plants to contaminants is not well understood and is less than that for algae, periphyton and animal species. This trend persists despite the fact that considerable information about the effects of toxics and nutrients on plants can be obtained easily by measuring biomass and chlorophyll *a*, and with further effort, the determination of bioaccumulation, productivity, and community composition can lead to additional useful information. The use of aquatic vegetation in environmental biomonitoring and laboratory chemical hazard assessments is increasing, however it remains difficult to differentiate the environmental factors that cause a change in the community dynamics of aquatic plants. Therefore, more information is needed before their value will be comparable to that of algae and higher level trophic organisms such as invertebrates and fish, which are more commonly used for these purposes.

A. Plant Types

Aquatic macrophytes have multi-cellular structure and specialized tissues. They include the mosses, liverworts and flowering plants. Their sizes range from the nearly microscopic to those easily visible. There is considerable variation in life cycles, morphology, physiology and reproduction of the various species which reflects their diversity and ability to adapt to a wide range of physical and chemical conditions. There are approximately 370,000 terrestrial species of angiosperms and, of these, there are about 7400 aquatic species (1). The freshwater aquatic species are found in lakes, reservoirs, ponds, wetlands and canals. Intertidal species include seaweeds, and estuarine forms are represented by marsh grasses and seagrasses. The various species can be classified as attached and free-floating. Attached plants can be emergent, floating-leaved and submergent such as those found in Figure 1. The floating plants have true leaves and floating roots (duckweeds). Submerged plants are attached to the substrate and may have floating leaves and aerial reproductive structures (water milfoil). These forms usually inhabit aquatic areas less than 10 m in depth. Emersed species occur in shallow water and in moist shoreline areas with depths less than 1 m. These plants can be either floated-leaved (water lilies) or have upright shoots (cattails). Epiphytes are associated with some aquatic vascular plants such as seagrasses.

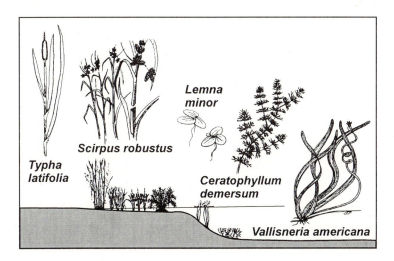

Figure 1. Representative aquatic vascular plants used in field biomonitoring programs and laboratory hazard assessments for contaminants.

Several species of vascular plants have been used for environmental phytoassessment in the laboratory (2), although no single species has been used more frequently than animal species. The selection criteria for determining the suitability of a plant test species has included considerations on the availability of culture techniques, its chemical sensitivity and the choice of species with seeds having a high germination percentage. Among aquatic plants, common duckweed has received more attention than most due to its ease of culturing, rapid growth, and sensitivity to several toxic substances (3-5). *Lemna minor* and *Lemna gibba* G3 are species recommended in standardized toxicity test methods in the United States (6,7).

Among the emergent plants, the effects of contaminants on millet has been studied commonly because it is a wetland species, the seeds are readily available, there is a relatively large historical toxicity database, the plant has a convenient seed size of 1 to 2 mm diameters, and the seeds have a high germination percentage (5). Other species used in phytoassessment include rice (8, 9), smartweed (*Polygonum coccineum*), Sesbania (*Sesbania macrocarpa*), duck wheat (*Fagopyrum tartaricum*), and chufa (*Cyperus esulentus*). The American Public Health Association et al. (6) recommends the use of rice cutgrass (*Leersia oryzoides*), American lotus (*Nelumbo lutea*), watercress (*Rorippa nasturtium-aquaticum*), and wild rice *Zizania aquatica*. More information on this topic is presented later in this chapter.

In some cases, terrestrial species have been used to assess the impact of contaminants in aquatic environments, such as industrial effluents. Many species have been recommended but cabbage, lettuce and oats are the more commonly used. Several terrestrial species are required for use in chemical assessments of soils by environmental regulatory agencies in the United States and other countries (10).

Although several plant species have been used in the hazard assessment process as required by regulatory agencies, the relative sensitivities of most vascular plant species to various toxicants and hazardous substances have not been studied systematically and are poorly understood. One of the reasons for this lack of information is that in the past, there was a school of thought that algae could be used as a surrogate species for vascular plants. Many published results have shown, however, that this assumption is false. Lewis (11) has shown that the sensitivities of different algal species can vary several orders of magnitude and it is very difficult therefore to quantify the typical response of "algae" unless several taxonomically different species are used, which is not the usual case. Hughes and Erb (12) reported that no species of algae and duckweed could be identified as "always being the most sensitive or always the least sensitive" after exposure to several pesticides. In another example, the relative sensitivities of ten species were compared based on their response to wastewater from a metal engraving industry (13). Using germination percentage and sensitivity as the selection criteria , rice, lettuce and tomato were considered the more promising test species.

It is important to note that some in the past have used toxicity for animal species as a surrogate for vascular plants (14-16). This practice is not supported technically and should be avoided. The perceived insensitivity of aquatic vascular plants relative to those for invertebrates and fish is due to a lack of an established data base describing their sensitivities and possibly also to the recognition that several of these plants are used in constructed wetlands to improve water quality. However, only a few of the many species of vascular plants have been used for this purpose. Based on the overall lack of comparative information, toxicity testing is essential in the future to measure the relative sensitivities of plants and animals to the same contaminants.

B. Importance and State of Phytoassessment

Phytoassessment is a diagnostic tool which can be applied to the aquatic and terrestrial environments. It is well known that vascular plants are essential components of these ecosystems and are important to their structural and functional integrity. This fact is more recognizable however, to scientists than to the public. Plants have many beneficial attributes but they also can be a detriment when excessive growth occurs (Table 1). For example, the combined negative effects of excessive growths of algae and vascular plants in Florida (757 mi^2 of estuaries) exceed the estuarine impact of chemical contaminants (245 mi^2) (17). Aquatic vascular plants influence many ecosystem processes and are important in oxygen and organic carbon production, soil and shore stabilization, nutrient recycling, and temperature moderation. Furthermore, plants provide food, shelter and substrates for many organisms including microalgae, invertebrates, fish, amphibians, birds and mammals. It should be noted that the ecological importance of submerged plants is considered to be greatest in small streams and rivers (18). The relationship of plant architecture and its role as a structural component in the survival of the biota has been evaluated for several species. Plants can also impact water and sediment quality, particularly the exchange of nutrients at the sediment-plant interphase (19-22) and their controlling influence on dissolved oxygen and pH concentrations are well-known. A good example of the importance of aquatic plants in stream ecology has been presented by Sand-Jensen et al. (23). It was found that the role of rooted macrophytes in regulating physical conditions and the biological structure in lowland streams was substantial and complex. Because of their ecological importance, it is essential to include plants in field biomonitoring programs as well as in any effort to provide toxicity profiles of potentially phytotoxic commercial chemicals, wastewaters and contaminated sediments (24).

Wastewaters, contaminated sediments and non-point source run-off contain phytostimulants and such phytotoxic chemicals as divalent metals and herbicides. Reviews of their phytotoxicities have been published (11, 25-27). Effects of acid precipitation, thermal effluents and energy development on plants have been reviewed (28-31). The phytotoxicities of herbicides, as would be expected, have been studied more than for any other group of chemicals. Herbicides are designed specifically to be phytotoxic. They are being used in increasing quantities on golf courses, lawns and roads and are an important part of modern agriculture for food and fiber production. Their use has caused widespread non-point source pollution in streams and rivers, especially during spring runoff events (32, 33).

Their effects on near-coastal areas are well documented and they may be responsible, in part, for the decline of submergent vegetation such as seagrasses in estuaries.

Table 1.   Beneficial and detrimental effects of vascular aquatic plants.

- Provide food for fish and waterfowl
- Shelter for biota
- Provide substrate for other biota
- Improve water quality
- Flood/erosion control
- Some commercially important
- Improve aesthetics
- Nuisance growths
- Impede navigation and recreational use
- Provide for vectors of disease
- Reduce fisheries

Since the early 1990s, phytoassessment studies have become more of an integral component of environmental monitoring and assessment. This is attributable to the concern for wetlands protection, use of vascular plants in the bioremediative process, management of excessive vegetative growths and the increase in environmental regulations. Excellent reviews are available describing the advantages and disadvantages of the use of algae and aquatic vascular plants to determine the environmental condition of various aquatic habitats and the toxicities of contaminants (12, 34-40). Furthermore, an increasing number of special symposia and test method standardization workshops have been conducted and several symposia proceedings have been published (41, 42). Phytoassessments are also part of two large-scale monitoring programs in the United States, the National Water Quality Assessment Program and the Biomonitoring of Environmental Status and Trends Program (43).

II. Toxicity Tests

Macrophytes, as stated earlier, are not commonly used in laboratory toxicity tests to estimate the toxicities of chemicals, wastewaters and contaminated sediments. Keddy et al. (44) evaluated the availability of toxicity test methods for 119 freshwater species and found only one reported methodology was available for a vascular plant.   Nevertheless, a few macrophytic species have been used in toxicity tests and reviews of their use are available (27, 34, 40, 45, 46). Additional sources of information on the subject have also been reported (3–5, 24, 41, 47-49).

A. Standardization

Phytotoxicity tests are required by several environmental regulations in the United States (Table 2). Most noteworthy of these is the U.S. Federal Insecticide, Fungicide, and Rodenticide Act (50) which requires that phytoassessment be used in the registration process of pesticides. Similar legislation is also present in Canada (51). The duckweeds, *Lemna minor* or *L. gibba* G3, are the more frequently used species for pesticide registrations (52, 53). Plants have also been required, on occasion, to assess the effects of another major source of environmental contaminants, wastewaters. Regulatory agencies in the United States recommend the use of freshwater and marine algae, invertebrates and fish, for the ecological hazard assessment of industrial and municipal effluents as mandated by the National Pollutant Discharge Elimination System (NPDES). Single species of algae are more commonly used for this purpose, although less frequently than animal species. Effluent toxicity tests can be conducted also with duckweed and with other macrophytic plants (4, 5, 48, 54, 55).

International regulatory agencies and standard writing organizations have promulgated or are developing phytoassessment test methods (24, 56-59). An OECD sponsored workgroup (Organization of Economic Cooperation and Development) has been attempting to harmonize published duckweed test methods so that they can be used by the member countries. The group consisted of representatives from Britain, Denmark, France, Italy, the Netherlands, Sweden, and the United States. The draft method will be evaluated by two rounds of inter-laboratory testing (calibration and validation) and is expected to be finalized in the near future.

Table 2. Phytotoxicity test methods required by legislation in United States.

Federal Insecticide, Fungicide, and Rodenticide Act (1982)	• Freshwater algal growth • Terrestrial seed germination, seedling emergence, vegetative vigor • Aquatic and terrestrial field testing • Rotational and irrigated crop residue uptake and metabolism (laboratory, greenhouse, field)
Food and Drug Act (1987)	• Freshwater algal growth • Terrestrial seed germination and root elongation
Toxic Substances Control Act (1985)	• Freshwater and marine algal growth • Duckweed growth • Terrestrial seed germination and root elongation • Terrestrial early seedling growth • Terrestrial uptake and translocation
Water Quality Act (1987) (Amended Clean Water Act)	• Freshwater algal growth Duckweed growth (optional)

Several of the phytotoxicity test methods using vascular plants in Table 2 have been standardized in the U.S. The available tests include those conducted with the duckweed, *Lemna gibba* G3, and several plant seedlings (7, 60, 61). Two joint task groups, each consisting of representatives from the American Public Health Association, American Water Works Association, and Water Environment Federation, have been developing and refining standardized test methods for the common duckweed (*Lemna minor*) and several emergent plants. These have been published as Parts 8211 and 8220, respectively in the Standard Methods for the Examination of Water and Wastewater (6).

B. Duckweed

The most commonly used aquatic vascular plant in toxicity tests in the U.S. are the duckweeds. An excellent review of their use has been presented (4), and additional information can be found in U.S. Environmental Protection Agency (59, 61), Cowgill and Milazzo (62) and Huebert and Shay (63). Toxicity tests with duckweeds are conducted in glass beakers or jars for 4 to 7 days. Plants with 2 to 4 frond colonies are used in each test. Twelve to 16 fronds are added to each test chamber which contains a nutrient-enriched media. The EC_{50} (concentration that causes 50% inhibitory effect on duckweed growth) and the NOEC (no-observed-effect concentration) values are calculated based on changes in biomass, root number, plant number, root length and chlorophyll content. Based on changes in these parameters, the duckweeds have been found sensitive to some chemicals and effluents but not to others. Therefore, additional information and experience is needed before they will become more widely used by the scientific and regulatory communities.

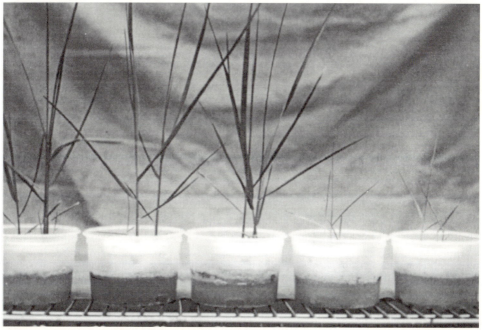

Figure 2. An example of a sediment phytotoxicity test and the response of *E. crusgalli* seedlings after exposure for two weeks to sediment samples collected below a wastewater discharge.

C. Rooted Species

Rooted vascular plants have been used less frequently than the duckweeds in toxicity tests due to their large size and the lack of standard test methodologies and culture techniques. Nevertheless, they are important species which are needed particularly to determine the phytotoxicity of sediments. Toxicity tests have been conducted with a variety of species which have included, among others, *Elodea canadensis, Myriophyllum spicatum, Vallisneria americana,* and *Potomogeton perfoliatus.* The test conditions have also varied considerably in these types of tests (Table 3). A photograph of a typical sediment phytotoxicity tests appears in Fig. 2. Most rooted test species used in laboratory testing are field collected but some have reported some culture methods. Test compounds in the laboratory studies have included herbicides and heavy metals (9, 64-72). In addition to single compounds, the effects of more complex contaminant matrices can also be evaluated using vascular plants. The following examples describe their use to monitor the quality of wastewater discharges, contaminated sediments and flood waters.

Early-seedling growth of several macrophytic species was used, in the first example, to determine the phytotoxicities of 10 wastewaters and the sediments in the corresponding receiving waters (73, 74). Effects were detectable using the seedlings for some undiluted effluents, and there were differences in the plant tissues affected (Table 4). It is important to note that the results differed from those for algae exposed to the same effluents. For example, the effluents in many cases were extremely phytostimulatory to single algal species after 96 hr exposure, but effects on seedlings of three species of vascular plants were less pronounced after 3 to 4 weeks exposure (Fig. 3). Some sediments collected below the outfalls were phytotoxic as shown for *Echinochloa crusgalli* after exposure for 2 weeks to the sediments (Fig. 2).

The sensitivity of seedlings exposed to estuarine sediments collected from an urbanized bayou near Pensacola, Florida, differed from that for a benthic amphipod exposed to the same sediments (Table 5). It can be seen that the effect of the sediment from Station 3 on biomass of *Spartina alterniflora* was inhibitory, it reduced whole plant biomass 56%. In contrast, the effect on animal survival was slight (3% mortality). Conversely, sediments from Station 1 were more toxic to the invertebrate than to the plant. This amphipod, *Ampelisca abdita*, is a recommended test species for sediment toxicity tests in the U.S. This study, in addition to showing species-specific differences in sensitivity, also found that the effects on shoots and roots also differed. Overall, these results reinforce the need to use both plants and animals in the hazard assessment process for sediments.

Another example of the use of rooted plants to monitor environmental quality is presented for floodwaters. Root growth of Japanese millet and rice were used as the indicators of effect following a standardized protocol (6). Episodic precipitation occurred in the Midcontinental United States during the spring and summer of 1993, when a hundred-year flood occurred in many parts of Iowa and neighboring states. During this period of record rainfall, stormwater samples were collected and analyzed for phytotoxicity. The results indicated that 12 of the 86 samples were phytotoxic to Japanese millet and that 5 of the 86 samples were phytotoxic to rice. The results also indicated that seasonal differences in phytotoxicity occurred. During the pre-planting season (in late March), 0 to 6% of the water samples were phytotoxic relative to 9 to 19% during the post-planting season. In addition, 19% of the samples collected during the post-planting season were toxic to duckweed.

D. Ecological Relevance

It was shown in the previous examples that aquatic vascular plants can be effective indicators of phytotoxicants in wastewater, contaminated sediments and flood-stage runoff. Unlike these "real-world" assessments, the ecological significance of laboratory-derived phytotoxicity data for single chemicals is not well understood. Single-species toxicity tests have been more frequently conducted

Table 3. Experimental conditions in several laboratory phytotoxicity tests conducted with primarily rooted vascular plants. Table adapted from Lewis [46].

Chemical	Plant Species	Effect Parameters	Test Duration	Test Chambers	References
●Atrazine Metribuzin Glyphosate	*Elodea canadensis* *Vallisneria americana* *Potomogeton perfoliatus* *Myriophyllum* sp.	Growth	3-6 weeks	Glass jars Plastic buckets	(65)
●Lead Cadmium Copper	*Eichhornia crassipes*	Growth Necrosis Chlorisis	6 weeks	Containers lined with plastic bags	(67)
●Chlorine	*Myriophyllum spicatum*	Biomass Chlorophyll *a*	4 days	Glass test chambers	(152)
●Mercury	*Hydrilla verticillata* *Pistia stratiotes* *Salvinia molesta*	Foliar injury Chlorophyll *a* Biomass	1-5 h	Aquaria	(153)
●Atrazine Linear alkylbenzene sulphonate 4-Nitrophenol Pentachlorophenol	*Elodea canadensis* *Chara hispida* *Myriophyllum spicatum*	Oxygen Production Membrane destruction	7-14 days	NS*	(154)
●Chromium	*Ceratophyllum demersum*	Chlorophyll *a*	1 week	NS*	(70)
●Copper Atrazine Lindane Chlordane	*Hydrilla verticillata*	Shoot length Chlorophyll *a* Dehydrogenase activity	4-14 days	Glass jars	(112)

* NS = not stated

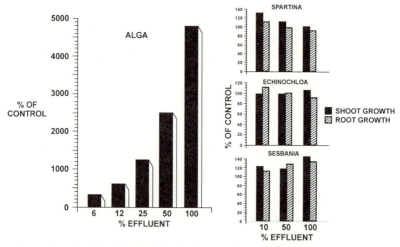

Figure 3. Comparative response of a cultured marine alga, *Dunaliella tertiolecta*, and three vascular plant seedlings after exposure to a treated effluent collected from an acrylic fiber manufacturing facility. The macrophyte test species were *Spartina alterniflora*, *Echinochloa crusgalli* and *Sesbania macrocarpa* .

Table 4. Effects of two treated wastewaters on early seedling growth of *Echinochloa crusgalli* exposed in an artificial sediment. Values represent the average ash free dry weight (mg) of six plants after exposure to 50% and 100% wastewater.

| | | WASTEWATER TYPE | |
EFFECT PARAMETER	WASTEWATER CONCENTRATION	PULP MILL	FIBER MANUFACTURER
Shoot Biomass	Control	185.1	185.1
	50	172.0	197.4
	100	124.0*	158.7*
Root Biomass	Control	72.4	72.4
	50	68.1	84.8
	100	38.4*	67.0
Whole Plant Biomass	Control	257.5	257.5
	50	240.1	282.2
	100	162.4*	225.7*

* Significant effect (P > 0.05)

than multispecies tests, mesocosms and other more realistic test designs. Reported attempts to field validate the available laboratory-derived data reported in the scientific literature are almost non-existent. Therefore, in addition to the further development of laboratory toxicity tests, the validation of the results is an important research need. Validated laboratory toxicity tests, are needed to derive water quality criteria for wetlands, confirm the value of vascular plants as sensitive indicator species in field biomonitoring programs and to derive sediment quality criteria for contaminated sediments.

Table 5. Comparative response of a benthic amphipod, *Ampelisca abdita*, and a rooted macrophyte, *Spartina alterniflora*, exposed to the same sediments collected from four locations in an urbanized Florida estuary. Test durations were 10 days (animals) and 3 weeks (*S. alterniflora*). Values for animal species represent mean percent survival and for plants, percent relative to control based on biomass.

SAMPLING STATION	TEST SPECIES			
	Ampelisca abdita	*Spartina alterniflora* Shoot	Root	Total
1	11	71	128	90
2	95	114	217	147
3	97	32	70	44
4	93	78	178	110

III. Biomonitoring Techniques

A. Population Surveys

Macrophyte population surveys are conducted to determine the location, frequency, cover and biomass of plant species in areas affected by different types of environmental contaminants. Additional studies usually follow in which functional aspects of the plants are investigated. Excellent reviews, methodology descriptions and site-specific examples of the use of the assessment methodologies are available (6, 75-81). In addition, some information on the relative effects of several types of environmental disturbances on submersed aquatic vegetation has been reported (Table 6). It is important to note that it is often difficult to determine or differentiate the causal factors of environmental impacts on plants in aquatic environments (18).

Macrophytic species are considered by some to be more useful as indicators of ecological condition than algae and periphyton since they can be sampled easily due to their large size and usual littoral habitats. Furthermore, they can be sampled manually or by using devices such as rakes and grab samplers and once collected, they can be identified to species in the field. Quantitative sampling is used to determine the extent and rate of growth per unit area and it can be accomplished using line intercepts, quadrants and belt transects. The choice of method depends upon sediment type, water depth, current, vegetation type and density. Remote sensing can also be used. These techniques include analog aircraft and satellite aerial photography, digital aircraft and satellite multi-species scanning in the visible, infrared and thermal bands; microwave techniques; side-looking airborne radar and shuttle imaging radar.

The most frequently used quantitative method for sampling vascular plants is the quadrant technique. The time-intercept method is generally quicker and more appropriate in some situations but it is less frequently used. Quadrant size is important since it must be sufficiently large to allow for measurement of all the desired parameters. The size of the plants influences the choice of the quadrant size. For example, square quadrants, 1.0 m^2 to 0.1 m^2, have been used for rooted vascular plants. More specifically, *Typha* may be best sampled with a 0.25 or 1.0 m^2 quadrant, whereas *Zostera marina* may be better sampled using a 0.1 m^2 quadrant. High replication using smaller quadrants results in lower variance than using fewer larger quadrants.

Table 6. Effects of environmental disturbances on submersed aquatic plants as reported by
Davis and Brinson (97).

	IMPACTS			
PERTURBATION	Suspended Sediments	Eutrophication	Toxicity	Sedimentation
Silviculture	Low	Low	Low	Low
Mining	Variable	Variable	Low	Variable
Clear Cutting	High	Low	Low	Variable
Logging Roads	High	Low	Low	High
Urban Construction	High	High	Low	High
Wastewaters	Low	High	Low/High	Low
Stormwaters	High	High	High	Medium
Agriculture	High	High	High	High
Stream Channelization	High	High	Variable	High
Dredging	High	High	Variable	Variable
Dams	High	Variable	Low	High

B. Biometrics

(i) *Percent cover, shoots, roots* - Percent cover is the amount of surface area covered by a taxa or
group of taxa, and its analysis provides a measurement of community structure. Percent cover can be
determined using photographic techniques or visually estimating the amount of area covered by each
taxon in the quadrant. Median percent cover is usually used when calculating the percent cover for a
taxon. In addition to percent cover, shoot density and length can be determined for each taxa and used
as an indicator of environmental condition. Shoot density is the number of shoots per unit area. Live
shoots are counted separately from dead shoots. Shoot length is determined by measuring the length
of all or a random sample of shoots within the quadrant using a tape measure or similar device. The
total number of flowers per shoot, per plant or per quadrant can also be determined. Biernacki et al.
(71) evaluated the leaf to root surface area of *Vallisneria americana* collected from areas impacted by
organochlorine contamination. It was concluded that this relatively simple ratio provided a rapid and
cost-effective metric of site quality.

(ii) *Biomass* - Biomass sampling in the field is labor intensive and involves destructive sampling.
However, it is the most commonly determined effect parameter for aquatic vascular plants. This
measurement of growth and productivity can be determined for the whole plant or for specific tissues,
i.e., roots, stems, leaves, shoots and nodes. If tissues are analyzed, some idea on the allocation of
resources can be determined. Differences in rhizomes, tubers and seeds have been used as indicators
of effects (82).

Biomass can be based on wet weight, dry weight or ash-free dry weight and can also be estimated
from chlorophyll and carbon content. Dry weight and ash-free dry weight are more frequently used,
and methods for their analysis are standardized (6). Species standing stock is a measure of the weight
of plants occupying a unit area of substrate. Standing stock is sampled by harvesting all plant material
either above and below the substrate or both. Above-ground species standing stock represents
biomass of plant material above the substrate; that below, including roots and rhizomes, is referred to
as the below-ground standing stock. The plants are usually separated by species before determination
of dry weight. For above-ground measurements, only the shoot or stem portion should be removed.
For below-ground biomass determinations, plant roots and rhizomes are removed from a core taken
from the study area.

(iii) *Productivity* - Productivity can be determined based on biomass and metabolic activity (Table 7). Methods to measure productivity based on biomass are standardized and described with other methods elsewhere (6). Photosynthetic activity of vascular plants can be determined based on oxygen evolution and consumption using an oxygen meter or a colorimetric technique. Measurements can be performed on plant segments in bottles (83) or on whole plants in larger test chambers. The use of radioisotopes such as ^{14}C is also common to determine primary productivity.

Table 7. Several methods used to determine production of different types of aquatic plants. A = plants with above-ground biomass year-round, B = plants without year-round above-ground biomass, ● = method used, ○ = method not used. Adapted from APHA et al. (6).

METHODS	PLANT TYPES					
	Emergent		Floating		Submerged	
	A	B	A	B	A	B
Biomass:						
Above-ground biomass	●	○	●	○	●	○
Below-ground biomass	●	●	●	●	●	●
Biomass tagging:						
Turnover, growth increment, and summed shoot maximum	●	●	●	●	●	●
Cohort	●	○	●	○	●	○
Oxygen measurement:						
Light and dark bottle	○	○	○	○	●	●
Radioisotopes	○	○	○	○	●	●
Inorganic carbon exchange:						
Continuous CO_2 exchange	●	●	●	●	●	●
Discrete inorganic C measurement	○	○	○	○	●	●

(iv) *Physiological Effects* - Toxic effects on plants usually occur first at the suborganismal level before causing whole-organismal effects. These sublethal effects can be biochemical and genetic and include changes in carbohydrate content, cytochrome f, ethylene/ethane, oxidative enzyme activity and protein concentration (84, 85). In addition to these, concentrations of phytochelatins can indicate plant stress, as well as measurements of cellular P and N and enzymes such as phosphatase (86-88). Although several physiological parameters of plants have been used as indicators of chronic contaminant exposure, few have been related to the growth and survival of the whole plant or to their effect on the composition of the indigenous plant community. Therefore, the use of these sublethal effect parameters and their current value in the environmental hazard assessment process has been limited. In particular, their ability to provide a consistent and a replicable indication of environmental stress is unproven and is in need of validation.

(v) *Pigment Content* - Pigment content is often measured to determine the physiological status of aquatic plants. One advantage in the use of pigment content as an indicator of stress is that the samples can be frozen and held for months before solvent extraction and analysis. There are several types of pigments that can be analyzed using HPLC and spectrophotometric methods. These include the anthocyanins, carotenoids and phytoene. The more commonly measured pigments, however, are the chlorophylls which have been determined for aquatic plants on numerous occasions in response to a variety of environmental perturbations (83, 89). Chlorophyll *a* content, in particular, is often used as a measure of plant biomass and can be used as an indicator of water quality as part of the autotrophic index (90). Reduction in chlorophyll can result in the inability to produce carbohydrates. Pheophytin is a degradation product of chlorophyll, and its proportion to chlorophyll can be used as an environmental indicator. Pheophytin is usually found in low concentrations in plant tissue, and the presence of large amounts indicates possible water quality degradation.

C. Indicator Species

Indicator species have been used in the past to provide information on water and sediment quality and the advantages and disadvantages of the concept have been discussed in detail (91, 92). The approach has been used more commonly for algae and aquatic invertebrates than for vascular plants due to a general lack of published information describing the sensitivities of vascular plants to most contaminants (93-95). However, some plants have been used for this purpose. For example, mosses have been used as indicators of changes in pH and water level (96) and *Phragmites* spp. are considered to be an indicator of disturbed wetlands. The relative sensitivities of several vascular species has been reported by Davis and Brinson (97) and appear in Table 8.

Table 8. Submersed macrophytes listed in order of their decreasing tolerance to environmental change. Adapted from Davis and Brinson (97).

Advective Species	•	*Myriophyllum spicatum*
	•	*Potamogeton crispus*
Tolerant Species (high biomass in disturbed systems)	•	*Ceratophyllum demersum*
	•	*Vallisneria americana*
	•	*Najas guadalupensis*
	•	*Potamogeton perfoliatus*
Tolerant Species (low biomass in disturbed systems)	•	*Elodea canadensis*
	•	*Najas flexilis*
	•	*Potamogeton pusillus*
Rosulate Species	•	*Eleocharis acicularis*
	•	*Isoetes macrocarpa*
	•	*Lobelia dortmanna*
	•	*Eriocaulon septangulare*

The basis for the indicator species concept is that plants and animals cannot survive indefinitely in areas not suitable for their physical, chemical and nutritional requirements. Therefore, the presence of a species indicates the suitability of the environment. There are problems with this approach. The absence of a species does not necessarily indicate "pollution" but could reflect other factors such as competitive exclusion. Some species may have wide tolerances for environmental conditions and exist under a variety of environmental conditions. Also, the sensitivity of a species may be influenced by other environmental factors and for plants, these are often poorly understood.

D. Numerical Indices

Researchers have correlated water and sediment quality to the diversity and density of the resident aquatic biota using numerical indices. These indices are based on considerations of the number of species and the distribution of the individuals representing the species. They have been used more commonly for invertebrates particularly those associated with the benthos. The basis for their calculation is the assumption that "healthy" communities support a diverse number of species, whereas contaminated areas are inhabited by only a few. This is due to the differences in sensitivities of the various species to contaminants that inhabit the area of interest. There are approximately 18 diversity indices that have been used by the scientific community, and reviews on their value are available (98, 99). The differences in these indices are based on the degree of "weighing" given to the number of species and their distribution.

Other indices are available that determine the similarity between two biotic samples (100). These coefficients of similarity do not indicate the degree of contaminant stress but are valuable for detecting differences in community composition due to spatial and temporal considerations. Any unusual differences in composition can be an indication that further research is needed to determine the cause(s). Related to these indices are statistical techniques for cluster analysis (101) and determining the probability of similarity (102).

The use of numerical indices alone as indicators of environmental quality is not encouraged. To some, they are considered simplistic (103). The types of species and their sensitivities to contaminants and functional attributes are rarely accounted for in the numerical index approach. Taxonomic error and difficulty in comparing values among different studies further complicate their use and value as a diagnostic tool.

IV. Bioaccumulation

Plants have the ability to adsorb contaminants and retain them in their tissues for long periods of time. These contaminant concentrations (bioresidues) are often greater than those in the surrounding water and sediments. They reflect the effect of all sources of contaminants present in the study area, including those from intermittent contaminant sources and those present but below the detection limit as measured by chemical methods. This ability of vascular plants to bioaccumulate contaminants has resulted in their widespread use as indicators of water and sediment quality and as bioremediative agents in constructed wetlands designed to improve wastewater quality.

Vascular plants are useful biomonitors (bioaccumulators) due to the fact that they are common and, unlike most other biota, are stationary. Many species have extensive root systems which facilitates in the absorption, accumulation, and sequestering of sediment contaminants in their tissues. In addition, plants are usually the predominant species in many ecosystems in terms of biomass and hold the dominant position in wetlands. Aquatic and wetland species have leaves and epiphytes which provide plants an expanded area to trap particulate matter, sorb, translocate, and sequester metal and organic substances (49).

Plant tissues have shown a differential ability to concentrate contaminants. Leaves and stems have been found to concentrate contaminants less than roots and shoots (104) but leaves and shoots are readily available to grazers and should be included in any study. It is important to note that, before analysis, epiphytes should be removed from field-collected submersed species.

The potential of vascular plants to bioaccumulate chemicals has been determined in the laboratory. Test methods, test durations and uptake rates reported for various species have been presented (25, 45, 105). The objectives of many studies have been to determine their bioremediation potential (constructed wetlands), identify useful bioindicator species and to determine their role in nutrient cycling. It has been found that the process of bioaccumulation is dependent upon pH, temperature, tissue type, tissue age, growth rate and species. Furthermore, the contaminant uptake, metabolism, and depuration affects the tissue quality.

Bioaccumulation of nutrients, specific metals and organic substances by plants has been reported in numerous reports and widely reviewed (82, 105-115). The low lipid content in vascular plants is thought to restrict the degree of uptake of organic contaminants (116); however, some have shown that plants have the ability to accumulate and biodegrade organic contaminants such as herbicides, polychlorinated biphenyls and polynuclear aromatic hydrocarbons (115, 117, 118). Levels of nitrogen and phosphorous in plant tissues provide estimations of growth (82, 119).

Significant species-specific differences in contaminant uptake have been reported. To some researchers, the species more useful as biomonitors are the submersed types (*Ceratophyllum, Elodea*) (39). The value of emergent species is thought to be reduced due to uptake from atmospheric sources; however, due to translocation of contaminants, others consider them to be as useful as submersed species (120). Rooted emergent species tend to accumulate more metals than mobile, floating species (114) like the duckweeds. Other species used in monitoring have included emergents such as *Pontederia cordata* (120), *Typha latifolia* and *Eichhornia crassipes*. Whitton et al. (88) recommended 10 species for monitoring metals in rivers, including *Elodea canadensis* and *Potamogeton pectinatus*.

Of particular note are the aquatic bryophytes which are good bioaccumulators. Aquatic mosses have several qualities which make them good biomonitors. These include their tolerance to high levels of contamination and long-term storage stability. Furthermore, their use is considered rapid, efficient and cost-effective. They have been commonly used as environmental indicators since at least 1969 in Belgium, Germany, France, Portugal, Sweden and Canada (121-123) and, to a lesser extent, in the United States (43, 106).

Vascular plants, indigenous and transplanted (124), have been used in the field to detect metal contamination resulting from a variety of sources. Two examples are provided that illustrate their ability to detect site-specific differences in environmental quality based on the bioaccumulative process at the time of the study. The first example is for a base-metal processing plant in Pori, Finland, which discharged, at the time of the study, an effluent containing copper into a river dominated by *Nuphar lutea* (125). Nuphar was collected at intervals of 300 m upstream and downstream from the discharge and analyzed for copper. The copper concentration in giant petioles was largest near the outfall, 110 mg/Kg. Leaves and rhizomes also showed peaks, although with smaller concentrations, 90 and 50 mg/Kg, respectively. Only plants collected 600 m upstream and 1,200 m downstream of the outfall had tissue bioresidues similar to the background level. The exception were flowers which showed a relatively small peak copper concentration at 300 m downstream (Fig. 4).

A second example is also provided to demonstrate the practical use of the bioaccumulative nature of vascular plants. The Natchaug River is located in Connecticut (United States) and flows through a predominantly forested watershed characterized by minimal urbanization (120). The Willimantic River also originates in a pristine area, but most of its watershed is more developed and the river receives an industrial effluent. The Shetucket River is formed at the confluence of the these two rivers. *Pontederia cordata* and *Potamogeton epihydrus* are abundant and were collected at regular intervals along the three rivers. Copper concentrations in shoots and roots of *Pontederia* spp. appear in Figure 5. Roots and rhizomes of *Pontederia* spp. had greater copper concentrations than the shoots. In the pristine Natchaug River, the difference was small, whereas in the Willimantic River, copper concentrations in the below-ground organs were two to four times greater than those in the shoots. Copper concentrations in rhizomes and roots were, on average, four and six times greater in the Shetucket and Willimantic Rivers, respectively, than those from plants collected from the Natchaug River.

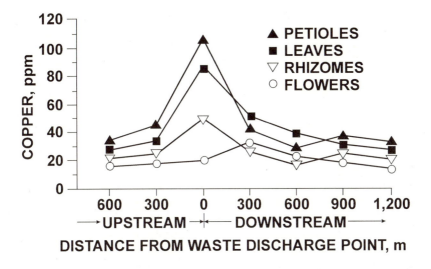

Figure 4. Copper accumulation in the yellow water lily (*Nuphar lutea*) collected at different distances from the wastewater outfall of a metal-processing plant in Finland. Adapted with permission from Aulio (125).

V. Eutrophication and Phytoassessment

Eutrophication is the result of excessive nutrient enrichment of a waterbody. It is a common condition in many freshwater and salt water habitats affected by anthropogenic activities and has been studied for many years. The rate of eutrophication is greatly accelerated by discharges of municipal sewage and non-point source runoff from agricultural and urban areas. As result of extreme eutrophication, excessive growth of algae and/or extensive mats of aquatic vascular plants can occur.
These excessive growths have resulted in fish kills, taste and odor problems, clogging of water filtration systems, impedance of boating activities, and generally decreasing aesthetic and recreation values of the waterbody (126). In some countries, they also can serve as vectors for disease. Excessive nutrient runoff from animal rearing facilities caused excessive growth of *Pfiesteria piscicida* along portions of the East Coast of the United States in 1997. This dinoflagellate was suspected of causing fish lesions, fish kills, memory loss and other ailments of workers in the field.

Eutrophication is considered the most important water-quality issue in lakes, reservoirs, ponds, and estuaries in the southeastern United States. As mentioned earlier, Florida estuaries are more impacted by excessive plant growth, which is a result in some cases of eutrophication, than toxic chemicals. In contrast to the usual proliferation in plant growth, some plant species, however, are sensitive to eutrophic conditions and may be absent. For example, there is considerable evidence that many broad-leaved Potamogetons do not tolerate eutrophicated conditions such as *P. foliosus, P. natans, and P. nodosus* (97). However, species with small leaves are often the only plants in eutrophicated waters. An excellent description of the relationship of eutrophication and freshwater vegetation can be found in Kohler and Labus (127).

Considerable economic resources have been used to eradicate and control excessive vascular plant growths world wide, such as for hydrilla, *Salvinia molesta* and *Eichhornia crassipes* in the United States. Eradication procedures have included mechanical harvesting, alteration of water levels, use of herbicides and plant pathogens and biological control (aquatic insects and fish). One scientific benefit

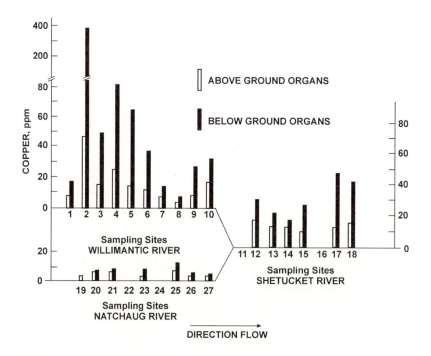

Figure 5. Copper accumulation in pickerel weed (*Pontederia cordata*) collected at
stations (indicated by number) in streams in Connecticut. Adapted with permission
from Heisey and Damman (120)

of the development and use of these methods has been a better understanding of the ecology and
physiology of vascular plants (128-131). For more detailed information on this topic and its
international scope, see the special edition of the Journal of Plant Management (132).

The algal growth potential test (AGPT) was developed in the early 1970s to determine the phytostimulatory properties of a waterbody. Field collected water samples are augmented with nitrogen or
phosphorus and then inoculated with the green alga, *S. capricornutum*. Growth of the alga is then
determined for a prescribed time period to determine if the water samples are either nitrogen and/or
phosphorus limited. The AGPT has been used by many regulatory agencies. For example, it is part of
the Clean Lakes Program for monitoring lake water quality by the U.S. Environmental Protection
Agency (6).

The duckweeds can also be used to estimate the eutrophication potential of natural waters (4, 55).
Water samples in the duckweed growth potential test (DGT) are spiked with nitrogen or phosphorus,
similar to the AGPT (6). Recently, a comparison of the AGPT and DGT was made using lake water
samples collected from the southeastern United States (Table 9). The results showed that duckweed
growth was stimulated in six samples spiked with nitrogen and in one sample spiked with phosphorus.
Algal growth in the AGPT was limited in four and five samples based on nitrogen and phosphorus
content, respectively. Since this comparative testing is the first reported in the literature, the results
should not be overly interpreted. However, the results do show that this vascular plant shows promise
as an indicator of nutrient enrichment.

Table 9. Comparison of the results of the duckweed growth potential test (DGPT)(mean ± standard deviation of net frond count increase) and algal growth potential test (AGPT) using lake water. N and P treatments were equivalent to the recommended levels by the American Public Health Association et al. (6). Duckweed growth solution is equivalent to 10-fold concentration of algal growth solution.

SAMPLE	DGPT (frond increase%)			AGPT	
	Control	N treated	P treated	Dry Weight (mg/sample)	Limiting Nutrient
1	17 ± 2.5	$31* \pm 1.2$	$26* \pm 2.1$	16.2	N
2	25 2.5	32* 1.1	29 3.2	34.2	N
3	27 0.6	31 1.7	27 1.2	25.7	N
4	20 3.2	30* 2.9	23 2.1	4.9	N
5	20 1.5	35* 5.0	22 1.5	2.4	P
6	21 0.6	31* 1.0	24 1.2	1.5	P
7	20 3.0	30 4.9	23 1.0	1.3	P
8	24 4.2	34 13.0	24 2.9	3.0	P
9	21 1.2	31* 1.5	24 1.5	1.7	P

* = significant difference, $P = < 0.05$.

VI. Wetlands and Phytoassessment

Scientific knowledge concerning the value of aquatic plants in environmental toxicology and the usefulness of the various phytoassessment methods has increased in recent years. This is attributable, in part, to increased regulatory activity and scientific interest related to preserving natural wetlands and the use of constructed wetlands as bioremediative agents. Wetland acreage is decreasing rapidly and, as a consequence, these plant-dominated habitats have undergone scientific scrutiny in an effort to define the chemical and biological requirements necessary for their sustainability. Water quality standards for wetlands have been discussed (133) and specific biological criteria are available (134) which attempt to maintain the baseline vegetation species. These criteria are based on an understanding of the sensitivities of plants to contaminants and nutrients such as those reported by Tripathi and Shukla (135). Other indicators of wetland condition have been proposed which include considerations of diversity, abundance and biomass (136). Leaf area, percent light transmittance and green light reflectance were also identified as "advanced developmental indicators."

Constructed wetlands are becoming more important in controlling water quality (bioremediation). They are being constructed to control flooding, restore coastal areas, enhance wildlife and fisheries and to treat wastewaters originating from diverse sources. There were approximately 140 wetlands constructed to treat mine-drainage in 1989 (137). There are over 300 constructed wetlands in North America and 500 in Great Britain treating storm water and municipal effluents (138). A variety of submersed and emergent vascular species have been used for this purpose (Table 10). It has been estimated that only approximately 1% of the available taxa have been evaluated (139). Constructed wetlands can effectively remove organic matter, heavy metals, toxic organics, BOD, suspended solids and nutrients. Some organic compounds can be biodegraded in the process. This is attributed, in part, to the plant's root system and associated microorganisms. Aquatic macrophyte based treatment systems use the ability of plants to translocate O_2 from leaves into roots producing an aerobic zone which is desirable to treat numerous wastes. This topic is only briefly introduced here and an extensive literature is available. See Hammer (139) for a good introduction.

Table 10 . Vascular plants used in constructed wetlands. From Hammer (139).

EMERGENT	SUBMERGED	FLOATING
•Scripus robustus	•Egeria densa	•Lagorosiphon major
• Scirpus lacustris	•Ceratophyllum demersum	•Salvinia rotundifolia
•Schoenoplectus lacustris	•Elodea nuttallii	•Spirodela polyrhiza
•Phragmites australis	•Myriophyllum aquaticum	•Pistia stratiotes
•Phalaris arundinacea		•Lemna minor
•Typha domingensis		•Eichhornia crassipes
•Typha latifolia		•Wolffia arrhiza
•Canna flaccida		•Azolla caroliniana
•Iris pseudacorus		•Hydrocotyleumbellata
•Scirpus validus		•Lemna gibba
•Scirpus pungens		•Lemna spp.
•Glyceria maxima		
•Eleocharis dulcis		
•Eleocharis sphacelata		
•Typha orientalis		
•Zantedeschia aethiopica		
•Colocasia esculenta		

VII. The Future/Recommendations

The use of vascular plants is becoming more important in the regulatory hazard assessment process and in environmental monitoring programs. This is particularly true when related to plant-dominated habitats such as wetlands and small streams and for species at risk such as sea grasses. It is obvious, however, that scientific opinion concerning their value is "mixed" which is due to a lack of scientific information. There are many research needs before the use of vascular plants will become as frequent as that of other flora such as algae and periphyton. For example, the interactive effects of physical and chemical modifying factors with the sometimes simultaneous presence of nutrients and toxicants needs to be determined. More research is needed to identify sensitive plant species and effect parameters and to standardize and validate laboratory and field assessment protocols. The sensitivities of different vascular plant species to contaminants need to be compared with those of animal species and other types of aquatic plants, particularly algae for which a relatively large historical data base exists. Furthermore, the relationship of bioresidues and physiological effects to whole plant performance and community effects needs investigation and related to the value of vascular plants as *in-situ* bioindicators of environmental impact.

One research area where the use of vascular plants is more needed than others is for the environmental assessment of contaminated sediments. Historically, plants have not been as commonly used as benthic animal species (140). Furthermore, when the phytotoxic effects of contaminated sediments have been determined, it is usually for elutriates and various algal species (141-144). Germany currently is using algae and higher-order plants for sediment evaluations (145) and although the invertebrate component of the benthos (7, 146) has been the focus in the U.S., in some cases rooted plants have been used also(144). Rooted plants, for reasons discussed in this Chapter, need to be included in the battery of toxicity tests used to determine the extent, sources, and causes of sediment contamination. In addition, these data are needed for the development of numerical sediment quality assessment guidelines and sediment quality criteria designed to protect the benthos. Some reported studies using vascular species have shown promise for this purpose (9, 147-151).

In closing, this Chapter has briefly summarized the availability and use of several methodologies used by the scientific community to assess the effects of nutrients and toxicants on aquatic vascular plants. The summary provides only an overview of a multifaceted subject which includes field and laboratory studies as well as the related topics of eutrophication and wetlands. Therefore, it is recommended that the readers further explore the references for more detailed information.

VIII. References

1. S. McComas: Integrating aquatic pesticides with other strategies: building on the terrestrial experience. Lakesline (17)15 (1997).

2. L.A. Kapustka: Selection of phytotoxicity tests for use in ecological risk assessment. IN: Plants for Environmental Studies (eds.), W. Wang, J.W. Gorsuch, J.S. Hughes. CRC Press/Lewis Publishers, Boca Raton, FL. (1997) 563 p.

3. B. Landolt: Biosystematic investigations in the family of duckweed (*Lemnaceae*). Vol. 2. Geobotanischen Inst. ETH, Stiftung Rrbel, Zurich, Switzerland (1986).

4. W. Wang: Literature review on duckweed toxicity testing. Environmental Research. (51)7-22 (1990).

5. W. Wang: Literature review on higher plants for toxicity testing. Water Air and Soil Pollution (59)381-400 (1991a).

6. American Public Health Association, American Water Works Association, and Water Environment Association: Standard Methods for the Examination of Water and Wastewater, 19th edition. Washington D.C. (1995).

7. American Society for Testing and Materials: Annual Book of ASTM Standards. Vol. 110,5. Biological Effect and Environmental Fate: Biotechnology; Pesticides. ASTM, Philadelphia, PA. (1997).

8. W. Wang: Comparative rice seed toxicity tests using filter paper, Growth Pouch-TM, and seed tray methods. Environmental Monitoring Assessment (24)257-266 (1993).

9. R.L. Powell, Kimerle, R.A. and Moses, E.M.: Development of a plant bioassay to assess toxicity of chemical stressors to emergent macrophytes. Environmental Toxicology and Chemistry (15)1570-1576 (1996).

10. Organization for Economic Cooperation and Development (OECD): Terrestrial plants: Growth test, Test Guideline No. 208. OECD Guidelines for Testing of Chemicals, Paris (1984).

11. M.A. Lewis: Are algal toxicity data worth the effort? Journal of Environmental Toxicology and Chemistry (9)1279-1284 (1990a).

12. J.S. Hughes: The use of aquatic plant toxicity tests in biomonitoring programs. Canadian Technical Report of Fisheries and Aquatic Sciences. No. 1863, 169-174 (1992).

13. W. Wang and Keturi, P.H.: Comparative seed germination tests using ten plant species for toxicity assessment of a metal engraving effluent sample. Water Air and Soil Pollution (52)369-376 (1990).

14. E. Kenaga and Moolenaar, R. J.: Fish and Daphnia toxicity as surrogates for aquatic vascular and algae. Environmental Science and Technology (3)1479-1480 (1979).

15. P.M. Gersich and Mayes, M.A.: Acute toxicity test with *Daphnia magna* Straus and *Pimephales promelas* Rafinesque in support of National Pollutant Discharge Elimination permit requirements. Water Research (20)939-941 (1986).

16. M.A. Mayes, Hopkins, D.L., and Dill, D.C.: Toxicity of picloram (4-amino-3,.5,6-trichloropicolinic acid to life stages of the rainbow trout. Bulletin of Environmental Contamination and Toxicology (38)653-660 (1987).

17. Florida Dept. of Environmental Protection: Water-Quality Assessment for the State of Florida Section 305(B) Main Report. Bureau of Surface Water Management, Tallahassee (1996).

18. D.F. Westlake: Aquatic macrophytes in rivers. A review. *Pol. Hydrobiol.* 20(1)31-40 (1973).

19. J.C. O'Kelley and Deason, T.R.: Degradation of pesticides by algae. Office of Research and Development, EPA-600/13-76-022, Athens, GA (1976).

20. P.A. Chambers and Prepas, E.E.: Nutrient dynamics in riverbeds: the impact of sewage effluent and aquatic macrophytes. Water Research (28)453-464 (1994).

21. M.A. Lewis and Wang, W.: Water quality and aquatic plants. In: Plants for Environmental Studies, (ed.) W. Wang, J.W. Gorsuch, J.S. Hughes. Lewis Publishers, Boca Raton, FL (1997).

22. C. Wigand, Stevenson, J., and Cornwell, J.: Effects of different submersed macrophytes on sediment biogeochemistry. Aquatic Botany (56)233-244 (1997).

23. K. Sand-Jensen, Jeppesen, E., Nelsen, K., Vander Bijl, L., Hjermind, L., Nielsen, L.W. and Iversen, T.M.: Growth of macrophytes and ecosystem consequences in a lowland Danish Stream. Freshwater Biology (22)15-32 (1989).

24. W. Wang and Freemark, K.: The use of plants for environmental monitoring and assessment. *Ecotox. Environ. Saf.* (30)289-301 (1995).

25. B.A. Whitton: Algae as monitors of heavy metals in freshwater. IN: Algae as Ecological Indicators, ed. L.E. Shubert, Academic Press, London, pp. 257-280 (1984).

26. G.W. Stratton: The effects of pesticides and heavy metals towards phototrophic microorganisms. IN: Reviews in Environmental Toxicology 3, ed., E. Hodgson, Elsevier, New York, p. 71-147 (1987).

27. S.M Swanson, Rickard, C.P., Freemark, K.E., and MacQuarrie, P.: Testing for pesticide toxicity to aquatic plants: recommendation for test species. IN: Plants for Toxicity Assessment: Second Volume, eds., J.W. Gorsuch, W.R. Lower, W. Wang, M.A. Lewis, pp. 77-97. ASTM STP 1115. American Society for Testing and Materials, Philadelphia, PA (1991).

28. T.E.L. Langford: Ecological Effects of Thermal Discharges, Elsevier Applied Science, New York (1987).

29. M.G. Kelly: Mining and the Freshwater Environment. Elsevier Applied Science, London, 231 pp.(1988).

30. P.M. Stokes, Howell, E.T. and Krantzberg, G.: Effects of acid precipitation on the biota of freshwater lakes. IN: Acid Precipitation, Vol. 2, Biological Effects, Springer-Verlag, New York, New York, p. 273-305 (1989).

31. S. Shales, Thake, B.A., Frankland, B., Khan, D.H., Hutchinson, J.D., and Mason, C.F.: Biological and ecological effects of oils. IN: The Fate and Effects of Oil in Freshwater, eds. Green, J., and Treet, W., Elsevier Applied Science, New York, p. 81-173 (1989).

32. D.A. Goolsby, Coupe, R.C., and Markovchick, D. J.: Distribution of selected herbicides and nitrate in the Mississippi River and its major tributaries, April through June 1991. U.S. Geological Survey Water Resources Investigations Report 91-4163. Denver, CO (1991).

33. W.E. Periera and Hostettler, P. D.: Nonpoint source contamination of the Mississippi River and its tributaries by herbicides. Environmental Science and Technology (27)1542-1552 (1993) .

34. O. Sortkjaer.: Macrophytes and macrophyte communities as test systems in ecotoxicological studies of aquatic systems. Ecological Bulletins Stockholm (36)75-80 (1984).

35. M.A. Lewis.: Chronic toxicities of surfactants and detergent builders to algae: A review and risk assessment. Ecotoxicological Environmental Safety (20)79-84 (1990b).

36. W.C. Dennison, Orth, R.J., Moore, K.A., Stevenson, J.C., Carter, V., Kollar, S., Bergstrom, P.W. and Batuk, R.A.: Assessing water quality with submersed aquatic vegetation. Bioscience (43)86-94 (1993).

37. P.V. McCormick and Cairns, J. JR.: Algae as indicators of environmental change. Journal of Applied Phycology (6)509-526 (1994).

38. J.L. Doust, Schmidt, M. and Doust, L.L.: Biological assessment of aquatic pollution: a review with emphasis on plants as biomonitors. Biological Review (69)147-186 (1994).

39. B.A. Whitton and Kelly, M.G.: Use of algae and other plants for monitoring rivers. Australian Ecology (20)45-56 (1995).

40. I. Kong, Bitton, G., Koopman, B., and Jung, K.H.: Heavy metal toxicity testing in environmental samples. Reviews of Environmental Contamination and Toxicology (142)119-147 (1995).

41. American Society for Testing and Materials.: Pants for Toxicity Assessment, Wang, W., Gorsuch. J., and Lower, W. [Eds.] ASTM STP 1091, ASTM, Philadelphia, PA. 363p. (1990).

42. American Society for Testing and Materials.: Plants for Toxicity Assessment, Gorsuch, J., Lower, W., Lewis, M., and Wang, W. [Eds]. ASTM STP 1115, ASTM, Philadelphia, PA. 401p. (1991).

43. U.S. Geological Survey.: Summary report from a workshop on selection of tier 1 bioassessment methods. Information and Technology Report 7, Denver, CO (1996).

44. C.J. Keddy, Greene, J.G. and Bonnell, M.: Review of whole-organism bioassays: soil freshwater sediment, and freshwater assessment in Canada. Ecotoxicology and Environmental Safety (30)221-251(1995).

45. P. Guillizoni: The role of heavy metals and toxic materials in the physiological ecology of submersed macrophytes. Aquatic Botany (41)87-109 (1991).

46. M.A. Lewis.: Use of freshwater plants for phytotoxicity testing: a review. Environmental Pollution (87)319-336 (1995).

47. W. Wang: Factors affecting metal toxicity and accumulation by aquatic organisms (overview). Environment International (13)437-457 (1987).

48. W. Wang.: Toxicity reduction of photo processing wastewaters. Environmental Science and Health A27, 1313-1328 (1992).

49. W. Wang, Gorsuch, J., and Hughes, J. [eds.]: Plants for Environmental Studies. CRC/Lewis Publishers, Boca Raton. FL, 563 p. (1997).

50. U.S. Environmental Protection Agency.: Plant tier testing: a workshop to evaluate nontarget plant testing in subdivision J pesticide guidelines. EPA /600/9-91/041. Corvallis, OR (1991).

51. K. Freemark, MacQuarrie, P., Swanson, S. & Peterson, H.: Development of guidelines for testing pesticide toxicity to nontarget plants for Canada, In: Plants for Toxicity Assessment, ASTM STP, 1091. Eds. W. Wang, J.W. Gorsuch and W.R. Lower. ASTM, Philadelphia, PA (1990).

52. J.S. Hughes, Alexander, M. M., and Balu, K.: An evaluation of appropriate expression of toxicity in aquatic plant bioassays as demonstrated by the effects of atrazine on algae and duckweed. ASTM STP 971, pp. 531-545. American Society for Testing and Materials, Philadelphia, PA (1988).

53. J.S. Hughes and Erb, K.: The relative sensitivity of five non-target aquatic plant species to various pesticides. Presented at 10th Annual Meeting of the Society of Environmental Toxicology and Chemistry, October 28, Toronto. Canada (1989).

54. J.E. Taraldsen and Norberg-King. T.: New method for determining effluent toxicity using duckweed (*Lemna minor*). Environmental Toxicology and Chemistry (9)761-767(1990).

55. W. Wang.: Ammonia toxicity to macrophytes (common duckweed and rice) using static and renewal methods. Environmental Toxicology and Chemistry (10)1173-1177 (1991b).

56. U.S. Food and Drug Administration.: Seed germination and root elongation. FDA Environmental Assessment Technical Guide No. 4.06. Center for Food Safety and Applied Nutrition and Center for Veterinary Medicine, U.S. Department of health and Human Services, Washington, DC (1987).

57. SIS.: Water quality - Determination of growth inhibition (7-d) *Lemna minor* duckweed. Swedish Standard Institution. SS 02 82 13. 5 p (I Swedish) (1991).

58. AFNOR: XP T 90-337. Determination de l'inhibition de croissance de *Lemna minor*. 10p (I French) (1996).

59. U.S. Environmental Protection Agency.: OPPTS 850.4400 Aquatic Plant Toxicity Test Using *Lemna* spp., "Public Draft". EPA 712-C-96-156. 8 p. (1996).

60. U.S. Environmental Protection Agency.: U.S. Environmental Protection Agency. Seed germination/root elongation toxicity test. EG-12. Office of Toxic Substances, Washington, DC (1982).

61. U.S. Environmental Protection Agency.: Lemna acute toxicity test. Fed. Reg. 50, 39331-39333. (1985).

62. U.M. Cowgill and Milazzo, D.P.: The culturing and testing of two species of duckweed. IN: Aquatic Toxicology and Hazard Assessment: 12th volume, eds: U.M. Cowgill, and L.R. Williams pp. 379-391. ASTM STP 1027. American Society for Testing and Materials, Philadelphia, PA (1989).

63. D.B. Huebert and Shay, J.M.: Considerations in the assessment of toxicity using duckweeds. Environmental Toxicology and Chemistry (12)481-483 (1993).

64. R.A. Stanley.: Toxicity of heavy metals and salts to Eurasian water milfoil (*Myriophllum spicatum* L.). Archives of Environmental Contamination and Toxicology (4)331-341 (1974).

65. D.R. Forney and Davis, D.E.: Effects of low concentrations of herbicides on submersed aquatic plants. Weed Science (29)677-85 (1981).

66. D.L. Correll and Lu, T.L.: Atrazine toxicity to submersed vascular plants in simulated estuarine microcosms. Aquatic Botany (14)151-158 (1982).

67. DS.H. Kay, Haller, W.T., and Garrard, L.A.: Effects of heavy metals on water hyacinths (*Eichhornia crassipes* [Mart.] Solms). Aquatic Toxicology (5)117-28 (1984).

68. W.J. Fleming, Ailstock, J.S., Momot, J.J. and Norman, C.M.: Response of sago pondweed, a submerged aquatic macrophyte, to herbicides in three laboratory culture systems. IN: Plants for Toxicity assessment: Second Volume, ASTM STP1115, J.W. Gorsuch, W.R. Lower, W. Wang and M.A. Lewis (eds.), American Society for Testing and Materials, Philadelphia, pp 267-275 (1991).

69. T.D. Byl and Klaine, S.J.: Peroxidase activity as an indicator of sublethal stress in the aquatic plant *Hydrilla verticillata* (Royle). IN: Plants for Toxicity Assessment: second volume. ASTM STP 1115, J.W. Gorsuch, W.R. Lower, W. Wang, and M.A. Lewis, ed., American Society for Testing and Materials, Philadelphia, PA, 101-106 (1991).

70. P. Garg and Chandra, P.: Toxicity and accumulation of chromium in *Ceratophyllum demersum* L. Bulletin of Environmental Contamination and Toxicology (44)473-478 (1990).

71. M. Biernacki, Lovett-Doust, J., and Lovett-Doust, L.: *Vallisneria americana* as a biomonitor of aquatic ecosystems: leaf-to-root surface area ratios and organic contamination in the Huron-Erie Corridor. Journal of Great Lakes Research (22)289-303 (1996).

72. M. Biernacki, Lovett-Doust, J., and Lovett-Doust, L.: Laboratory assay of sediment phytotoxicity using the macrophyte, *Vallisneria americana*. Environmental Toxicology and . Chemistry (16)472-478 (1997).

73. M.A. Lewis, Weber, D., Stanley, R., Quarles, R. and Roush, T.: Sediment toxicity and chemical quality of Gulf of Mexico Coastal sediments below ten wastewater discharges. Submitted: Water Research (1998a).

74. M.A. Lewis, Weber, D.E. and Stanley, R.S.: Comparative toxicities of municipal, industrial, forestry product, and power generation effluents discharged to near-coastal areas of the Gulf of Mexico. Water Environment Research. In Press (1998b).

75. G.R.D. Wood.: Hydrobotanical Methods, University Park Press, Baltimore, MD, 173 pp. (1970).

76. K.S. Long.: Remote sensing of aquatic plants: Tech. Rep. Waterways Experiment Station U.S. Army Corps of Engineers, Vicksburg, MS, MS A79-2 (1979).

77. D.S. Andrews, Webb, D.H. and Bates, A.L.: The use of aersol remote sensing in quantifying submersed aquatic macrophytes. In: W.M. Dennis, B.G. Isom (eds.) Ecological Assessment of Macrophyton: Collection, Use and Meaning of Data, STP843, ASTM, Philadelphia, PA (1984).

78. W.M. Dennis and Isom, B.G.: Ecological Assessment of Macrophyton: Collection, Use and Meaning of Data. ASTM STP843, American Society for Testing and Material, Philadelphia (1983).

79. S.I. Kozuharov.: Plants as bioindicators. In: Biological Monitoring of the State of the Environment: Bioindicators (ed. J. Salanki), pp 17-25. The International Union of Biological Sciences (1985).

80. L.J. Britton and Greeson, P.E.: Techniques of Water-Resources Investigations of the United States Geological Survey. Chapter A-4 Methods for collection and analysis of aquatic, biological and microbiological samples. U.S. Geological Survey, Denver (1987).

81. A.M. Small, Adey, W.H., Lutz, S.M., Reese, E.G. and Roberts, D.L.: A macrophyte-based rapid biosurvey of stream water quality: restoration at the watershed scale. Restoration Ecology (4)24-145 (1986).

82. M.D. Systsma and Anderson, L.W.J.: Biomass, nitrogen, and phosphorous allocation in parrot feather (*Myriophyllum aquaticum*). Journal of Aquatic Plant Management (31)244-248 (1993).

83. M.D. Netherland and Lembi, C.A.: Gibberllin synthesis inhibition effects on submersed aquatic weed species. Weed Science (40)29-36 (1992).

84. S.L. Sprecher and Netherland, M.D.: Methods for monitoring herbicide-induced stress in submersed aquatic plants: a review. Aquatic Plant Control Research Program, Miscellaneous Paper A-95-1, U.S. Army Corps of Engineers, Vicksburg, MS (1995).

85. A. Ramanathan, Ownby, J.D. and Burkes, S.L.: Protein biomarkers of phytotoxicity in hazard evaluation. Bulletin of Environmental Contamination and Toxicology (56)926-934 (1996).

86. G.P. Fitzgerald and Nelson, T.C.: Extractive and enzymatic analyses for limiting or surplus phosphorus in algae. Journal of Phycology (2)232-37 (1996).

87. B.A. Whitton, Grainger, S.L.J., Hawley, G.R. and Simon, J.W.: Cell-bound and extracellular phosphatase activities of cyanobacterial isolates. Microbial Ecology (21)85-98 (1991a).

88. B.A. Whitton, Kelly, M.G, Harding, J.P.C. and Say, P.J.: Use of plants to monitor heavy metals in freshwaters. IN: Methods for the Examination of Waters and Associated Materials. HMSO, London. 43 pp. (1991b).

89. R.L. Doong, MacDonald, G.E., and Schilling, D.G.: Effect of fluridone on chlorophyll, carotenoid and anthocyanin content of hydrilla. Journal of Aquatic Plant Management (31)55-59 (1993).

90. C.I. Weber (ed).: Biological field and laboratory methods for measuring the quality of surface waters and effluents. Program Element 1BA027, EPA/670/4-783-001. Cincinnati, OH, U.S. EPA (1973).

91. J.M. Hellawell.: Biological surveillance of rivers- a biological monitoring handbook. Water Research Center (1978).

92. J.M. Hellawell.: Biological indicators of freshwater pollution and environmental management. Elseveir Applied Science Publishers, London (1986).

93. B.A. Whitton.: Algae and higher plants as indicators of river pollution. IN: Biological Indicators of Water Quality, ed. A. James and L. Evison, John Wiley, Chichester (1979).

94. S.M. Haslam.: A proposed method for monitoring river pollution using macrophytes. Environmental Technology Letters (3)19-34 (1982).

95. L.E. Shubert (ed.).: Algae as Ecological Indicators. Academic Press, London, England (1984).

96. A.M. Farmer.: The effects of lake acidification on aquatic macrophytes – a review. Environmental Pollution (65)219-240 (1990).

97. G.J. Davis and Brinson, M.M.: Responses of submersed vascular plant communities to environmental change. Biological Services Program, U.S. Fish and Wildlife Service FWS/OBS-79/33 (1980).

98. H.G. Washington.: Diversity biotic and similarity indices: a review with special reference to aquatic ecosystems. Water Research (18)653-694 (1984).

99. P.D. Abel.: Water Pollution Biology, John Wiley and Sons, New York, NY (1989).

100. D.A. Brock.: Comparison of community similarity indexes. Journal of Water Pollution Control Federation (49)2488-2495 (1977).

101. E.C. Pielou.: The Interpretation of Ecological Data. John Wiley & Sons, New York (1984).

102. M.G. Kendall.: Rank correlation methods. Griffin and Co., London (1962).

103. L. Maltby and Calow, P.: The application of bioassays in the resolution of environmental problems: past, present and future. Hydrobiologia (188/189)65-76 (1989).

104. H. Brix and Lyngby, J.E.: The distribution of cadmium, copper, lead, and zinc in eelgrass (*Zostera marina* L.). Science of the Total Environment (24)51-63 (1982).

105. W. Wang and Lewis, M.A.: Metal accumulation by aquatic macrophytes. In: Plants for Environmental Studies (eds.) W. Wang, J. Gorsuch, J. Hughes, CRC Press, Boca Raton, FL (1997).

106. N.E. Whitehead and Brooks, R.R.: Aquatic bryophytes as indicators of uranium mineralization. Bryologist (72)501-507 (1969).

107. B.T. Hart, Jones, M.J. and Breen, P.: *In-situ* experiments to determine the uptake of copper by the aquatic macrophyte, *Najas tenuifolia*. Environmental Technology Letters. (4)217-222 (1983).

108. D.C. Mortimer.: Freshwater aquatic macrophytes as heavy metal monitors - the Ottawa River experience. Environmental Monitoring Assessment (5)311-323 (1985).

109. M. Kovacs and Podani, K.: Bioindication: a short review on the use of plants as indicators of heavy metals. Acta Biologica Hungarica (37)19-29 (1986).

110. G.N. Mhatre and Chapekar, S.B.: The effect of mercury on some aquatic plants. Environmental Pollution Series. A. (39)207-216 (1985).

111. P.M. Outridge and Noller, B.N.: Accumulation of toxic trace elements by freshwater vascular plants. Reviews of Environmental Contamination and Toxicology (121)1-63 (1991).

112. M.L. Hinman and Klaine, S.J.: Uptake and translocation of selected organic pesticides by the rooted aquatic plant *Hydrilla verticillata* Royle, Environmental Science and Technology (36)609-613 (1992).

113. P. Reimer and H.C. Duthie.: Concentrations of zinc and chromium in aquatic macrophytes from the Sudbury and Muskoka regions of Ontario, Canada. Environmental Pollution (79)261-265 (1993).

114. A. Crowder.: Acidification, metals and macrophytes. Environmental Pollution (71)171-203 (1991).

115. T.A. Anderson, Holyman. A. M.. Edwards, N. T., Walton, B. T.: Uptake of polycyclic aromatic hydrocarbons by vegetation: a review of experimental methods. In: Plants for Environmental Studies, W. Wang, J. Gorsuch, and J. Hughes [Eds.] CRC Press/Lewis Publishers, Boca Raton, FL, 563 p. (1997).

116. T.P. Boyle.: The effect of environmental contaminants on aquatic algae. In: Algae as Ecological Indicators, ed. L.E. Shubert, pp. 237-256. Academic Press, New York (1984).

117. R.K. Puri, Ye, Q., Kapila. S. Lower, W. R.. and Puri, V.: Plant uptake and metabolism of polychlorinated biphenyls. In: Plants for Environmental Studies, eds. W. Wang, J. Gorsuch, and J. Hughes, CRC Press/Lewis Publishers, Boca Raton, FL (1997).

118. O.J. Schwartz and Jones, L. W.: Bioaccumulation of xenobiotic organic chemicals by terrestrial plants. In: Plants for Environmental Studies, W. Wang, J. Gorsuch, and J. Hughes [eds.] CRC Press/Lewis Publishers, Boca Raton, FL, 563 p. (1997).

119. S.A. Nichols and Keeney, D.R.: Nitrogen nutrition of *Myriophyllum spicatum*: variation of plant tissue nitrogen concentration with season and site in Lake Wingra. Freshwater Biology (6)137-144 (1976).

120. R.M. Heisey and Damman, A.W.H.: Copper and lead uptake by aquatic macrophytes in eastern Connecticut, U.S.A. Aquatic Botany (14)213-229 (1982).

121. C. Mouvet.: Accumulation of chromium and copper by the aquatic moss *Fontinalis antipyretica* Lex. HEDW transplanted in a metal-contaminated river. Environmental Technology Letters. (5)541-548 (1984).

122. E. Goncalves, Boaventura, R. and Mouvet, C.: Sediments and aquatic mosses as pollution indicators for heavy metals on the Ave River Basin (Portugal). Science of the Total Environment (114)7-24 (1992).

123. C. Mouvet, Morhain, E., Sutter, C. and Couturieux, N.: Aquatic mosses for the detection and follow-up of accidental discharges in surface waters. Water Air and Soil Pollution (66)333-348 (1993).

124. S.B. Chaphekar.: An overview on bioindicators. Journal of Environmental Biology (12)163-168 (1991).

125. K. Aulio.: Accumulation of copper in fluvial sediments and yellow water lilies (*Nuphar lutea*) at varying distances from a metal processing plant. Bulletin of Environmental . Contamination and Toxicology (25)713-717 (1980).

126. W. Novotny and Olem, H.: Water quality -- prevention, identification, and management of diffuse pollution. Von Nostrand Reinhold, NY, 1054 p. (1994).

127. A. Kohler and Labus, B.C.: Eutrophication processes and pollution of freshwater ecosystems including waste heat. IN: Physiological plant ecology. IV Encyclopedia of Plant Physiology. (eds.) O.L. Lange et al. New Series Vol. D, 413-464 (1983).

128. G. MacDonald, Shilling, D.G., and Berwick, T.A.: Effects of endothall and other aquatic herbicides on chlorophyll fluorescence, respiration and cellular integrity. Journal of Aquatic. Plant Management (31)50-55 (1993).

129. D. Francko, Delay, L., and Al-Hamdani, S.: Effect of hexavalent chromium on photosynthetic rates and petiole growth in *Nelumbo lutea* seedlings. Journal of Aquatic Plant Management (31)29-33 (1993).

130. S. Sprecher, Stewart, A.B., and Brazil, J.M.: Peroxidase changes as indicators of herbicide-induced stress in aquatic plants. Journal of Aquatic Plant Management (31)45-50 (1993).

131. M. Smart and Doyle, R.: Ecology theory and the management of submersed aquatic plant communities. U.S. Army Corp of Engineers: Vol. A-95-3, Vicksburg, MS (1995).

132. Journal of Aquatic Plant Management.: International Symposium on the Biology and Management of Aquatic Plants. Vol. 31. Publication of the Aquatic Plant Management Society, Washington (1993).

133. R.W. Nelson and Randall, R.: National guidance on water quality standards for wetlands. Phase II: Long-term development, USEPA Office of Wetlands Protection, Washington (1990).

134. Florida Department of Environmental Regulation.: Chapter 17-6, Regulations for Wastewater Facilities. Tallahassee, FL (1988).

135. B.D. Tripathi and Shukla, S.C.: Biological treatment of wastewater by selected aquatic plants. Environmental Pollution (69)69-78 (1991).

136. M. Brown, Brandt, K. and Adamus, P.: Potential Indicators of Ecological Conditions in Inland Wetlands (Draft). Prepared for Environmental Monitoring Assessment Program (EMAP), U.S. Environmental Protection Agency, Office of Research and Development, Washington, DC, by University of Florida Center for Wetlands and NSI Technology Services Corporation (October), 77 pp. (1989).

137. R.C. Weider.: A survey of constructed wetlands for acid-coal mine drainage treatment in the eastern United States. Wetlands (9)299-315 (1989).

138. R.L. Knight.: Treatment wetlands database now available. Water Environment Technology P. 31, February (1994).

139. D.A. Hammer.: Constructed Wetlands for Wastewater Treatment. Lewis Publishers, Chelsea, MI. (1989).

140. W. Traunspurger and C. Drews.: Toxicity analysis of freshwater and marine sediments with meio and macrobenthic organisms: a review. Hydrobiologia (328)215-261 (1996).

141. M. Munawar and I.F. Munawar (1987): Phytoplankton bioassays for evaluating toxicity of in-situ sediment contaminants. Hydrobiologia (149)87-105 (1987).

142. Y.S. Wong, N.F.Y. Tam, P.S. Lau and X.Z. Xue.: The toxicity of marine sediments in Victoria Harbour, Hong Kong. Marine Pollution Bulletin (31)464-470 (1995).

143. C. Nalewajko.: Effects of cadmium and metal-contaminated sediments on photosynthesis, heterotrophy and phosphate uptake in Mackenzie River delta phytoplankton. Chemosphere (30)1401-1414 (1995).

144. M.L. Wildhaber and C.J. Schmitt.: Assessment and Remediation of Contaminated Sediments (ARCS) Programs. Hazard ranking of contaminated sediments based on chemical analysis laboratory toxicity tests and benthic community structure: method of prioritizing sites for remedial action. USEPA 905-R94-024, Chicago, Illinois (1994).

145. W. Calmano, Ahlf and U. Förstner.: Sediment Quality Assessment: chemical and biological approaches. Sediments and Toxic Substances, W. Calmano and U. Förstner [eds.], Springer-Verlag, Berlin (1996).

146. U.S. Environmental Protection Agency.: Methods for assessing the toxicity of sediment-associated contaminants with estuarine and marine amphipods. USEPA 600/R-94/025. Office of Research and Development, Narragansett, RI (1994).

147. B.L. Folsom, Preston, K.M. and Lee, C.R.: Plant bioassay of contaminated dredged material. In: Dredging and Dredged Material Disposal. R.L. Montgomery and J.W. Leach (eds.), American Society of Civil Engineers, NY, NY (1984).

148. G.E. Walsh., Weber, D. E., Brashers, L. K. and Simon, T. L.: Artificial sediments for use in tests with wetland plants. Environmental and Experimental Botany (30)391-396 (1990).

149. G.E. Walsh, Weber, D.E. Simon, T.L. and Brashers, L.K.: Toxicity tests of effluents with marsh plants in water and sediment. Environmental Toxicology and Chemistry (10)517-25 (1991a).

150. G.E. Walsh, Weber D.E., Nguyen, M.T. and Esry, L.K.: Responses of wetland plants to effluents in water and sediment. Environmental Experimental Botany (32)351-8 (1991b).

151. G.E. Walsh, Weber, D. E., Simon, T. L., Brashers, L. K., and Moore, J. C.: Use of marsh plants for toxicity testing of water and sediment. IN: Plants for Toxicity Assessment: second volume, ASTM STP 1115, pp. 341-354. American Society for Testing and Materials, Philadelphia, PA (1991c).

152. C.H. Watkins and Hammerschlag, R.S.: The toxicity of chlorine to a common vascular aquatic plant. Water Research. (18)1037-1043 (1984).

153. G.N. Mhatre.: Bioindicators and biomonitoring of heavy metals. Journal of Environmental Biology (20)201-209 (1991).

154. W. Huber, Zieris, F.J., Feind, D. and Neugebaur, K.: Ecotoxicological evaluation of environmental chemicals by means of aquatic model. Bundesministerium fuer Forschung und Technologie, Research Report 03-7314-0, Bonn. (1987).

Environmental Science Forum Vol.96 (1999) pp. 275-302
© 1999 Trans Tech Publications, Switzerland

The Use of Aquatic Macrophytes in Monitoring and in Assessment of Biological Integrity

P.M. Stewart[1], R.W. Scribailo[2] and T.P. Simon[3]

[1] Lake Michigan Ecological Research Station, Biological Resources Division, U.S. Geological Survey, 1100 N. Mineral Springs Road, Porter, IN 46304, USA

[2] Biological Sciences and Chemistry Section, Purdue University North Central, 1401 S. U.S. 421, Westville, IN 46391 USA

[3] U.S. Environmental Protection Agency, Water Division, 77 W. Jackson Street, WW-16J, Chicago, IL 60604, USA

Keywords: Biomonitoring, Bioindicators, Aquatic Toxicity Testing, Index of Biotic Integrity

Abstract — Aquatic pl... ...ns, and communities should be used as indicators of the aquatic environment ... ...vstem response to different stressors. Plant tissues bioaccumulate a... ...than what is present in the sediments; and this appears to be r... ...ification, and buffering capacity. The majority of toxicity stu... ...ls, have been done with several *Lemna* species and *Vallisner...* ...viewed include pesticides and herbicides, polycyclic aromatic ... ...enyls, and other industrial contaminants. The use of aquati... ...ınd... of environmental quality was evaluated for specific characte... ...v as ...iological integrity. Indices such as the floristic quality ... ...ervatism (C) are pioneering efforts to describe the quality of ... ...iversity. Our case study in the Grand Calumet Lagoons foun... ...ie greatest aquatic plant species richness, highest FQI and C v... ...ce. Lastly, we introduce the concepts necessary for the develop... ...grity. Development of reference conditions is essential to ... ...munity structure, function, individual health, condition, ...guild development and tolerance definition are also integral to ... ...etric index.

Introduction

Aquatic organisms are used routinely as indicators that reflect the degree of pollution and disturbance [1,2,3,4,5] and allow the detection of ecosystem-level affects resulting from anthropogenic stress [6]. Indicators of environmental quality need to be appropriate to the situation, however, there are a limited array of indicators available. Endpoints measured include a loss of sensitive species, decline in species rich ness, trophic state changes [7], and shifts in growth habits [8].

Species diversity of an aquatic plant community may be a direct reflection of water and substrate quality. The loss or predominance of certain species may indicate the presence of pollutants. The use of bioindicators offers a rapid assessment of potential problem areas based on knowledge of the critical habitat requirements of one or more indicator species. Biomonitors involve the measurement of target toxin concentrations and can provide specific information on tolerance to contaminants. Toxicity and bioaccumulation studies using aquatic vascular plants can provide information about the effects of contaminants, biomagnification, and generate predictions about accumulation levels.

Objectives of this review

The objective of this paper is to provide an overview on the use of aquatic macrophytes as monitors and indicators of biotic integrity. By compiling this literature and providing an example from our own research, we are attempting to focus attention on the usefulness of aquatic macrophytes in the assessment of biotic integrity. For the purpose of this review, we will limit our discussion to the aquatic vascular plants (aquatic macrophytes) or aquatic angiosperms, although the majority of biomonitoring studies for aquatic organisms still use algae as the primary source of information for water quality determinations [9,10,11,12,13,14,15]. To facilitate future discussion on the biotic integrity of aquatic plant communities, we include preliminary information on the development of a multimetric index using aquatic plant information.

Biomonitoring and Bioassays

Terms such as biomonitors, bioassays, and bioindicators have been used interchangeably in the literature resulting in considerable confusion (16). Biomonitoring has historically been used synonymously to define both bioassays and toxicity testing. We interpret biomonitoring to only include *in situ* studies of toxicity. The closely related topic of bioassays is herein defined as laboratory based studies of toxicity to determine effect concentrations under controlled dose and time.

Toxin concentrations in sediments or water are not necessarily equivalent to bioaccumulation levels in aquatic organisms. Bioaccumulation may result in concentrations in plant tissues of ten times or more than those found in the surrounding environment. This may result in ambient toxin concentrations that are below detectable limits despite having deleterious effects on the biota. This is particularly evident where contaminated sediments may be leaching a constant source of pollutants into the water column over an extended period of time as with many polycyclic organic compounds (16,17,18,19,20) and pentachlorophenols (21).

What makes a biomonitor?

The usefulness of a species as a biomonitor is dependent upon being able to establish a strong correlation between tissue concentrations of a particular toxin with the concentration in the sediments or water column. The use of aquatic plants (freshwater macrophytes) as *in situ* biomonitors for water quality assessment has been reviewed (16,22,23,24). Aquatic plants offer many advantages as biomonitors over animal species. Their sedentary habit makes them ideal for use *in vivo*, clonal growth allows for multiple sampling and ease of propagation, and minimal care requirements make them inexpensive to maintain (25). In addition, plants often have greater sensitivities to many contaminants than those reported for animal groups (26). This includes metals, alcohols, pesticides, surfactants, effluents, and organic chemicals (Table 1).

Table 1. Test substances that are more toxic to freshwater plants than animals[1].

Metals
Nickel
Cadmium
Zinc
Copper
Chromium

Alcohols
Butanol
Hexanol
Heptanol
Octanol
Diethylene glycol
Proporgyl alcohol
Isooctanol

Pesticides
Diquat
Atrazine
Chlordane
2,4-D
Endrin
Dieldrin
Aldrin
Glyphosate
Tebuthiuron

Surfactants
Ditallo dimethyl ammonium chloride
Trimethyl ammonium chloride
Sodium dodecyl sulfate

Effluents
Paper mill
Textile
Oil refinery
Herbicides

Organics
Organotin
Chloramine
Phenol
Acrylates
Acridine
Potassium chlorate
Potassium dichromate
4-Chlorphenol
Chloronapthalene
4-Nitrophenol
Dibenzofuran
Sodium tetraborate
Sodium fluoride
Dinitrotoluene
Chloroacetaldehyde
Nitrobenzene
2,4,6-Trinitophenol
1,3-Dichlorpropene

[1](26) — Reprinted from Environmental Pollution, 87, M.A. Lewis, Use of freshwater plants for phytotoxicity testing, pp. 2-8, 1995, with permission from Elsevier Science.

Review of "in situ" testing

Our literature review found that aquatic plants have been relatively ignored in general works on pollution assessment. Many of these deal with aspects of biological assessment, water quality, aquatic toxicology, and profiles of biological species used in reporting chemical toxicity, but none of these discuss aquatic macrophytes (4,27,28,29,30). It is only in the last five years that reviews of ecotoxicology mention the use of aquatic macrophytes (16).

Most biomonitoring, using aquatic plants, has involved the use of a single clone of a species as in *Lemna* (31). In contrast, since the response of individuals within a species often varies markedly according to genotype, multiple genotypes of *Vallisneria americana* were used as a biomonitor (32). Furthermore, prolonged use of the same clone for extended periods may allow evolutionary selection for resistance or genetic drift away from wild populations. In another form of testing, the validity of using seed germination bioassays as indicators of toxicity to adult plants is questioned, since the expression of genes during seed germination is often very different than those expressed by adult plants. For aquatic plants that reproduce by vegetative means such as turions, rhizomes, and tubers rather than by seed formation, use of this test may be invalid (33).

Few studies have evaluated the effects of organic compounds on aquatic plants. *Vallisneria americana*, as a biomonitor for organochlorine contamination (16,32), showed a relationship between exposure to contaminants with roots accumulating the highest concentrations of organochlorine. A study on the effects of organic substances on starch and soluble carbohydrates showed that increased starch and sugar levels were indicators of eutrophication (34).

Heavy metal uptake

Most studies using aquatic plants as biomonitors involve heavy metal uptake (35,36,37,38,39,40,41,42,43). The distribution of heavy metals in plant tissues was examined in eight aquatic macrophytes collected from Okefenokee Swamp. Lower tissue metal concentrations were found in swamp plants than those taken from outside the swamp, reflecting its unpolluted and ombrotrophic state; however, differences in plant organs explained more of the variation in metal concentrations than did location (44). Likewise, in the fluvial lakes adjacent to the St. Lawrence River, Quebec, leaf concentrations of Cu, Ni, Pb, and Zn in *Vallisneria americana* reflected spatial variations in environmental contamination (45). *V. americana* concentrated trace metals to higher levels than did *Potamogeton richardsonii* for all metals except Pb (45). The heavy metal load of river ecosystems was studied and it was found that generally the concentrations of aquatic plant tissue metal levels followed in the order: $Cu \gg Pb > Hg \geq Cd$ (46). Nickel toxicity was tested in surface waters of different hardness (47). The nickel ion at 1 ppm caused a 30% inhibition of duckweed growth in most surface waters and 70% inhibition in extremely soft waters (47). A significant positive correlation was found between heavy metal (Pb, Cd, Fe, Hg) concentrations and organic matter content in *Eriocaulon septangulare* (48). Presence of high concentration of organic matter in sediments significantly reduced metal levels in plant tissues. Metal concentrations were consistently higher in roots than in shoots and were never less than sediment concentrations. The disappearance of aquatic plant species due to acidification processes associated with acid rain deposition in lakes led investigators to postulate that acid intolerant soft-water macrophytes would disappear after acidification due to aluminum toxicity (49). However, aluminum additions had no detectable effect on the vitality of aquatic macrophytes in poorly buffered environments (38).

Tolerance of aquatic plants to heavy metals varies widely (50). Metal tolerance and the role of metal-binding proteins and metal-regulated gene expression is thought to be related to free ion activity (51), which chelators such as EDTA reduce, and not total metal concentration (52,53,54). Measurement of metal toxicity may be confounded by the presence of EDTA or other chelators in the medium; however, removing chelators from the growth medium will cause precipitation. The presence of a chelator is also necessary for optimal growth (53,54,55) and insuring agreement among replicate flasks in a toxicity series (56).

Sediment effects of rooted aquatic vegetation in a contaminated area includes the withdrawal of heavy metals and phosphorus from the sediment and their subsequent release into the littoral water in a process termed "pumping" (57,58,59). This results in resuspension of metals into the aqueous environment. The accumulation and uptake of metals in 20 different species of aquatic macrophytes in lower Austrian Rivers was studied (34). Increased accumulation rates of Mn, Fe/Zn, Cu, Cd, and Ni were found in the field and these validated laboratory results of increasing Cu, Ca, and Ni concentrations. Species showing the greatest accumulation rates for the largest number of heavy metals included *Fontinalis* sp., *Elodea canadensis, Ranunculus trichophyllus, R. fluitans, Myriophyllum verticillatum, P. crispus,* and *Lemna trisulca.*

In further studies on metals in aquatic macrophytes (60), concentrations of Cd, Cu, Fe, Mn, Ni, Pb, V, and Zn in *Ceratophyllum demersum, Potamogeton lucens,* and *Vallisneria spiralis* were found to be lower than comparable published studies (61), suggesting that tissue concentrations are a site-specific phenomenon. In polluted water, the affects of acidification on metal accumulation in wetlands, ponds, and lakes in Maine and Maryland showed that six genera of aquatic plants accumulated metals in proportion to water concentrations (Table 2, 62). Evidence of relationships between metal concentrations in water and concentrations in plants was limited for unpolluted waters. Low pH levels appeared to have little effect on the accumulation of metals by aquatic plants.

Bioassays
Bioassays are synonymous with phytotoxicity testing (26,63,64). Toxicity testing, where one species is placed in a static system with one contaminant or effluent, has become an accepted technique (65) and there are published protocols for many forms of testing (e.g., 66). The specific goal of phytotoxicity testing is to develop water quality criteria that reflect maximum allowable limits of organic and inorganic chemicals in natural water bodies and in effluents.

While the use of aquatic plants in toxicity testing has been advocated, in practice it has been used much less often than aquatic animals. Only 7% of 528 phytotoxicity tests examined used aquatic macrophyte species (67). A recent review has summarized many of the problems associated with the standard use of algae rather than aquatic vascular plants for toxicity testing (68). Aquatic plants such as *Vallisneria americana* may show toxicity effects to herbicides at concentrations two thousand times lower than the algae species typically used in phytotoxicity testing. A standardized protocol for aquatic macrophytes is lacking, although there is work in progress on several species and test systems (26). There is a standard guide for conducting static toxicity tests with *Lemna gibba* (69) and information available for conducting flow-through tests on *Lemna minor* (70).

Duckweeds *(Lemna)* are by far the most commonly used aquatic plant species for aquatic toxicity testing (55,71). Their free-floating habit avoids the complications of introducing sediments to the system, while their rapid growth rates with a simple measurable parameter of plant health (number

Table 2. Element concentrations (μg/g dry weight) for bur-reed (n=7), bladderwort (n=22), and pondweed (n=5) collected in Maryland and Maine in 1987[1].

Elements	Bur-reed Mean	Bur-reed Range	Bladderwort Mean	Bladderwort Range	Pondweed Mean	Pondweed Range
Al	74.6	31-150	1,740	130-6,800	296	77-650
Cd	86% <LOD (<0.2-0.3)a		55% <LOD (0.2-4.6)a		0.46	0.3-0.7
Ca	7,890	3,600-14,000	8,550	4,800-17,000	8,620	6,000-9,900
Cu	7.81	2.9-15	14.3	3.7-71	4.2	3.5-51
Fe	759	270-1,800	11,100	1,600-25,000	687	190-900
Pb	100% <LOD (<4)a		9.18	<4-25b	7.6	<4-13b
Mg	2,820	2,100-3,500	4,540	1,200-7,500	3,510	2,400-5,000
Mn	696	190-1,100	5,650	91-19,000	92	57-130
Ni	6.14	<1-13b	12.3	<1-74b	60% <LOD (1-2)a	
P	1,930	1,200-2,700	1,740	710-3,400	1,860	1,500-2,500
Zn	28.3	15-43	92.6	37-240	65.4	48-82

[1](62) — aMean not calculated. Less than 50% of samples exceeded the limit of detection (LOD). Limit of detection values are in parentheses. bValues equal to 0.5 LOD were used when the analytical results were reported as <LOD. Reprinted with permission from Environmental Toxicology and Chemistry, (1993). Effects of acidification on metal accumulation by aquatic plants and invertebrates, by P.H. Albers and M.B. Camardese, Volume 12. Copyright Society of Environmental Toxicology and Chemistry (SETAC), Pensacola, FL, (1993).

of leaves) and ease of culture make them ideally suited for toxicity testing. Unfortunately, there is considerable reason to question whether results using duckweed are truly indicative of toxicity for the majority of aquatic vascular plant species where uptake is primarily from sediments.

In vitro phytotoxicity testing measures toxic concentrations of pollutants through the use of a simplified system involving serial dilutions or dosages with one or a small number of toxins. Unfortunately, the impairment response curves generated for individual toxins are seldom representative of the interactive effects among toxins that occur in natural systems. In addition, tests run in sterile laboratory settings may not accurately predict "real-world" results. The wide array of stochastic elements present in aquatic habitats such as the dynamics of the sediment-water column interface can alter toxin availability and further complicate response (28,29). An excellent use of aquatic macrophytes is for *in vitro* toxicity testing of ambient water or sediments from a contaminated site. Unlike animals, their rooted sedentary habit directly samples sediment contaminants. Although tests of this type are more "realistic" than single species laboratory bioassays, these tests have the disadvantage that the toxicity profile of the water or sediment being assessed is often not fully characterized.

Many contaminated sites have mixtures of toxic chemicals with a wide array of potential synergistic toxicity effects. One novel approach towards testing the overall impact of multiple toxins on ecosystem health may be to document developmental instability in aquatic plants as an early warning sign of possible contaminant problems (72). Measure of developmental instability, or asymmetry, in the form of repeating units such as leaves, is a measure of stress in plants (73). There were greater within plant variance in leaf symmetry (including characteristics such as within whorl variance of leaf length, leaf width, and the ratio of leaf length to leaf width) at sites with higher pollution concentrations. Species studied included the aquatic vascular plants *Ceratophyllum demersum, Elodea canadensis,* and the marine algae, *Fucus furcatus* (73). Another novel approach is the use of measurements of root elongation as an index of toxicity for single and mixtures of compounds on the Japanese millet (*Echinochloa crus-galli*) (47,63).

Phytotoxicity testing for metals and organic contaminants
Trace metals are important micronutrients in plant and animal nutrition. Optimal ranges for these micronutrients are often quite narrow; concentrations outside of these ranges can cause death, decreased growth, and reduced fecundity (74). A metal is considered essential if without it the plant cannot complete its life cycle or if the element is part of a molecule used as a constituent or metabolite (75). Excess metals can be contributed from weathering of soils, landfills, industrial sources, and acid mine drainage (29).

The response of clones of two species of duckweed, *Lemna gibba* and *L. minor,* to additions of selenium, tin, cobalt, and vanadium was tested (76). They found varied toxic response to metals in different *Lemna* clones (76,77). The availability, uptake rates, and accumulation of different metals by submersed and floating macrophytes showed that differences in metal enrichment could be attributed to species tested, seasonal growth rate changes, tissue age, and metal type (50). *Lemna trisulca,* axenically grown with a regular replacement of a complete filter-sterilized medium containing various amounts of zinc and cadmium, showed that concentrations of Zn greater than 3.0 to 4.5 μM decreased the reproduction rate (78). The EC_{50} was estimated at 5.0 μM Zn for final yield and at 14 μM Zn for reproduction rate. There was a significant Zn by Cd interaction; increased Zn levels mitigated the inhibitory effect of Cd on the multiplication and final yield of duckweed. Increased Zn (3.06 μM) decreased Cd uptake, however, further Zn increases (6.12 and 12.2 μM) promoted plant uptake of Cd from the medium. The combined effects of cadmium and linear alkyl benzene sulfonate (LAS) on *L. minor* was found to decrease the ^{14}C incorporated into proteins, DNA, RNA, and phospholipids (79). They found the presence of LAS increased the uptake of [109]Cd in the plants. *Wolffia globosa* exhibited a linear accumulation rate for Cr and Cd with increasing concentration and exposure rate (80). A level of tolerance was exhibited by *W. globosa* to Cr enabling a stimulation in growth at 0.05 ppm Cr, while either no significant change or slight inhibition was seen at 1 ppm Cd and 2 ppm Cr. The concentration factor, which indicates the efficiency of a plant in accumulating the metal, has been found to decrease with an increase in the metal concentration. Concentration factors as high as 5000 were observed for duckweeds to various metals including Ag, Cd, Cu, Ni, and Zn (81,82,83,84,85).

Organic chemical toxicity literature on aquatic macrophytes includes information on pesticide and herbicide leaching and runoff from agricultural sources (86), the effects of polycyclic aromatic hydrocarbon (PAH) (20,87), polychlorinated biphenyls (PCB) (88), and other organics from industrial sources (21,89,90). A comprehensive review of the effects of atrazine (currently the most extensively used herbicide in the United States) on surface waters, included a substantial section on aquatic macrophytes (86). Atrazine produces a reversible inhibition of photosynthesis but is subject to biotic and abiotic breakdown. While some degradates are phototoxic, they are usually much less so than atrazine itself. Most accumulation is episodic and coincides with spring and fall applications of the herbicide and subsequent transport into water bodies. Diquat is an aquatic herbicide that inhibits photosynthesis and destroys membranes. An evaluation of the effects of diquat on the nuisance aquatic weed *Hydrilla verticillata* found that dissolved oxygen concentration was the most sensitive measure of response followed by membrane permeability (91). The lethal concentration of diquat effecting *H. verticillata* was approximately 600 mg/kg dry weight.

Polycyclic aromatic hydrocarbons (PAH) are a group of toxic and mutagenic contaminants that readily accumulate in the membrane systems of aquatic organisms (92). These compounds have high affinities for organic matter and very low water solubilities, making them extremely persistent in aquatic sediments (88). Exposure of PAH to light results in production of photoproducts with enhanced toxicity relative to the parent compound (20,87). *Lemna gibba* accumulated three PAHs (i.e., anthrazene, phenanthrene, and benzo-pyrene), preferentially in thylakoids and microsomes, suggesting that these subcellular compartments are the most at risk from PAH damage. Phenathroquinone, the photoproduct of phenanthrene, was toxic to *Lemna gibba* (87).

One of the most harmful classes of organic compounds are the polychlorinated biphenols (PCBs). Concentrations several orders of magnitude higher than in water have been reported in sediments. PCB levels can be 3-4 times higher in plants than animals and 6,000-9,000 times higher in plants than in water (93).

Chlorinated phenols have been used extensively as fungicides, herbicides, and insecticides; and in pulp and paper mills for bleaching (94,95). Pentachlorophenol (PCP) uptake in *Eichhornia crassipes* was studied by exposing plants to differing solutions in aquaria for 48 hours (21). Analysis of plant extracts indicated higher PCP concentrations in roots than shoots and metabolism of PCP to conjugates, partially dechlorinated derivatives, and reactive intermediates that generate free radicals. There were significant increases in antioxidative enzymes such as peroxidases after PCP exposure, particularly in roots. Assays using these enzymes could provide useful markers for indicating environmental stress in aquatic plants. Toxicity of phenols increased with the number of chlorine substitutions on the phenol ring (90). *Lemna gibba* conjugated 2,4,5-trichlorophenol with d-glucose to reduce toxicity and these effects depended on growth conditions, however, up to 95% of the 2,4-DCP was metabolized over a 6-d growth period (96). Assays involving measurements of elevated levels of antioxidative enzymes could become a quick and accurate method for assessing stress levels in aquatic plants.

Phthalate esters are widely used in the manufacturing of plastics and are polar narcotics. These polar narcotics do not react in a predictive manner but nevertheless cause toxicity. In a review of phthalate esters, that excluded effects on aquatic plants, a broad characterization of toxicity was provided (89). Low molecular weight phthalate esters were acutely and chronically toxic through an unspecified mechanism to many species. In contrast, *Lemna* species showed reduced toxicity (55), suggesting that they may not be a good representative indicator of potential sensitivity in aquatic plants as a group.

Constructed wetlands

Constructed wetlands are similar to flow-through bioassays and mesocosms and approximate a subset of the physical, chemical, and biological functions of a natural aquatic ecosystem (97,98,99,100). They have been used to treat a variety of waste streams, but have been most effectively used in Europe for the treatment of sewerage wastes for small communities (101). These systems are typically composed of a series of natural ponds with a diversity of emergent and floating vegetation. Regulations governing the design and operation of constructed wetlands typically require a detailed knowledge of the waste stream entering the system and a constant monitoring of toxin concentrations leaving the system (e.g., 102).

The majority of studies on constructed wetland systems have documented the removal rates of effluents for variables such as nitrogen, iron, or BOD. A primary function of aquatic plants in constructed wetlands is to act as pumps that provide oxygen to an otherwise anaerobic sediment and thus greatly enhance microbial metabolism rates for organic matter, nitrogen and phosphorus loss, and oxidation of metals (103,104,105,106).

Aquatic plant species used in constructed wetlands are those that can tolerate high BOD and are more tolerant of eutrophication. Emergent species are typically rapidly growing 'weedy' species with high rates of organic carbon assimilation such as cattails, reed, and bulrush. Reduction rates for BOD were generally similar for all species investigated including arrowhead, bulrush, reed, and cattail (105). Nitrogen removal followed the order: bulrush > reed > cattails (107). In certain systems, a noxious aquatic weed species (108), *Eichhornia crassipes*, has been effective for BOD reduction (109,110). *Lemna* has also been quite effective in increasing the efficiency of treatment in sewage lagoons (111). Further study is needed to determine which plants most efficiently break down organic matter, the mechanism that plants use to degrade toxins, and to prevent the escape of invasive species, such as *Phragmites*, into neighboring natural wetlands.

Bioindicators

Bioindicators are species whose presence indicates a certain habitat type or quality. Bioindicators of good habitat quality may be species of concern because of their rarity, often are very sensitive to disturbance or the presence of pollutants, and are often the first to disappear with habitat degradation. A good bioindicator species shows fidelity to a specific habitat. The absence of a species, without historical data, is not nearly as informative; the species may have once been there and is now extirpated, or it may never have reached the site but could survive there if it did. Species diversity of an aquatic plant community may be a direct reflection of water quality and the loss or predominance of certain species may indicate the presence of pollutants. Papers that discuss the role of aquatic plants as biomonitors are often referring to species as bioindicators of degraded habitat quality (e.g., 112).

The national ranking of wetland species assigns species to categories based on the probability that they will be found in wetlands (113). The current approach to wetland delineation by the U.S. Army Corps of Engineers involves assessment on the basis of finding positive indicators of three parameters: hydrophytic vegetation, hydric soils, and wetland hydrology (114). This approach has evolved over the past 25 years from an evaluation based solely on the presence of hydrophytic vegetation (115). Delineation of many sedge meadows and wet prairies, in particular, rely on accurate assessment of species to verify that a habitat is a wetland. Unfortunately, a great deal of habitat of this type has been lost (116). Natural habitats have been highly modified and fragmented since presettlement times in North America and Europe. Wetlands have been particularly vulnerable to destruction because their soils provide excellent farmland once drained (116). Lakes and wetlands also act as watershed repositories for sediments, excess nutrients, and pollutants (117).

As level of disturbance or modification increases, many sensitive native species are replaced by weedy exotics. Given the tremendous pressure on the remaining natural areas in the world, it is necessary to recognize remnants of presettlement vegetation and assess the floristic quality of habitats. One of the few available indices using this type of information, the floristic quality index (FQI), was developed (118) and subsequently modified and used in Michigan, USA (119). In order to use the FQI for assessing the quality of habitats, a comprehensive list of plant species must be assembled and assigned values that are indicative of the likelihood that they represent a species occurring in a habitat undisturbed from presettlement times. Values of this type, termed coefficients of conservatism (C), have been assigned for all plant species known to occur in the Chicago region (118) and Michigan (119). Species that are considered to be presettlement typically have higher coefficients of conservatism with threatened or endangered species having the highest value. Sixty percent of state listed species for Michigan have a C value of 10 or above (119). To derive the FQI, C values are summed and divided by the total number of plant taxa (n) giving a mean coefficient of conservatism. The mean C is then multiplied by the square root of the total number of plant species (n) to yield the floristic quality index.

An examination of the number of rare species in wetlands from the Great Lakes to the maritime provinces of Canada found that rare species occurred almost exclusively in sites with biomass of less than 100 g/0.25 m^2 (120). Studies of this type have shown that eutrophication leads to reductions in native species and rare species in particular.

Aquatic macrophytes in water quality assessment

Aquatic macrophytes have been largely ignored in the assessment of water quality. For instance, a study concerned with the loss of biota as a result of lake acidification found that a pH reduction in lakes of approximately 1.5 standard units led to a 31% reduction in the number of species in various taxonomic groups, other than aquatic macrophytes, which were omitted from the study (121). The use of aquatic macrophytes in ecotoxicological testing includes research performed in the field on the impact of a road salt pile on the plant community of a northern ombrotrophic bog dominated by *Sphagnum* mosses (122). The area nearest the salt pile had been invaded by *Vaccinium oxycoccus* and *Typha angustifolia* (123,124). After the salt pile was removed, succession was monitored for seven years. When chloride levels declined to below 300 ppm, the bog began returning to a more typical species assemblage.

Aquatic macrophytes should be considered in any study concerned with determination of the trophic status of a body of water (125,126). Organic pollution has been identified as one of the major factors that changes species composition (127), increases extension of areal coverage of aquatic macrophytes (7), and decreases or replaces aquatic plant growth forms (8). Nutrient uptake by macrophytes was significantly correlated with total sediment N and available P (128). Aquatic plants allow for increased sedimentation, uptake of nutrients and contaminants, litter decomposition, soil retention, and increased microbial processes. This has led to the consensus that wetlands improve water quality and their use in wastewater disposal (129,130,131).

Macrophyte and algal communities were used as bioindicators of water pollution by examining the relationship between species and sampling sites (132). Cluster analysis revealed groupings of biological species that were important in understanding patterns that could not have been revealed otherwise (132). In another study, correlations existed between the main components of an aquatic macrophyte community and trophic status as measured by chemical variables including ammonia, nitrate, phosphate, and others (133). Discriminant analysis was used to compare the successional sequence that was found with one based on statistical analysis (133). Further work showed that root length of *Lemna* was controlled by concentrations of phosphorus and oxygen in the water (134). From this a bioindication scale was derived based on the macrophyte community and the degree of eutrophication. Further work was termed the "bioindication method of the degree of eutrophication" (135), done in an area periodically flooded with high permutations of nutrient concentrations from groundwater fed streams. This flooding and nutrient enrichment caused the development of hypereutrophic and pollution tolerant macrophyte communities.

Increased eutrophication, mortality from pesticides, and fish and plankton interactions results in loss of submerged aquatic plants (136,137,138). Fish and zooplankton alter plant communities and it was found that the loss of aquatic plant species was not due to increased shading by phytoplankton alone (137). Pesticides may influence plant communities by selectively eliminating certain zooplankton species. A reduction in the number of aquatic plant growth forms (floating, submergent, emergent) was related to increased levels of nutrients and contaminants near an industrial landfill in the Grand Calumet Lagoons (see below). In a separate study, at several stream sites draining a large wetland area, aquatic plant diversity appeared more closely related to the number of habitat types available than to either land use or water quality (unpublished data).

Case Study -- The Grand Calumet Lagoons, Southern Lake Michigan Drainage

Introduction

We present data from a study of the Grand Calumet Lagoons in Northwest Indiana, part of the Great Lakes watershed, as an example of how aquatic plant communities can be used to assess biological degradation. The Grand Calumet Lagoons are located in a Great Lakes Area of Concern (AOC), which is one of 42 areas of the Great Lakes watershed identified by the International Joint Commission as having severe environmental contamination. This is the only AOC with all fourteen designated uses impaired (139).

One of the Great Lake Region's most important environmental issues is the contamination of water, air, soil, and biota by persistent toxic substances. These include heavy metals (copper, zinc, etc.), organic contaminants (organochlorines, pesticides, polychlorinated biphenyls (PCBs), and polycyclic aromatic hydrocarbons (PAHs). Lead concentrations in precipitation are higher in northwestern Indiana than in any other part of the Great Lakes region (140). A national park, Indiana Dunes National Lakeshore (INDU), contains part of the Lagoons and has the highest wet deposition levels of sulfate and nitrate of any monitored park in the country (141). Much of the western section of the Grand Calumet Lagoons is surrounded by a large industrial landfill that has received millions of tons of steel slag and other industrial waste. The purpose of this research was to compare the distribution and growth habit of aquatic plant species in the Grand Calumet Lagoons from least-impacted to most-impacted sites.

Methods

Lake and area description -- The Grand Calumet Lagoons (Fig. 1) are palustrine wetlands formed by siltation damming of the former outlet to Lake Michigan of the Grand Calumet River. The Grand Calumet River now flows westward, making our study area part of its headwaters. The 32.6 ha Grand Calumet Lagoons system drains a 3.5 km^2 watershed and is located on the eastern edge of the AOC. The Lagoons are divided into three similarly sized sections: the East, Middle, and West Lagoons. Most of the land use in the East and Middle Lagoons includes residential, natural areas, and several small non-point pollution sources. Land use adjacent to the West Lagoon includes some natural areas that yields to an industrial area which includes slag disposal, industrial storage, refuse dumping, scrap preparation, basic oxidation sludge processing, and coal, coke, and rail car storage, and a historic hazardous waste dump.

Two small ponds are located north of the lagoons: East Pond and West Pond (Fig. 1), which were historically backwaters of the Grand Calumet River, and are separated from each other by a high dune ridge. The West Pond is bordered by the landfill along its western shore and was partially filled by the industrial landfill.

Plant collection -- Aquatic plant species in the Grand Calumet Lagoons were identified in the field and enumerated with an abundance scale (observed=1, rare=2, rare/common=3, common=4, very common=5, abundant=6), which is a modification of commonly used techniques. Difficult species were collected and identified in the laboratory using standard floristic manuals (142,143). Aquatic plant species were also assigned to a growth habit (floating, emergent, submergent) for further evaluation of patterns in aquatic plant response to disturbance.

Statistics -- The aquatic plant communities found in the West Lagoons and West Pond (closest to the industrial landfill) were compared to those found near and far-field (Middle Lagoon) or across the dune ridge (East Pond) from the industrial area by one-way ANOVA (144). Various indices, including the number of species, coefficient of community by cluster analysis (145,146), and the coefficient of conservatism and floristic quality index (118) were used to analyze community structure.

Results

We found 40 aquatic plant species in our survey, with several species having a coefficient of conservatism of 8 or above (Table 3). The Middle Lagoon sites had the greatest number of aquatic plant taxa, however, this was not significantly different among lagoons (F = 4.139, p =

0.087, df = 2). The mean coefficient of conservatism (F = 82.385, p < 0.001, df = 2) and the floristic quality index (F = 15.775, p = 0.007, df = 2) were significantly different among lagoons.

Figure 1. Map of the Grand Calumet Lagoons and ponds showing proximity to the Indiana Dunes National Lakeshore and the industrial area.

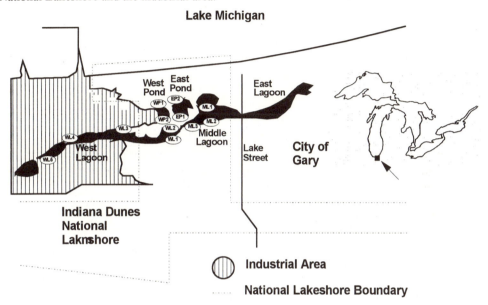

Floating plant species were absent from the West Lagoon at the time of our sampling (Table 4), yet these had been observed on previous trips to the Lagoons. An exotic plant species, *Myrophyllum spicatum*, was found throughout the Middle Lagoon and at several sites in the West Lagoon earlier in the season, but not at the time that we collected abundance data. Submergent plant species were found at all sites except WL 3-5. All growth habits were represented at the West Pond but not in the East Pond, which lacked floating macrophytes.

Relationships among sites based on their aquatic plant communities using cluster analysis (an assessment of similarity) showed two main groupings (Fig. 2). The Middle Lagoon sites and the East and West Ponds formed one cluster and all West Lagoon sites formed another cluster. The most similar sites in each of the two main clusters were sites ML2 and ML3 and sites WL1 and WL4.

Discussion
Aquatic plant communities at sites in the Grand Calumet Lagoons showed decreased species richness, less abundance, and fewer sensitive species based on their proximity to industrial areas. Sites closest to the landfill (near-field) had fewer species, lower mean coefficient of conservatism, and lower floristic quality than did the sites located further from the landfill (far-field). Growth habits disappeared at most West Lagoon sites with no submergent plants and no floating plants found during our survey. Our data gathering efforts supported the need for multiple site visits, several plant taxa were found at a site during one visit and not the next.

Table 3. Aquatic plant species and relative abundances at sites in the Grand Calumet Lagoons and ponds. WS = Wetland status, C = Coefficient of conservatism, ex=exotic, and a = algae.

Species	WS	C	ML1	ML2	ML3	WL1	WL2	WL3	WL4	WL5	EP	WP
Asclepias incarnata	obl	4							3	1		
Carex bebbii	obl	6					1					
Carex comosa	obl	5					4		4			4
Carex stricta	obl	5				6	3					
Ceratophyllum demersum	obl	5				2						
Chara globularis	a	a	4	5	5		2				6	
Cladium mariscoides	obl	10			3						4	4
Eleocharis ovata	obl	10						3	3	5		
Elodea canadensis	obl	5	4		3	6						
Iris virginica v. shrevei	obl	5	2		4		3	3	4	2		
Juncus acuminatus	obl	6	6		4							
Juncus balticus v. littoralis	facw	6			4	2	2					
Juncus brachycephalus	obl	9										3
Juncus torreyi	facw	4			4						3	
Lycopus virginicus	obl	9				2	3	3	3	2		
Myriophyllum exalbescens	obl	7					3					
Myriophyllum spicatum	obl	ex	3	4	6	3	4	6				
Najas flexilis	obl	6	4	4	6						6	
Nitella sp.	a	a			3							
Nuphar advena	obl	7	5	4	2							
Nymphaea tuberosa	obl	7	4		3							
Panicum clandestinum	facw	6									4	5
Phragmites australis	facw+	1					1	4	1	4		
Polygonum hydropiperoides	obl	7			2							
Polygonum lapathifolium	facw+	0							3	5		
Pontederia cordata	obl	10	6		3							
Potamogeton crispus	obl	ex		3								3
Potamogeton foliosus	obl	7									1	
Potamogeton illinoensis	obl	7	6	2	3						6	6
Potamogeton pectinatus	obl	5	4	3	4	4	2				4	4
Potamogeton zosteriformis	obl	8	3	4	2		2					2
Scirpus atrovirens	obl	4									2	
Scirpus pungens	obl	5	4	3	6		3	4	4		6	6
Scirpus validus v. creber	obl	5	4								6	4
Triglochin maritima	obl	10		2								
Typha angustifolia	obl	1				6			3			
Typha latifolia	obl	1				6	6	3	3	2	6	
Utricularia minor	obl	10										5
Utricularia vulgaris	obl	9					6					6
Vallisneria americana	obl	7		4	5							
Number of species			14	11	19	11	13	7	10	7	12	12
Number of rated species			12	8	16	10	11	6	10	7	11	11
Coefficient of Conservatism (mean)			6.3	6.9	6.6	5.3	5.5	4.5	4.1	4.3	5.5	7.2
Floristic Quality Index			21.9	19.4	26.3	16.8	18.1	11.0	13.0	11.3	18.1	23.8

Table 4. Number of aquatic plant species that occupy submergent, floating, and emergent growth habits at each site in the Grand Calumet Lagoons and ponds.

Growth habit	Site									
	ML1	ML2	ML3	WL1	WL2	WL3	WL4	WL5	EP	WP
Submergent	7	8	9	6	4	0	0	0	5	6
Floating	2	1	2	0	0	0	0	0	0	1
Emergent	5	2	8	5	9	7	10	7	7	5

Figure 2. Cluster analysis of aquatic plants species collected from the Grand Calumet Lagoons and ponds.

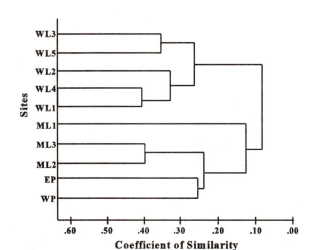

An impaired aquatic plant community existed within and surrounding the Grand Calumet Lagoons and ponds. There were high levels of contaminants in some areas of the Grand Calumet Lagoons and in the ponds located near the landfill and of these, PAHs elicited the greatest concern (147). Heavy metal concentrations were elevated at some sites, this may be an additional cause for concern (148).

Development of an Index of Biotic Integrity

One of the primary objectives of the Clean Water Act is the preservation of biological integrity. Biological integrity is defined as "... the ability of an aquatic ecosystem, to support and maintain a balanced, integrated, adaptive community of organisms having a species composition, diversity, and functional organization comparable to that of natural habitats of a region" (149). Since most areas of the world have had human impact to some extent, it becomes necessary to search for "least-impacted" areas for the development of reference conditions (150).

The index of biotic integrity (IBI) was used to evaluate fish communities of small streams in the Midwestern United States (149,151). The IBI has evolved into a family of multimetric indices that are modified for regional use, resource type, and organismal composition. The index uses twelve characteristics, termed "metrics", to provide assessment of community stability. Karr et al. (152) used three broad-based groups of metrics including species richness and composition, trophic guilds, and individual health, condition, and abundance. When metrics or attributes are being considered for modification, it is important that the original purpose of the metric be considered.

Primary producer indices of biotic integrity

Most biological assessment efforts using primary producers have focused on algal and periphyton communities. Rosen (153) discussed the use of periphyton communities as environmental indicators and described the use of multimetric indices. In Montana (USA), the index includes a diversity metric, several pollution metrics (most tolerant, less tolerant, and sensitive groups), a similarity index metric, and a siltation metric (154). The State of Kentucky (USA) has constructed the diatom bioassessment index (DBI) using attributes of diatom species richness, species diversity, percent community, a pollution tolerance index, and percent sensitive index (155). Scores for each metric range between 1 and 5 and are cumulatively summed to provide a narrative description of excellent, good, fair, and poor. The State of Oklahoma (USA) (156) developed a rapid bioassessment protocol for diatoms modeled after that used by Plafkin et al. (157). Metrics used for this index include the number of taxa and a sequential comparison index (158).

Structural attributes of aquatic macrophyte communities

In palustrine habitats, littoral and profundal habitats possess unique plant communities. Around the margin of these wetlands are another suite of plants that are wetted seasonally, yet maintain an important affinity to differing water levels and qualities. Species richness and composition metrics include species diversity and the presence of indicator species to describe the impacts to various habitats in the area being evaluated. For aquatic macrophyte communities, specific metric substitutions may include proportions of floating, emergent, and submergent species, or may focus on specific littoral or profundal zone species and emergent riparian zone aquatic macrophytes (Table 5).

Table 5. Metrics to be considered in the index of biotic integrity for aquatic plant communities.

Metric	Possible score		
	1	3	5
Species richness	low	moderate	high
Species diversity	low	moderate	high
Abundance	low	moderate	high
Dominance	high	moderate	low
Proportion exotics	high	moderate	low
Indicator species	absent	present	numerous
Endangered species	absent	present	numerous
Coefficient of conservatism	low	moderate	high
Floristic quality index	low	moderate	high
Growth habit	one	few	many
Percent tolerant species	high	moderate	low
Percent exhibiting stress	high	moderate	low

Additional metrics should describe the presence of tolerant and sensitive species. High abundances of tolerant species may indicate degraded conditions and reflect a decline in biological integrity. Substitution metrics for aquatic macrophytes may include the proportion of exotic and invasive taxa. For aquatic macrophytes, a species list of various indicator species that would be extremely sensitive to changes in habitat should be produced. Karr et al. (152) suggested that this list should be less than 5-10% of the total species possible so that only the initial loss of these taxa would indicate the beginning of a decline in biological integrity. The floristic quality index (FQI) could be substituted as a type of tolerance and sensitivity measure (118).

The introduction of exotic species may be causing some of the greatest reductions in native species that have been seen in recent times (159,160,161). Native species are often outcompeted and often, as in the examples of purple loosestrife (*Lythrum salicaria*), Eurasian water milfoil (*Myriophyllum spicatum*), glossy buckthorn (*Rhamnus frangula)*, and common reed (*Phragmites australis*) produce dense monospecific stands. Exotic species provide poor food and shelter for native animal species and often produce large numbers of seeds or reproduce vegetatively by fragmentation and are often difficult to control (e.g., glossy buckthorn) (162).

Functional aspects of aquatic macrophyte communities
Various life styles and affinities to specific habitat requirements can be used to quantify community function. Galatowitsch and McAdams (163) grouped plant species along the Mississippi River into various plant guilds to begin addressing functional attributes of community stability. Species that are sensitive to habitat changes, such as erosion, turbidity, and increases in total dissolved solids, could be grouped. Another metric may focus on changes in water levels, i.e., including soil saturation for emergent species, presence of pioneering species that are first to colonize an area after disturbance, and species unable to tolerate water depth changes, which could be an indicator of stability. In wetlands, the extent of inundation governs the gradient from marsh (flooded) to sedge meadow (periodically flooded) to prairie (periodically saturated soil) with a change in emergent aquatic species.

Individual health, condition, and abundance
Karr et al. (152) used abundance, the proportion of hybrids, and the presence of disease to evaluate individual health, condition, and abundance. Plant communities may require a relative abundance rating, dominance, disease, and pioneer species as substitute metrics. For aquatic macrophyte communities, one proposed metric may be the ratio of rhizome reproduction, clonal growth, and stolon and turion production compared to sexual reproduction. The presence of increased cloning and monotypic stands would indicate a reduction in biological integrity.

The lowest level of biological integrity is identified by the proportion of disease metric (152). Diseases identified by Karr et al. (152) and the Ohio Environmental Protection Agency (164) were those described as deformities, lesions, and tumors. For aquatic macrophyte communities, these could include the presence of deformities, pitted leaves and stalks, and lesions caused by insects, rust, or acidic air deposition.

Predicted metric response to increased biological integrity
Hughes and Noss (165) provide a summary of responses to environmental degradation. Decline in biological integrity would result from the loss of native species and reduction of specialized taxa or guilds, an increase in the percentage of exotic or introduced species, and a decline in intolerant or sensitive species. Conversely, the percentage of the community comprising generally tolerant or insensitive species would increase and the percentage of trophic and habitat specialists declines, while trophic and habitat generalists increase. A decline in biotic integrity will cause a reduction in number of individuals and an increase in disease or anomalies, while spatial and temporal fluctuations are more pronounced.

Future Research Needs

Autecological studies of native aquatic plant species are limited in North America. Species potential as a bioindicator is difficult to discern when little is known about the basic life history, including studies of habitat requirements, reproductive strategy, distribution, and resistance to stress. More research is needed on abundance, distribution, and tolerance of native wetland species. There is a need to develop biotic indices that allow estimates of ecological integrity (e.g., 166). Establishment of critical habitat requirements of species can provide important information about water quality. For example, within a watershed composed of several lakes, the presence of bioindicator aquatic plant species in headwater lakes versus their absence in the lower lakes may be directly related to degraded water quality.

Little attention has been given to the effects of organic contaminants on aquatic plants. The development of toxicity testing protocols should be encouraged and additional species tested to derive a wider range of phytotoxicity results for writing regulations to protect the health of the aquatic environment. Post-construction monitoring on the effectiveness of constructed wetlands for mitigation should include data collection on man-made wetlands as an option in restoration and reclamation efforts. Not only would it increase our understanding of tolerance levels and critical habitat requirements of aquatic plant species, but it would lead to the design and creation of constructed wetlands of greater efficiency (167).

Development and testing of metrics designed to assess the quality of the aquatic environment should incorporate aquatic plant information into the criteria. A multimetric index of biotic integrity should be developed for wetland plant communities. Regional reference conditions will need to be established (168), tested, and modified to fit a variety of wetland types and conditions (169).

Acknowledgments

We thank Noel Pavlovic and Richard Whitman for reading earlier versions of this manuscript. We thank Jason Butcher for preparing the figures and tables and helping with the references. Ann Zimmerman provided a literature search and helped with references. Although this article may have been funded wholly or in part by the U.S. Environmental Protection Agency, opinions expressed herein do not necessarily represent those of the agency. This chapter is contribution 1035 of the USGS Great Lakes Science Center.

References Cited

[1] M.S.A. Campbell (1939): Biological indicators of stream pollution. Sewage Works J., 11(1), 123-127.
[2] K.P. Krishnamoorthi, M.K. Abdulappa, and A. Gadkari (1978): Biological indicators of pollution. Proc. Indian Natl. Sci. Acad., 44(B), 98-110.
[3] P.M. Stewart and D.J. Robertson (1992): Aquatic organisms as indicators of water quality in suburban streams of the lower Delaware River Region, USA. J. Pa. Acad. Sci., 65(3), 135-141.
[4] R. Patrick, F. Douglass, D.M. Palavage, and P.M. Stewart (1992): *Surface Water Quality: Have the Laws Been Successful?* Princeton University Press, Princeton, N.J., USA.
[5] K.A. Moser, K.M. McDonald, and J.P. Smol (1996): Applications of freshwater diatoms to geographical research. Prog. Phys. Geography., 20(1), 21-52.

[6] D.W. Schindler (1987): Detecting ecosystem responses to anthropogenic stress. Can. J. Fish. Aquat. Sci., 44, 6-25.

[7] O. Ravera (1983): Assessment of the trophic state of a water body. Ann. Limnol., 19, 229-234.

[8] C.M.L. Mesters (1995): Shifts in macrophyte species composition as a result of eutrophication and pollution in Dutch transboundary streams over the past decades. J. Aquat. Ecosyst Health, 4, 295-305.

[9] B.J.Cholnoky (1960): The relationship between algae and the chemistry of natural waters. C. S. I. R., Reprint R.W., 129, 215-229.

[10] P. Huttunen and J. Meriläinen (1986): Application of multivariate techniques to infer limnological conditions from diatom assemblages, 201-212. In J.P. Smol, R.W. Battarbee, R.B. Davis, and J. Meriläinen: *Diatoms and Lake Acidity: Reconstructing pH from Siliceous Algal Remains in Lake Sediments,* Developments in Hydrology 29, W. Junk Publishers, The Netherlands.

[11] H.J.B. Birks, J.M. Line, S. Juggins, A.C. Stevenson, and C.J.F. ter Braak (1990): Diatoms and pH reconstruction. Philos. Trans. R. Soc. Lond., 327(B), 263-278.

[12] M.D. Agbeti (1992): Relationship between diatom assemblages and trophic variables: a comparison of old and new approaches. Can. J. Fish. Aquat. Sci., 49, 1171-1175.

[13] S.S. Dixit, J.P. Smol, J.C. Kingston, and D.F. Charles (1992): Diatoms: powerful indicators of environmental change. Environ. Sci. Technol., 26, 23-33.

[14] C.E. Christie and J.P. Smol (1993): Diatom assemblages as indicators of lake trophic status in southeastern Ontario lakes. J. Phycol., 29(5), 575-586.

[15] P.M. Stewart (1995): Use of algae in pollution assessment. Nat. Areas J., 15(3), 234-239.

[16] J. Lovett Doust, M. Schmidt, and L. Lovett Doust (1994): Biological assessment of aquatic pollution: a review, with emphasis on plants as biomonitors. Biol. Rev., 69, 147-186.

[17] M.S. Simmons (1984): PCB contamination in the Great Lakes, 287-309. In J.O. Nriagu and M.S. Simmons: *Toxic Contaminants in the Great Lakes*, John Wiley and Sons, New York, N.Y., USA.

[18] A.J. Stewart, G.J. Haynes, and M.I. Martinez (1992): Fate and biological effects of contaminated vegetation in a Tennessee stream. Environ. Toxicol. Chem., 11, 653-664.

[19] A.H. Nielson (1994): *Organic Chemicals in the Aquatic Environment - Distribution, Persistence and Toxicity*, Lewis Publications, Boca Raton, Fla., USA.

[20] C.L. Duxbury, D.G. Dixson, and B.M. Greenburg (1997): Effects of simulated radiation on the bioaccumulation of polycyclic aromatic hydrocarbons by the Duckweed *Lemna gibba.* Environ. Toxicol. Chem., 16, 1739-1748.

[21] S. Roy and O. Hänninen (1994): Pentachlorophenol: uptake/elimination kinetics and metabolism in an aquatic plant, *Eichhornia crassipes.* Environ. Toxicol. Chem., 13, 763-773.

[22] K.J. Killgore, E.D. Dibble, and J.J. Hoover (1993): *Relationship Between Fish and Aquatic Plants: A Plan of Study.* U. S. Army Corps of Engineers, Misc. Paper A-93-1, Vicksburg, Miss., USA.

[23] B.A. Whitton (1979): Plants as indicators of water quality, (5)1-(5)34. In: A. James and L. Evison: *Biological Indicators of Water Quality*, John Wiley and Sons, Inc., Chichester, UK.

[24] L.E. Shubert (1984): *Algae as Ecological Indicators*, Academic Press, London, UK.

[25] S. Roy and O. Hänninen (1993): Biochemical monitoring of the aquatic environment: possibilities and limitation, 119-135. In: M. Richardson: *Ecotoxicology Monitoring*, VCH, London, UK.

[26] M.A. Lewis (1995): Use of freshwater plants for phytotoxicity testing. Environ. Pollut., 87, 319-336.

[27] J. Cairns, Jr. and K.L. Dickson (1973): *Biological Methods for the Assessment of Water*

Quality, ASTM Special Publication 528, American Society for Testing and Materials, Philadelphia, Pa., USA.

[28] G.M. Rand and S.R. Petrocelli (1985): *Fundamentals of Aquatic Toxicology. Methods and Applications*, Hemisphere Publishing Co., Washington, D.C., USA.

[29] E.A. Laws (1993): *Aquatic Pollution. An Introductory Text*, 2nd edition, John Wiley and Sons, Inc., New York, N.Y., USA.

[30] S. Ramamoorthy and E.G. Baddaloo (1995): *Handbook of Chemical Toxicity Profiles of Biological Species*, Volume I, *Aquatic Species*, Lewis Publishers, Boca Raton, Fla., USA.

[31] D.B. Huebert and J.M. Shay (1993): Considerations in the assessment of toxicity using duckweeds. Environ. Toxicol. Chem., 12, 481-483.

[32] L. Lovett Doust, J. Lovett Doust, and M. Biernacki (1994): American wildcelery, *Vallisneria americana*, as a biomonitor of organic contaminants in aquatic ecosystems. J. Gt. Lakes Res., 20(2), 333-354.

[33] C.D. Sculthorpe (1967): *The Biology of Aquatic Vascular Plants*, Edward Arnold Ltd., London, UK.

[34] G.A. Janauer (1985): Heavy metal accumulation and physiological effects on Austrian macrophytes, 21-28. In: J. Salánki: *Heavy Metals in Water Organisms*, Symposia Biologica Hungarica 29, Akadémiai Kiadó, Budapest, Hungary.

[35] H. Gonzalez, H., M. Lodenius, and M. Otero (1989): Water hyacinth as indicator of heavy metal pollution in the tropics. Bull. Environ. Contam. Toxicol., 43, 910-914.

[36] P. Reimer and H.C. Duthie (1993): Concentrations of zinc and chromium in aquatic macrophytes from the Sudbury and Maskoka regions of Ontario, Canada. Environ. Pollut., 79, 261-265.

[37] B.A. Manny, S.J. Nichols, and D.W. Schloesser (1991): Heavy metals in aquatic macrophytes drifting in a large river. Hydrobiologia, 219, 333-344.

[38] M. Maessen, J.G.M. Roelofs, M.J.S. Bellemakers, and G.M. Verheggen (1992): The effects of aluminum, aluminum/calcium ratios and pH on aquatic plants from poorly buffered environments. Aquat. Bot., 43, 115-127.

[39] M. Greger and L. Kautsky (1991): Effects of Cu, Pb, and Zn on two *Potamogeton* species grown under field conditions. Vegetatio 97, 173-184.

[40] M. Greger and L. Kautsky (1993): Use of macrophytes for mapping bioavailable heavy metals in shallow coastal areas, Stockholm, Sweden. Appl. Geochem. Suppl., 2, 37-43.

[41] U. Vedagiri and J. Ehrenfeld (1992): Partitioning of lead in pristine and runoff-impacted waters from acidic wetlands in the New Jersey pinelands. Water Air Soil Pollut., 64, 511-524.

[42] L. Yurukova and K. Kochev (1994): Heavy metal concentrations in freshwater macrophytes from the Aldomirovsko swamp in the Sofia district, Bulgaria. Bull. Environ. Contam. Toxicol., 52, 627-632.

[43] M. Gupta, M., S. Sinha, and P. Chandra (1994): Uptake and toxicity of metals in *Scirpus lacustris* L. and *Bacopa monnieri* L. J. Environ. Sci. Health, A29(10), 2185-2202.

[44] R.W. Bosserman (1985):Distribution of heavy metals in aquatic macrophytes from Okefenoke Swamp, 31-39. In: J. Salánki: *Heavy Metals in Water Organisms*, Symposia Biologica Hungarica 29, Akadémiai Kiadó, Budapest, Hungary.

[45] L. St-Cyr and P.G.C. Campbell (1994): Trace metals in submerged plants of the St. Lawrence River. Can. J. Bot., 72, 429-439.

[46] B. Wachs (1985): Bioindicators for the heavy metal load of river ecosystems, 179-188. In: J. Salánki: *Heavy Metals in Water Organisms*, Symposia Biologica Hungarica 29, Akadémiai Kiadó, Budapest, Hungary.

[47] W. Wang (1987): Root elongation method for toxicity testing of organic and inorganic

pollutants. Environ. Toxicol. Chem., 6, 409-414.

[48] M. Coquery and P.M. Welbourn (1995): The relationship between metal concentration and organic matter in sediments and metal concentration in the aquatic macrophyte *Eriocaulon septangulare*. Wat. Res., 29(9), 2094-2102.

[49] G.H.P. Arts, J.G.M. Roelofs, and M.J.H. de Lyon (1990): Differential tolerance between softwater macrophyte species to acidification. Can. J. Bot., 68, 2127-2134.

[50] P. Guilizzoni (1991): The role of heavy metals and toxic materials in the physiological ecology of submersed macrophytes. Aquat. Bot., 41, 87-109.

[51] A.B. Tomsett, A.K. Sewell, S.J. Jones, J.R. deMiranda, and D.A. Thurman (1992): Metal binding proteins and metal-regulated gene expression in higher plants, 1-24. In: J.L. Wray: *Society for Experimental Biology Series 49: Inducible Plant Proteins*, Cambridge University Press, Cambridge, UK.

[52] D.B. Huebert and J.M. Shay (1992): The effect of EDTA on cadmium and zinc uptake and toxicity in *Lemna trisulca*. Arch. Environ. Contam. Toxicol., 22, 313-318.

[53] U. Borgmann (1983): Metal speciation and toxicity of free metal ions to aquatic biota, 47-72. In: J.O. Nriagu: *Aquatic Toxicology*, John Wiley and Sons, New York, N.Y., USA.

[54] Y.M. Nor and H.H. Cheng (1986): Chemical speciation and bioavailablity of copper: uptake and accumulation by *Eichornia*. Environ. Toxicol. Chem., 5, 941-947.

[55] W. Wang (1990): Literature review on duckweed toxicity testing. Environ. Res., 52, 7-22.

[56] J.S. Greene (1988): *Comments on the E47.01.07 Algal Test Standard Practice for Conducting Static 96-H Toxicity Tests with Microalgae (Draft #12)*. Presented to ASTM Task Force Group E47.01.07, April 27, 1999, Sparks, Nevada, USA.

[57] I. Kufel and L. Kufel (1985): Heavy metals and mineral nutrient budget in *Phragmites australis* and *Typha angustifolea*, 61-66. In: J. Salánki: *Heavy Metals in Water Organisms*, Symposia Biologica Hungarica 29, Akadémiai Kiadó, Budapest, Hungary.

[58] C.P. McRoy, R.J. Barsdate, and M. Nebert (1972): Phosphorus cycling in an eelgrass, *Zostera marina*, ecosystem. Limnol. Oceanogr., 17, 58-67.

[59] R.J. Reimold (1972): The movement of phosphorus through the salt marsh cord grass *Spartina alterniflora* Loisel. Limnol. Oceanogr., 17, 606-611.

[60] J.K. Abaychi and S.Z. Al-Obaidy (1987): Concentrations of trace elements in aquatic vascular plants from Shatt Al. Arab River, Iraq. J. Biol. Sci. Res., 18(2), 123-129.

[61] R.P.H Welsh and P. Denny (1980): The uptake of lead and copper by aquatic macrophytes in two English lakes. J. Ecol., 68, 443-455.

[62] P.H. Albers and M.B. Camardese (1993): Effects of acidification on metal accumulation by aquatic plants and invertebrates. 2. Wetlands, ponds and small lakes. Environ. Toxicol. Chem., 12(6), 969-976.

[63] W. Wang (1986): Comparative toxicity of phenolic compounds using root elongation method. Environ. Toxicol. Chem., 5, 891-896.

[64] X. Huang, B.J. McConkey, T.S. Babu, and B.M. Greenberg (1997): Mechanisms of photoinduced toxicity of photomodified anthracene to plants: inhibition of photosynthesis in the aquatic higher plant *Lemna gibba* (duckweed). Environ. Toxicol. Chem., 16(8), 1707-1715.

[65] J. Cairns, Jr. (1980): Beyond single species toxicity testing. Mar. Environ. Res., 3, 157-159.

[66] C.I. Weber, W.H. Peltier, T.J. Norberg-King, W.B. Horning II, F.A. Kessler, J.R. Menkedick, T.W. Neiheisel, P.A. Lewis, D.J. Klemm, Q.H. Pickering, E.L. Robinson, J.M. Lazorchak, L.J. Wymer, and R.W. Freyberg (1989): *Short-term Methods for Estimating the Chronic Toxicity of Effluents and Receiving Waters to Freshwater Organisms*, 2nd edition, Environmental Monitoring Systems Laboratory, Cincinnati, Ohio, USA.

[67] H. Blanck, G. Wallin, and S. Wangberg (1984): Species dependent variation in algal

sensitivity to compounds. Ecotoxicol. Environ. Saf., 8, 339-351.

[68] S.M. Swanson, Rickard, C.P., Freemark, K.E., and P. MacQuarrie (1991): Testing for pesticide toxicity to aquatic plants: recommendations for test species, 77-97. In J.W. Gorsuch, W.R. Lower, W. Wang, and M.A. Lewis: *Plants for Toxicity Assessment*, American Society for Testing and Materials, Philadelphia, Pa., USA.

[69] American Society for Testing and Materials (1991): *Standard Guide for Conducting Static Toxicity Tests with Lemna gibba G3. E141591,* Annual Book of ASTM Standards, Vol. 11.04, Philadelphia, Pa., USA.

[70] C.T. Walbridge (1977): *A Flow-through Testing Procedure with Duckweed (Lemna minor L.).* EPA 600/3-77-108, U.S. Environmental Protection Agency, Duluth, Minn., USA.

[71] W. Wang (1991): Literature review on higher plants for toxicity testing. Wat., Air, Soil Pollut., 59, 381-400.

[72] M. Tracy, D.C. Freeman, J.M. Emlen, J.H. Graham, and R.A. Hough(1995): Developmental instability as a biomonitor of environmental stress: an illustration using aquatic plants and microalgae, 313-337. In: F.M. Butterworth, L.D. Corkum, and J. Guzmán-Rincón: *Biomonitors and Biomarkers as Indicators of Environmental Change. A Handbook*, Plenum Press, New York, N.Y., USA.

[73] D.C. Freeman, J.H. Graham, and J.M. Emlen (1993): Developmental stability in plants: symmetries, stress and epigenesis. Genetica, 89, 97-119.

[74] H.V. Leland and J.S. Kuwabara (1985): Trace metals, 374-415. In: G.M. Rand and S.R. Petrocelli: *Fundamentals of Aquatic Toxicology*, Hemisphere Publishing Corporation, New York, N.Y., USA.

[75] E. Epstein (1972): *Mineral Nutrition of Plants: Principles and Perspectives*, John Wiley and Sons, Inc., New York, N.Y., USA.

[76] U.M. Cowgill and D.P. Milazzo (1989): The culturing and testing of two species of duckweed, 379-391. In: U.M. Cowgill and L.R. Williams: *Aquatic Toxicology and Hazard Assessment*, 12th volume, ASTM STP 1027, Philadelphia, Pa., USA.

[77] R.F.M. Van Steveninck, M.E. Van Steveninck, and D.R. Fernando (1992): Heavy-metal (Zn, Cd) tolerance in selected clones of duck weed (*Lemna minor*). Plant Soil, 146, 271-280.

[78] D.B. Huebert and J.M. Shay (1992): Zinc toxicity and its interaction with cadmium in the submerged aquatic macrophyte *Lemna trisulca* L. Environ. Toxicol. Chem., 11, 715-720.

[79] J. Singh, G. Chawla, S.H.N. Naqvi, and P.N. Viswanathan (1994): Combined effects of cadmium and linear alkyl benzene sulfonate on *Lemna minor* L. Ecotoxicology, 3, 59-67.

[80] P. Garg and P. Chandra (1994): The duckweed *Wolffia globosa* as an indicator of heavy metal pollution: sensitivity to Cr and Cd. Environ. Monit. Assess., 29, 89-95.

[81] T.C. Hutchinson and H. Czyrska (1975): Freshwater macrophytes as indicators of organic pollution. Int. Ver. Theor. Angew. Limnol. Verh., 19(3), 693-697.

[82] J.B. Lisiecki and C.D. McNabb (1975): Heavy metal toxicity and synergism to floating aquatic weeds. Verh. Int. Ver. Limnol. 19, 2102-2111.

[83] M. Van der Werff and M.J. Pruyt (1982): Long term effects of heavy metals on aquatic plants. Chemosphere, 11, 727-739.

[84] R.P. Glandon and C.D. McNabb (1985): The uptake of boron by *Lemna minor.* Aquat. Bot., 4, 53-64.

[85] J.H. Rogers, Jr., D.S. Cherry, and R.K. Guthrie (1978): Cycling of elements in duckweed (*Lemna perpusilla*) in an ash settling basin and swamp drainage system. Water Res., 12, 765-770.

[86] K.R. Solomon, D.B. Baker, R.P. Richards, K.R. Dixon, S.J. Klaine, T.W. LaPoint, R.J. Kendall, C.P. Weisskopf, J.M. Giddings, J.P.Giesy, L.W. Hall, Jr., and W.M. Williams (1996): Ecological risk assessment of atrazine in North American surface waters. Environ. Toxicol.

Chem., 15(1), 31-76.

[87] B.J. McConkey, C.L. Duxbury, D.G. Dixon, and B.M. Greenberg (1997): Toxicity of a PAH photooxidation product to the bacteria *Photobacterium phosphoreum* and the duckweed *Lemna gibba*: effects of phenanthrene and its primary photoproduct, phenanthrenequinone. Environ. Toxicol. Chem., 16, 892-899.

[88] D.J. Phillips (1978): Use of biological indicator organisms to quantitate organochlorine pollutants in aquatic environments - a review. Environ. Pollut., 16, 167-229.

[89] C.A. Staples, W.J. Adams, T.F. Parkerton, J.W. Gorsuch, G.R. Biddinger, and K.H. Reinert (1997): Aquatic toxicity of eighteen phthalate esters. Environ. Toxicol. Chem., 16(5), 875-891.

[90] H.A. Sharma, J.T. Barber, H.E. Ensley, and M.A. Polito (1997): A comparison of the toxicity and metabolism of phenol and chlorinated phenols by *Lemna gibba*, with special reference to 2,4,5 -trichlorophenol. Environ. Toxicol. Chem., 16, 346-350.

[91] K. Cassidy and J.H. Rodgers, Jr. (1989): Response of *Hydrilla* (*Hydrilla verticillata* (L.f.) Royle) to diquat and a model of uptake under nonequilibrium conditions. Environ. Toxicol. Chem., 8, 133-140.

[92] M. Cooke and A.J. Dennis (1983): *Polycyclic Aromatic Hydrocarbons: Formation, Metabolism, and Measurement*, Battelle, Columbus, Ohio, USA.

[93] S. Painter (1990): Ecosystem health and aquatic macrophytes: status and opportunities, Proceedings of the Aquatic Ecosystem Health Symposium, Waterloo, Ontario, Canada.

[94] J.W. Owens (1991): The hazard assessment of pulp and paper effluents in the aquatic environment: a review. Environ. Toxicol. Chem., 10, 1511-1540.

[95] V.P. Kozak, G.V. Simsiman, G. Chesters, D. Stensby, and J. Harkin (1979): *Reviews of the Environmental Effects of Pollutants, 11, Chlorophenols*, EPA 600/1-79-012, U.S. Environmental Protection Agency, Cincinnati, Ohio, USA.

[96] H.E. Ensley, J.T. Barber, M.A. Polito, and A.I. Oliver (1994): Toxicity and metabolism of 2,4-dichlorophenol by the aquatic angiosperm *Lemna gibba*. Environ. Toxicol. Chem., 13(2), 325-331.

[97] E.P. Odum (1984): The mesocosm. BioScience, 34, 558-562.

[98] J.R. Voshell, Jr. (1989): *Using Mesocosms for Assessing the Aquatic Ecological Risk of Pesticides: Theory and Practice*. Entomological Society of America, Miscellaneous Publications, 75.

[99] J.R. Voshell, Jr. (1990): Introduction and overview of mesocosms, 1-3. In: *Experimental Ecosystem Applications to Ecotoxicology*, North American Benthological Society, Technical Information Workshop, Virginia Polytechnic Institute and State University, Blackburg, Va., USA.

[100] J.H. Kennedy, Z.B. Johnson, P.D. Wise, and P.C. Johnson (1995): The use of microcosm and mesocosm as surrogate aquatic ecosystems in ecotoxicological research: a review, 117-162. In: D.J. Hoffman, B.A. Rattner, G.A. Burton, Jr., and J. Cairns, Jr.: *Handbook of Ecotoxicology*, Lewis Publishers, Boca Raton, Fla., USA.

[101] C. Etnier and B. Guterstam (1997): *Ecological Engineering for Wastewater Treatment*, 2nd edition, Lewis Publishers, Boca Raton, Fla., USA.

[102] R.H. Kadlec and R.L. Knight (1996): *Treatment Wetlands*. Lewis Publishers, Boca Raton, Fla., USA.

[103] C.M. Finlayson, I. von Oertzen, and A.J. Chick (1990): Treating poultry abattoir and piggery effluents in gravel trenches, 559-562. In: P.F. Cooper and B.C. Findlater: *Constructed Wetlands in Water Pollution Control*, Pergamon Press, Oxford, U.K.

[104] K.R. Reddy, E.M. D'Angelo, and T.A. DeBusk (1990): Oxygen transport through aquatic macrophytes: the role in wastewater treatment. J. Environ. Qual., 19(2), 261-267.

[105] P.S. Burgoon, K.R. Reddy, T.A. DeBusk, and B. Koopman (1991): Vegetated submerged

beds with artificial substrates. 2. N and P removal. J. Environ. Eng., 117, 408-424.

[106] H. Brix (1993): Macrophyte-mediated oxygen transfer in wetlands: Transport mechanisms and rates, 391-398. In: G.A. Moshiri: *Constructed Wetlands for Water Quality Improvement*, Lewis Publishers, Boca Raton, Fla., USA.

[107] R.M. Gersberg, B.V. Elkins, S.R. Lyons, and C.R. Goldman (1986): Role of aquatic plants in wastewater treatment by artificial wetlands. Water Resour., 20, 363-367.

[108] National Academy of Sciences (1976): *Making Aquatic Weeds Useful: Some Perspectives for Developing Countries*, National Academy of Sciences, Washington, D.C., USA.

[109] B.C. Wolverton and R.C. McDonald (1977): *Proceedings of the 1977 National Conference on Treatment and Disposal of Industrial Wastewaters and Residue, Houston, Texas*, Information Transfer Inc., Rockville, Md., USA.

[110] T.A. DeBusk and K.R. Reddy (1987): BOD removal in floating aquatic macrophyte-based treatment systems. Water Sci. Technol., 19, 273-279.

[111] S.J. Hancock and L. Buddhavarapu (1993): Control of algae using duckweed (*Lemna*) systems, 399-406. In: G. A. Moshiri: *Constructed Wetlands for Water Quality Improvement*, Lewis Publishers, Boca Raton, Fla., USA.

[112] S.M. Haslam (1982): A proposed method for monitoring river pollution using macrophytes. Environ. Technol. Lett., 3, 19-34.

[113] P.B. Reed (1988): *National List of Plant Species that Occur in Wetlands*, Biological Report 88,24, U.S. Fish and Wildlife Service, Washington, D.C., USA.

[114] U.S. Army Corps of Engineers (1987): *Corps of Engineers Wetland Delineation Manual*, U.S. Army Corps of Engineer Waterways Experiment Station, Vicksburg, Miss., USA.

[115] R.W. Tiner (1993): Using plants as indicators of wetland. Proc. Acad. Nat. Sci. Phila., 144, 240-253.

[116] National Research Council (1992): *Restoration of Aquatic Ecosystems*, National Academy Press, Washington, D.C., USA.

[117] R.K. Olson (1993): *Created and Natural Wetlands for Controlling Nonpoint Source Pollution*, C.K. Smoley, Boca Raton, Fla., USA.

[118] F. Swink and G. Wilhelm (1994): *Plants of the Chicago Region*, Indiana Academy of Sciences, Indianapolis, Ind., USA.

[119] K.D. Herman, L.A. Masters, M.R. Penskar, A.A. Reznicek, G.S. Wilhelm, and W.W. Brodowicz (1997): Floristic quality assessment: development and application in the state of Michigan (USA). Nat. Areas J., 17(3), 265-276.

[120] D.R.J. Moore, P.A. Keddy, C.L. Gaudet, and I.C. Wisheu (1989): Conservation of wetlands: do infertile wetlands deserve a higher priority? Biol. Conserv., 47, 203-217.

[121] D.W Schindler, S.E.M. Kasian, and R.H. Hesslein (1989): Losses of biota from American aquatic communities due to acid rain. Environ. Monit. Assess., 12, 269-285.

[122] D.A. Wilcox and R.E. Andrus (1987): The role of *Sphagnum fimbriatum* in secondary succession in a road salt impacted bog. Can. J. Bot., 65(11), 2270-2275.

[123] D.A. Wilcox (1986): The effects of deicing salts in Pinhook Bog, Indiana. Can. J. Bot., 64, 865-874.

[124] D.A. Wilcox and R.W. Buchholz (1986): Vegetation restoration in a road salt impacted bog (Indiana). Restor. Manag. Notes, 4(1), 28.

[125] G. Chawla, P.N. Viswanathan, and S. Viswanathan (1986): Aquatic flora in relation to water pollution, 99-128. In: P.K.K. Nair: *Aspects of Energy and Environment Studies, Glimpses in Plant Research*, Volume 7, Vikas Publishing House, Delhi, India.

[126] C. Steinberg and S. Schiefele (1988): Biological indication of trophy and pollution of running waters. J. Water Wastewater Res., 21(6), 227-234.

[127] Š. Husák, V. Sládeček, and A. Sládečková (1989): Freshwater macrophytes as indicators of organic pollution. Acta hydrochim. hydrobiol., 17(6), 693-697.

[128] J.M. Klopatek (1978): Nutrient dynamics of freshwater riverine marshes and the role of emergent macrophytes, 195-216. In: R.E. Good, D.F. Whigham, and R.L. Simpson: *Freshwater Wetlands. Ecological Processes and Management Potential*, Academic Press, Inc., San Diego, Calif., USA.

[129] C.A. Johnston (1991): Sediment and nutrient retention by freshwater wetlands: effects on surface water quality. Crit. Rev. Environ. Control, 21(5,6), 491-565.

[130] E. Chan, T.A. Bursztynsky, N. Hantzsche, and Y.J. Litwi (1982): *The Use of Wetlands for Water Pollution Control*, Municipal Environmental Research Laboratory, Office of Research and Development, U.S. Environmental Protection Agency, Cincinnati, Ohio, USA.

[131] E.C. Stockdale (1990): *Freshwater Wetlands, Urban Stormwater, and Nonpoint Pollution Control: A Literature Review and Annotated Bibliography*, Washington State Department of Ecology, Olympia, Wash., USA.

[132] J.A. Stratis, V. Simeonov, G. Zachariadis, T. Sawidis, P. Mandjukov, and S. Tsakovski (1996): Chemometrical approaches to evaluate analytical data from aquatic macrophytes and marine algae. Fresenius' J. Anal. Chem., 355(1), 65-70.

[133] R. Carbiener, M. Trémolières, J.L. Mercer, and A. Ortscheit (1990): Aquatic macrophyte communities as bioindicators of eutrophication in calcareous oligosaprobe stream waters (Upper Rhine plain, Alsace). Vegetatio, 86, 71-88.

[134] J. Elster, J. Kvet, and V. Hauser (1995): Root length of duckweeds (Lemnaceae) as an indicator of water trophic status. Ekológia (Bratislava), 14(1), 43-59.

[135] A. Schnitzler, I. Eglin, F. Robach, and M. Trémolières (1996): Response of aquatic macrophyte communities to levels of p and n nutrients in an old swamp of the upper Rhine plain (eastern France). Écologie, 27(1), 51-61.

[136] H. Balls, B. Moss, and K. Irvine (1989): The loss of submerged plants with eutrophication. I. Experimental design, water chemistry, aquatic plant and phytoplankton biomass in experiments carried out in ponds in the Norfolk Broadland. Freshwater Biol., 22(1), 71-87.

[137] K. Irvine, B. Moss, and H. Balls (1989): The loss of submerged plants with eutrophication. II. Relationships between fish and zooplankton in a set of experimental ponds, and conclusions. Freshwater Biol., 22(1), 89-107.

[138] J. Stansfield, B. Moss, and K. Irvine (1989): The loss of submerged plants with eutrophication. III. Potential role of organochlorine pesticides: a palaeoecological study. Freshwater Biol., 22(1), 109-132.

[139] M.O. Holowaty, M. Reshkin, M.J. Mikulka, and R.D. Tolpa (1992): Working toward a remedial action plan for the Grand Calumet River and Indiana Harbor Ship Canal, 211-233. In: J.H. Hartig and M.A. Zarull: *Under RAPs: Toward Grassroots Ecological Democracy in the Great Lakes Basin*, The University of Michigan Press, Ann Arbor, Mich., USA.

[140] D.F. Gatz, V.C. Bowersox, and J. Su (1989): Lead and cadmium loadings to the Great Lakes from precipitation. J. Gt. Lakes Res., 15, 246-264.

[141] National Atmospheric Deposition Program (1995): National Atmospheric Deposition Program, Colorado State University, Fort Collins, Colo., USA.

[142] N.C. Fassett (1957): *A Manual of Aquatic Plants*, 2nd edition, The University of Wisconsin Press, Madison, Wis., USA.

[143] G.S. Winterringer and A.C. Lopinot (1977): *Aquatic Plants of Illinois*, Illinois State Museum, Springfield, Ill., USA.

[144] J.H. Zar (1984): *Biostatistical Analysis*, 2nd edition, Prentice Hall, Englewood Cliffs, N.J., USA.

[145] J.G. Pearson and C.F.A. Pinkham (1992): Strategy for data analysis in environmental surveys emphasizing the index of biotic simlarity and BIOSIM1. Water Environ. Res., 64(7), 901-909.

[146] D.A. Gonzales, J.G. Pearson, and C.F.A. Pinkham (1993): *Users Manual: Biosim1, Beta Version 1.0*, EPA 600-R-93-219, U.S. Environmental Protection Agency, Environmental Monitoring and Support Laboratory, Las Vegas, Nev., USA.

[147] P.M. Stewart and J.T. Butcher (1997): Grand Calumet Lagoons, F230-F266. In: *Grand Calumet River - Indiana Harbor and Ship Canal Sediment Cleanup and Restoration Alternatives Project*, U.S. Army Corps of Engineers, Chicago District, Chicago, Ill., USA.

[148] U.S. Army Corps of Engineers (1997): Appendix B, Attachment 7. In: *Grand Calumet River - Indiana Harbor and Ship Canal Sediment Cleanup and Restoration Alternatives Project*, U.S. Army Corps of Engineers, Chicago District, Chicago, Ill., USA.

[149] J.R. Karr and D.R. Dudley (1981): Ecological perspective on water quality goals. Environ. Manage., 5, 55-68.

[150] R.M. Hughes (1995): Defining acceptable biological status by comparing with reference conditions, 31-47. In: W.S. Davis and T.P. Simon: *Biological Assessment and Criteria: Tools for Water Resource Management and Decision Making*, Lewis Publishers, Boca Raton, Fla., USA.

[151] J.R. Karr (1981): Assessment of biotic integrity using fish communities. Fisheries, 6, 21-27.

[152] J.R. Karr, K.D. Fausch, P.L. Angermeier, P.R. Yant, and I.J. Schlosser (1986): *Assessing Biological Integrity in Running Waters: A Method and its Rationale*, Illinois Natural History Survey Special Publication 5, Champaign, Ill., USA.

[153] B.H. Rosen (1995): Use of periphyton in the development of biocriteria, 209-215. In: W.S. Davis and T.P. Simon: *Biological Assessment and Criteria: Tools for Water Resource Planning and Decision Making*, Lewis Publishers, Boca Raton, Fla., USA.

[154] L. Bahls (1993): *Periphyton Bioassessment Methods for Montana Streams*. Water Quality Bureau, Department of Health and Environmental Science, Helena, Mont., USA.

[155] Kentucky Division of Water (1993): *Methods for Assessing Biological Integrity of Surface Waters*. Kentucky Department of Environmental Protection, Frankfort, Ky., USA.

[156] Oklahoma Conservation Commission (1993): *Development of Rapid Bioassessment Protocols for Oklahoma Utilizing Characteristics of the Diatom Community*. Oklahoma Conservation Commission, Okla., USA.

[157] J.L. Plafkin, M.T. Barbour, K.D. Porter, S.K. Gross, and R.M. Hughes (1989): *Rapid Bioassessment Protocols for Use in Streams and Rivers. Benthic Macroinvertebrates and Fish*, EPA 440-4-89-001, Office of Water Regulations and Standards, U.S. Environmental Protection Agency, Washington, D.C., USA.

[158] J. Cairns, Jr., D.W. Albaugh, F. Busey, M.D. Chanay (1968): The sequential comparison index: a simplified method for non-biologists to estimate relative differences in biological diversity in stream pollution studies. J. Water Pollut. Control Fed., 40(9), 1607-1613.

[159] C.S. Elton (1958): *The Ecology of Invasions by Animals and Plants*, Methuen Publishers, London, UK.

[160] M.B. Bain (1993): Assessing impacts of introduced aquatic species: grass carp in large systems. Environ. Manage., 17(2), 211-224.

[161] C.D.N. Barel, R. Dorit, P.H. Greenwood, G. Fryer, N. Hughes, P.B.N. Jackson, H. Kawanabe, R.H. Lowe-McConnell, M. Nagoshi, A.J. Ribbink, E. Trewavas, F. Witte, and K. Yamaoka (1985): Destruction of fisheries in Africa's lakes. Nature, 315, 19-20.

[162] J.A. Reinartz (1997): Controlling glossy buckthorn (*Rhamnus frangula* L.) with winter herbicide treatments. Nat. Areas J., 17(1), 38-41.

[163] S.M. Galatowitsch and T.V. McAdams (1994): *Distribution and Requirements of Plants on*

the Upper Mississippi River: Literature Review. Iowa Cooperative Fish and Wildlife Research Unit, Ames, Iowa, USA.

[164] Ohio Environmental Protection Agency (1989): *Biological Criteria for the Protection of Aquatic Life. Volume II: Users Manual for Biological Field Assessment of Ohio Surface Water*, Ohio EPA, Division of Water Quality Planning and Assessment, Ecological Assessment Section, Columbus, Ohio, USA.

[165] R.M. Hughes and R.F. Noss (1992): Biological diversity and biological integrity: current concerns for lakes and streams. Fisheries, 17(3), 11-19.

[166] S. Woodley, J. Kay, and G. Francis (1993): *Ecological Integrity and the Management of Ecosystems*, St. Lucie Press, Heritage Resources Centre, University of Waterloo and Canadian Parks Service, Ottawa, Canada.

[167] D.A. Hammer (1997): *Creating Freshwater Wetlands*, 2nd edition, Lewis Publishers, Boca Raton, Fla., USA.

[168] T.P. Simon (1998): Modification of an index of biotic integrity and development of reference condition expectations on dunal, palustrine wetland fish communities along the southern shore of Lake Michigan. Aq. Ecosys. Health Manage., 1, 49-62.

[169] T.P. Simon and P.M. Stewart (1998): Application of an index of biotic integrity for dunal, palustrine wetlands: emphasis on assessment of nonpoint source landfill effects on the Grand Calumet Lagoons. Aq. Ecosys. Health Manage., 1, 63-74.

AUTHOR INDEX

Behrens, A. 33

Bowers, N.J. 141

Braunbeck, T. 33

Cleveland, L. 195

Eckwert, H. 33

Fairchild, J.F. 195

Gerhardt, A. 1, 95

Janssens de Bisthoven, L. 65

Köhler, H.-R. 33

Konradt, J. 33

Lewis, M.A. 243

Little, E.E. 195

Mösslacher, F. 119

Müller, E. 33

Notenboom, J. 119

Pawert, M. 33

Pratt, J.R. 141

Rinderhagen, M. 161

Ritterhoff, J. 161

Roux, D.J. 13

Schramm, M. 33

Schwaiger, J. 33

Scribailo, R.W. 275

Segner, H. 33

Simon, T.P. 275

Stewart, P.M. 275

Tremp, H. 233

Triebskorn, R. 33

Wang, W. 243

Zauke, G.-P. 161

KEYWORD INDEX

Acidification 233
Amoebae 141
Amphibians 65
Amphipods 161
Aquatic
~ Plants 243
~ Systems 195
~ Toxicity Testing 275
Automated Biomonitors 95

Behaviour 161
Bioaccumulation 1, 161
~ Monitoring 119
Bioindicator(s) 1, 233, 275
Biological Indices 1
Biomarker 33
Biomonitoring 1, 13, 65, 243, 275
Biotic Indices 161
Biotransformation Enzymes 33
Birds 65

Cellular Pathology 33
Chironomidae 65
Ciliates 141
Community Response Biomonitoring 1
Contaminants 243
Continuous Biotests 95

Early Warning Systems 95
Ecological Integrity 13
Ecology 233
Ecosystem Monitoring 119
Ecotoxicology 195
Environmental
~ Assessment 13
~ Impacts 141
Eutrophication 233
Exposure Monitoring 1

Feeding 161
Fish 33, 65, 195

Gills 33
Groundwater Contamination 119
Growth 161

Index of Biotic Integrity 275
Indicators 243

Integrated Environmental Monitoring 1
Invertebrates 65
Isopods 161

Kidney 33

Lake Zooplankton 161
Life History 161
Liver 33

Metabolic Enzymes 33
Metazoa 119
Microbial Communities 119
Monitoring 33
~ Programme Design 13
Morphological Deformities 65

Pollution-Induced Stress 195
Population Studies 161
Protozoa 141

Reproduction 161
Running Waters 233

Salmo Trutta f. Fario 33
Saprobian System 141
Sediment Toxicity 161
South African River Health Programme ....
13
Stress Protein 33
Submerged Bryophytes 233
Surface/Subsurface Water Interactions 119

Toxic Effects 1
Toxicity Monitoring 119
Trout 33

Ultrastructure 33

Vertebrates 65

Waste Heat 233
Water Quality Monitoring 95
WET 1